Microbial Insights into Wastewater Treatment and Environmental Sustainability

Edited By

Harshita Jain

Amity Institute of Environmental Sciences, Amity University, Noida, 201303, India

&

Maulin P. Shah

*Department of Research Impact and Outcome
Research and Development Cell
Lovely Professional University, Phagwara, Punjab, India*

Microbial Insights into Wastewater Treatment and Environmental Sustainability

Editors: Harshita Jain and Maulin P. Shah

ISBN (Online): 979-8-89881-066-5

ISBN (Print): 979-8-89881-067-2

ISBN (Paperback): 979-8-89881-068-9

need for a court order if at any point you breach any terms of this License Agreement. In no event will any delay or failure by Bentham Science Publishers in enforcing your compliance with this License Agreement constitute a waiver of any of its rights.

3. You acknowledge that you have read this License Agreement, and agree to be bound by its terms and conditions. To the extent that any other terms and conditions presented on any website of Bentham Science Publishers conflict with, or are inconsistent with, the terms and conditions set out in this License Agreement, you acknowledge that the terms and conditions set out in this License Agreement shall prevail.

Bentham Science Publishers Pte. Ltd.
No. 9 Raffles Place
Office No. 26-01
Singapore 048619
Singapore
Email: subscriptions@benthamscience.net

CONTENTS

PREFACE

The quality of many environmental compartments (water, soil, air) is being compromised by the growing negative effects of human-generated pollution. Although physical and chemical-based remediation techniques are available, care must be taken due to their possible long-term environmental concerns. In-depth and current research involving microbiological processes essential to environmental protection is presented in Microbial Insights into Wastewater Treatment and Environmental Sustainability. Relevant processes include sequestering, mitigating, and managing water-based pollutants. The goal of this work is to enhance our knowledge of microbial populations responsible for pollution detoxification by focusing on their detection, observation, and avoidance. This intensive investigation will encourage the creation of novel strategies and ideas that advance the growing subject of environmental microbiology. There has been a noticeable trend in environmental cleanup technology in recent years towards biologically driven systems. These technologies provide a number of advantages over conventional techniques, including reduced maintenance, cost-effectiveness, reusability, energy efficiency, and efficient detoxification procedures.

Biologically driven technologies reduce secondary contamination hazards by reducing the volumes of residual by-products commonly generated by classical processes. Worldwide, environmental restrictions are becoming more stringent, thus increasing the demand for sustainable technologies and accelerating the adoption of biologically-based solutions. Such technologies are preferred over alternatives as they are more closely aligned with environmental safety, regulatory compliance, and sustainable development goals. This book emphasizes the importance of understanding bio-based technologies in order to manage modern global pollution properly. This work presents microbial sequestration of pollutants, including micropollutants, heavy metals, xenobiotics, and pollutants derived from petroleum. Great detail on co-metabolism, nutrient recycling, water treatment, energy production, and waste management are provided. Explored are recent developments in the fields of geomicrobiology, aeromicrobiology, biocontrol, plant-microbe interactions, and microbial energetics.

The book highlights emerging technologies for environmental management, including DNA microarrays, metagenomics, proteomics, green nanotechnology, and biosensor-based techniques. Alongside advances in hazard assessment and environmental monitoring, environmentally benign technologies such as waste valorization, biomining, biosolids utilization, and microbial metabolites are examined. By learning more about fundamental microbiology, readers will be able to comprehend biochemical processes in bioremediation and biocontrol technologies on a deeper level. From a scientific perspective, the book addresses important elements and field application issues while conducting a thorough evaluation of prospective and existing biotechnology techniques. This compilation provides a thorough overview of cutting-edge environmental microbiology technology, highlighting innovative green avenues to handle a range of environmental contamination issues successfully.

Harshita Jain
Amity Institute of Environmental Sciences
Amity University, Noida, 201303
India

&

Maulin P Shah
Department of Research Impact and Outcome
Research and Development Cell
Lovely Professional University, Phagwara, Punjab, India

List of Contributors

Aditi Bagade Department of Chemistry, Division of Biochemistry, Savitribai Phule Pune University (Formerly University of Pune), Pune, 411007, Maharashtra, India

Anamika Shrivastava Amity Institute of Environmental Sciences, Amity University, Noida, Gautam Budh Nagar, Uttar Pradesh, 201313, India

Anish Kumar Sharma School of Sciences, P P Savani University, Surat, Gujarat, India

Divya Bharti Amity Institute of Environmental Sciences, Amity University, Noida, Gautam Budh Nagar, Uttar Pradesh, 201313, India

Geetansh Sharma School of Bioengineering and Food Technology, Shoolini University, Solan, Himachal Pradesh, 173229, India

Harshita Jain Amity Institute of Environmental Sciences, Amity University, Noida, Gautam Budh Nagar, Uttar Pradesh, 201313, India

Jaya Dayal Department of Microbiology, S.S. Jain Subodh P.G. College, Jaipur, India

Kisan Kodam Department of Chemistry, Division of Biochemistry, Savitribai Phule Pune University (Formerly University of Pune), Pune, 411007, Maharashtra, India

Kinzey M. Abohussein Biotechnology Program, New Programs Administration, Faculty of Agriculture, Ain Shams University, Cairo, Egypt

Maya Kumari Amity School of Natural Resources and Sustainable Development, Amity University Uttar Pradesh, Noida, Uttar Pradesh, India

Mayank Chaudhary Ladakh Ecological Development Group (LEDeG), Leh, 194101, India

Maria A. Farag Biotechnology Program, New Programs Administration, Faculty of Agriculture, Ain Shams University, Cairo, Egypt

Maulin P. Shah Department of Research Impact and Outcome Research and Development Cell, Lovely Professional University, Phagwara, Punjab, India

Mohammad H. El-Zmrany Biotechnology Program, New Programs Administration, Faculty of Agriculture, Ain Shams University, Cairo, Egypt

Muhammad Aslam Khan Department of Biological Sciences, Faculty of Sciences, International Islamic University (IIU), Islamabad, Pakistan

Mohamed Ebrahim Department of Plant Pathology, Faculty of Agriculture, Ain Shams University, Cairo, Egypt

Nilanjan Chakraborty Scottish Church College, Kolkata, 700006, India

Niketa Bhati Amity Institute of Environmental Sciences, Amity University, Noida, Gautam Budh Nagar, Uttar Pradesh, 201313, India

Paurabi Das Crop Production and Protection Division, CSIR- Central Institute of Medicinal and Aromatic Plants, Lucknow, 226015, India

Priyanka Chauhan School of Sciences, P P Savani University, Surat, Gujarat, India

Renu Dhupper Amity Institute of Environmental Sciences, Amity University, Noida, Gautam Budh Nagar, Uttar Pradesh, 201313, India

Shanvi Rana School of Bioengineering and Food Technology, Shoolini University, Solan, Himachal Pradesh, 173229, India

Samah H. Abu-Hussien — Department of Agricultural Microbiology, Faculty of Agriculture, Ain Shams University, Cairo, Egypt

Shivali Pal — Amity Institute of Environmental Sciences, Amity University, Noida, Gautam Budh Nagar, Uttar Pradesh, 201313, India

Savita Tapase — Department of Chemistry, Division of Biochemistry, Savitribai Phule Pune University (Formerly University of Pune), Pune, 411007, Maharashtra, India

Sougata Ghosh — Department of Physics, Faculty of Science, Kasetsart University, Bangkok, 10900, Thailand
Department of Microbiology, School of Science, RK University, Rajkot, 360020, Gujarat, India

Vidiksha Singla — School of Bioengineering and Food Technology, Shoolini University, Solan, Himachal Pradesh, 173229, India

Vinod Nandre — Department of Chemistry, Division of Biochemistry, Savitribai Phule Pune University (Formerly University of Pune), Pune, 411007, Maharashtra, India

Vrushali Desai — School of Sciences, P P Savani University, Surat, Gujarat, India

Ziad Samy — Biotechnology Program, New Programs Administration, Faculty of Agriculture, Ain Shams University, Cairo, Egypt

CHAPTER 1

Introduction to Environmental Pollution and Microbial Technologies

Harshita Jain[1,*], Maulin P. Shah[2] and Renu Dhupper[1]

[1] *Amity Institute of Environmental Sciences, Amity University, Noida, Gautam Budh Nagar, Uttar Pradesh-201313, India*

[2] *Department of Research Impact and Outcome Research and Development Cell Lovely Professional University, Phagwara, Punjab, India*

Abstract: Environmental pollution poses intricate problems to ecosystems, adversely affecting air, water, and soil quality due to toxins like heavy metals, organic pollutants, micropollutants, and petroleum derivatives. This chapter provides a comprehensive introduction to these significant pollutants, emphasising their origins, durability, and environmental effects. This chapter examines the essential function of bacteria in natural environmental processes, emphasising their capacity to transform, decompose, or immobilise toxic chemicals. Microorganisms, through their varied metabolic pathways, facilitate the detoxification of contaminants by processes including biodegradation, biosorption, and bioaccumulation. This chapter provides an overview of contemporary and developing microbial technologies that utilise these natural processes to mitigate pollution. Significant progress in microbial-based pollution management, including bioremediation, bioaugmentation, and the application of microbial consortia, is examined, along with cutting-edge technologies such as metagenomics, proteomics, and microbial biosensors. This chapter combines insights on microbial activity with technical advancements to investigate sustainable, bio-based remedies for environmental degradation.

Keywords: Bioremediation, Bioaccumulation, Environmental pollution, Microbial biotransformation, Microbial degradation.

INTRODUCTION

Environmental pollutants arise from several sources, including both natural and anthropogenic causes. Anthropogenic pollutants, such as heavy metals, Persistent Organic Pollutants (POPs), micropollutants, and petroleum-derived contaminants, are prevalent in industrial, agricultural, and urban waste streams [1]. Heavy metals, such as lead, cadmium, and mercury, predominantly infiltrate the

---

* **Corresponding author Harshita Jain:** Amity Institute of Environmental Sciences, Amity University, Noida, Gautam Budh Nagar, Uttar Pradesh-201313, India; E-mail: hjain@amity.edu

environment *via* industrial operations, including mining, metal processing, and waste incineration [2]. POP, including polychlorinated biphenyls (PCBs) and pesticides, survive in ecosystems owing to their resistance to biodegradation and are frequently disseminated *via* air, aquatic, and terrestrial pathways. Micropollutants, encompassing pharmaceuticals, personal care items, and endocrine-disrupting substances, often stem from wastewater discharge and are challenging to eliminate using traditional treatment procedures [3]. Petroleum-derived contaminants, including hydrocarbons from oil spills, constitute a significant environmental issue owing to their toxicity and potential for bioaccumulation. These contaminants, derived from diverse industrial and agricultural activities, present considerable threats to ecological integrity and biodiversity [4].

The longevity of pollutants in ecosystems substantially depends on their chemical makeup, ambient conditions, and interactions with biological species. Pollutants, including POPs and heavy metals, are recognised for their enduring stability, frequently remaining underground in soil and aquatic systems for decades with minimal degradation [5]. These enduring pollutants bioaccumulate in organisms and biomagnify along the food chain, resulting in detrimental ecological and health effects. Heavy metals disturb enzymatic activities in both aquatic and terrestrial organisms, compromising metabolic functions and adversely affecting reproductive health. POPs, owing to their lipophilic characteristics, accumulate in adipose tissues and induce toxicological consequences in both wildlife and humans, encompassing carcinogenicity and endocrine disruption [6]. Micropollutants, even in minimal amounts, have demonstrated the capacity to impact aquatic organisms by interfering with hormonal systems and fostering antibiotic resistance. The pervasive and enduring presence of these contaminants destabilises ecosystems, diminishes biodiversity, and hinders vital ecosystem services, including water purification, soil fertility, and carbon sequestration [7].

Mitigating environmental pollution presents intricate issues, chiefly owing to the varied chemical characteristics of contaminants and their interactions with environmental matrices [8]. Traditional physical and chemical treatment approaches, although efficient in specific situations, frequently produce secondary contaminants or residues that necessitate additional control. Moreover, conventional methods may be inadequate in addressing low-concentration contaminants, such as micropollutants, that endure in wastewater discharges [7]. Formulating cost-effective, efficient, and sustainable pollution management solutions is further limited by varying global regulatory standards and the necessity to reconcile economic and environmental concerns. The complex dynamics of pollutant bioaccumulation and biomagnification further complicate risk assessment, rendering long-term effects on ecosystems and human health challenging to anticipate. Emerging biologically-driven technologies are promising but necessitate strong frameworks for large-scale implementation and thorough safety assessments to mitigate any ecological disturbances [9]. The incorporation of microbial technology presents a means for sustainable pollution management; however, these methods require optimisation for diverse pollutants and environmental conditions to guarantee effectiveness and adherence to regulations.

ROLE OF MICROBES IN ENVIRONMENTAL PROCESSES

Microorganisms are essential in regulating environmental processes, especially for contaminant reduction and ecosystem vitality. They participate in multiple biogeochemical cycles, including carbon, nitrogen, sulphur, and phosphorus cycling, thereby aiding in nutrient recycling, detoxification, and ecosystem stabilisation. Microbes catalyse the decomposition and conversion of organic matter, and their metabolic adaptability allows them to thrive in many environmental situations, rendering them crucial for preserving ecosystem integrity and fostering sustainability [10]. In contaminated settings, bacteria serve as natural bioremediation agents by metabolising dangerous compounds, detoxifying contaminants, and restoring ecological equilibrium. Their function in pollution control beyond mere degradation, encompassing intricate metabolic transformations that affect the disposition of pollutants in soils, water, and sediments (Table **1**).

Table 1. Role of microbes in environmental processes and pollution mitigation [11].

Role	Microbial Mechanism	Key Microbial Groups Involved	Implications for Ecosystem Health	Examples of Pollutants	Environmental Impact
Nutrient Cycling	Microbes play a key role in cycling essential elements like carbon, nitrogen, sulfur, and phosphorus.	Bacteria, fungi, archaea, and algae.	Nutrient cycling is critical for ecosystem productivity and maintaining biodiversity.	Organic matter, ammonia, nitrates, and phosphates.	Essential for soil fertility, water quality, and ecosystem stability.
Biodegradation of Organic Matter	Degradation of complex organic compounds into simpler molecules, often leading to mineralization.	Bacteria (*e.g.*, *Pseudomonas*, *Bacillus*) and fungi (*e.g.*, *Trichoderma*).	Prevents the accumulation of waste materials, enhancing soil and water quality.	Petroleum hydrocarbons, plastics, and pesticides.	Reduces pollutant load, preventing the contamination of water bodies and soil.

(Table 1) cont.....

Role	Microbial Mechanism	Key Microbial Groups Involved	Implications for Ecosystem Health	Examples of Pollutants	Environmental Impact
Pollutant Detoxification	Transformation of harmful substances into less toxic forms through enzymatic reactions.	*Bacillus, Pseudomonas, Desulfovibrio,* and fungi.	Detoxification minimizes the toxic effects of contaminants on organisms and ecosystems.	Heavy metals, xenobiotics, and POPs.	Reduces toxicity, mitigating bioaccumulation and ecosystem degradation.
Bioremediation (Natural)	Microorganisms metabolize and transform pollutants to restore environmental balance.	Bacteria, fungi, algae, and actinomycetes.	Natural remediation through microbial activity helps in the cleanup of contaminated sites.	Petroleum, solvents, and pharmaceuticals.	Restores contaminated environments, ensuring ecosystem services like water purification.
Co-metabolism	Some microbes degrade pollutants as a secondary process while metabolizing other compounds.	*Pseudomonas, Sphingomonas,* and *Rhodococcus.*	Enhances pollutant breakdown by providing additional metabolic pathways.	Chlorinated compounds and hydrocarbons.	Facilitates the removal of pollutants that are otherwise resistant to degradation.
Symbiosis and Plant-Microbe Interactions	Symbiotic relationships between microbes and plants enhance pollutant uptake and degradation.	Rhizobacteria, mycorrhizal fungi, *Bradyrhizobium, Azospirillum.*	Enhances plant growth and pollutant degradation in contaminated soils or waters.	Heavy metals and hydrocarbons.	Improves soil quality, promotes phytoremediation, and restores ecosystem health.
Heavy Metal Remediation	Microbial reduction or precipitation of metals into less toxic or insoluble forms.	*Geobacter, Shewanella, Bacillus,* and *Pseudomonas.*	Prevents metal toxicity in ecosystems and reduces bioaccumulation.	Lead, mercury, chromium, and cadmium.	Reduces environmental toxicity, aiding in the restoration of contaminated water and soils.
Phytoremediation Enhancement	Microbes in plant root zones enhance the degradation of pollutants through plant-microbe interactions.	*Mycorrhizal fungi, Pseudomonas,* and *Bacillus.*	Strengthens phytoremediation by aiding in pollutant uptake and detoxification.	Pesticides, heavy metals, and petroleum products.	Enhances soil fertility and reduces contamination, improving overall ecosystem health.

(Table 1) cont.....

Role	Microbial Mechanism	Key Microbial Groups Involved	Implications for Ecosystem Health	Examples of Pollutants	Environmental Impact
Gaseous Pollutant Removal	Microbes break down gaseous pollutants like methane and nitrogen oxides through specialized pathways.	Methanotrophs, *Nitrosomonas*, *Nitrobacter*, and denitrifiers.	Reduces atmospheric pollutants, improving air quality and mitigating climate change.	Methane, nitrogen oxides, and sulfur dioxide.	Contributes to global climate regulation by reducing greenhouse gas emissions.
Wastewater Treatment	Microbial consortia in activated sludge systems degrade organic waste, nutrients, and contaminants in wastewater.	*Nitrosomonas*, *Pseudomonas*, *Acinetobacter*, and *Flavobacterium* .	Critical for maintaining water quality and preventing eutrophication in aquatic ecosystems.	Organic pollutants, nitrogen, and phosphorus.	Reduces pollution load in water bodies, preventing eutrophication and improving water quality.
Microbial Electrochemical Systems (MES)	Microbes in electrochemical systems reduce pollutants *via* electron transfer reactions in bioreactors.	*Geobacter*, *Shewanella*, and *Desulfovibrio*.	Offers an innovative, energy-efficient solution to treat pollutants in wastewater.	Heavy metals, organic compounds, and dyes.	Provides an energy-efficient approach to wastewater treatment and pollutant removal.

Microbial Diversity and Metabolic Pathways

Microbial diversity includes a variety of species, such as bacteria, fungi, archaea, and algae, each with specialised metabolic pathways that allow them to interact with different environmental contaminants [11]. Diversity is essential in contaminated environments, as many microbial species have unique enzyme systems that may degrade or change a wide range of pollutants. Certain bacteria employ aromatic hydrocarbons as carbon sources through dioxygenase enzymes, whereas others decompose complex chemical molecules using peroxidases or laccases [12]. The metabolic versatility of microorganisms enables them to utilise various electron acceptors (*e.g.*, oxygen, nitrate, sulphate) and energy sources (*e.g.*, organic substrates, inorganic substances), hence promoting pollutant transformation in both aerobic and anaerobic environments. These metabolic pathways are essential for pollutant detoxification and for sustaining microbial community dynamics, which are crucial for ecosystem health and resilience [13].

Mechanisms of Microbial Degradation, Biotransformation, and Bioaccumulation

Microbial degradation, biotransformation, and bioaccumulation are essential processes by which bacteria alleviate the effects of environmental contaminants. Microbial degradation denotes the decomposition of contaminants by microbial enzymes, converting them into less dangerous or non-toxic compounds. Bacteria such as *Pseudomonas* and *Rhodococcus* are recognised for their ability to break down petroleum hydrocarbons *via* oxidative mechanisms [14]. Biotransformation entails the conversion of contaminants into various chemical forms, which may exhibit increased or decreased toxicity depending on microbial metabolic activity. An exemplary case is the microbial conversion of dangerous heavy metals, such as mercury, wherein bacteria like *Desulfovibrio* and *Geobacter* change mercury ions into elemental mercury, which, while volatile, is less accessible [13]. Bioaccumulation refers to the absorption and retention of contaminants by microorganisms, potentially resulting in detoxification or, conversely, secondary pollution if the microorganisms themselves become a source of pollutants. Certain fungi and bacteria sequester heavy metals within their cellular structures, offering an efficient method for detoxifying contaminated environments, although they also present a potential hazard if ingested by higher trophic levels [14].

Microbial Interactions in Polluted Environments: Symbiosis, Competition, and Co-metabolism

In contaminated environments, microbial interactions can profoundly affect pollutant breakdown and ecosystem functionality. Symbiotic connections, such as those between plants and nitrogen-fixing bacteria or between fungi and algae in lichen, can augment pollution detoxification. Rhizoremediation, the process whereby plant roots release chemicals that enhance microbial activity, is an effective method for remediating soils contaminated with heavy metals or organic contaminants [15]. Competition among microbial species for scarce resources, including nutrients and oxygen, can determine microbial community topologies and affect the efficacy of pollutant breakdown. Microbial competition can impede pollutant breakdown when less efficient degraders surpass more specialised organisms. Co-metabolism, in which one microbe destroys a pollutant while another supplies an essential co-factor or metabolic product, is a prevalent relationship that facilitates pollutant clearance. An instance of co-metabolism is the conversion of chlorinated solvents by specific bacteria that employ simple organic molecules as carbon sources, wherein the breakdown of the solvent is a secondary process occurring concurrently with the main metabolism of other substances [16].

Case Studies: Microbial Responses to Specific Pollutants

Several case studies highlight the diverse and critical roles that microbes play in responding to specific environmental pollutants. One such example is the bioremediation of petroleum hydrocarbons in marine environments, where bacteria such as *Alcanivorax borkumensis* have been shown to degrade aliphatic and aromatic hydrocarbons rapidly. These microbes utilize oil as a primary carbon source, significantly reducing the environmental impact of oil spills. Another case study involves the treatment of heavy metal contamination, where *Shewanella* and *Geobacter* species have been employed to reduce toxic metals such as chromium and uranium to less toxic forms, often through microbial electrochemical systems [17]. In agricultural soils contaminated with pesticides, microbes like *Pseudomonas putida* have been shown to degrade organophosphates and other chemical residues, reducing the long-term persistence of these pollutants. In wastewater treatment facilities, microbial consortia have been employed to decompose intricate organic contaminants, such as pharmaceuticals and endocrine-disrupting compounds, *via* advanced biological treatment methods. These case studies highlight the versatility of microbial communities and their capacity to remediate various pollutants in different contexts, illustrating their potential for extensive application in environmental management and pollution mitigation [18].

CURRENT MICROBIAL TECHNOLOGIES IN POLLUTION CONTROL

Microbial technologies have arisen as efficient instruments for addressing environmental pollution, especially in reducing pollutants in water, soil, and air. These methods utilise the metabolic variety of microbes to decompose, detoxify, and convert contaminants into less dangerous or non-toxic compounds. In the realm of pollution control, various advanced microbial approaches are emerging, such as bioremediation, bioaugmentation, biostimulation, and the application of microbial consortia.

Bioremediation: Principles and Applications

Bioremediation is the process by which microorganisms decompose or convert dangerous compounds into less toxic forms, thereby restoring environmental equilibrium [19]. This natural process utilises the capacity of microbes to metabolise organic contaminants, including hydrocarbons, pesticides, and heavy metals, into innocuous by-products such as carbon dioxide, water, and biomass. Bioremediation can take place *in situ*, where pollutants are addressed at the contaminated location, or *ex-situ* when contaminated materials are extracted for treatment in regulated settings. Nutrient availability, temperature, pH, and the presence of oxygen or electron acceptors substantially affect microbial activity in bioremediation. In recent decades, bioremediation has effectively addressed contaminated soils, groundwater, and industrial wastewater, achieving significant success in petroleum spill remediation and the treatment of hazardous metals and synthetic compounds [11] (Fig. **1**).

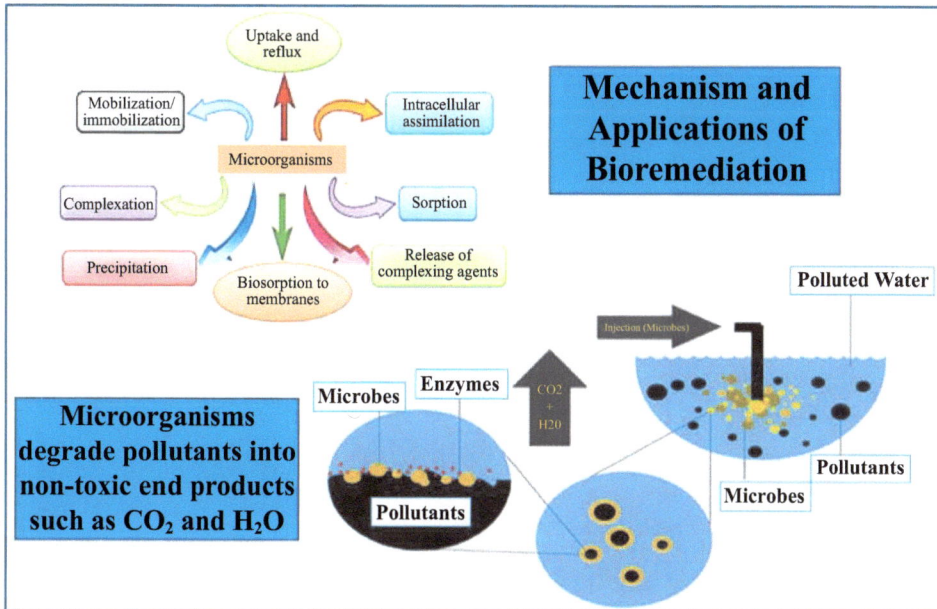

Fig. (1). Mechanism and applications of bioremediation [19, 20].

Bioaugmentation and Biostimulation Techniques

Bioaugmentation and biostimulation are two synergistic methods that improve the efficacy of bioremediation (Fig. **2**). Bioaugmentation entails the introduction of certain strains or consortia of microorganisms to a contaminated location to enhance the microbial population capable of decomposing specific contaminants. The introduced bacteria may have specialised enzymes or metabolic pathways that allow them to break down contaminants more effectively than indigenous microbial populations [21]. Conversely, biostimulation is a method that entails altering the environment to enhance the proliferation and function of native bacteria already existing in the contaminated region. This may involve the incorporation of nutrients, oxygen, or other growth factors that augment microbial metabolic activities. Both procedures have been employed successfully to remediate organic contaminants such as petroleum hydrocarbons, chlorinated solvents, and agricultural pesticides. Bioaugmentation is utilised when indigenous bacteria are inadequate in quantity or deficient in essential metabolic pathways, whereas biostimulation is predominantly applied to enhance the activity of naturally occurring microbes [22].

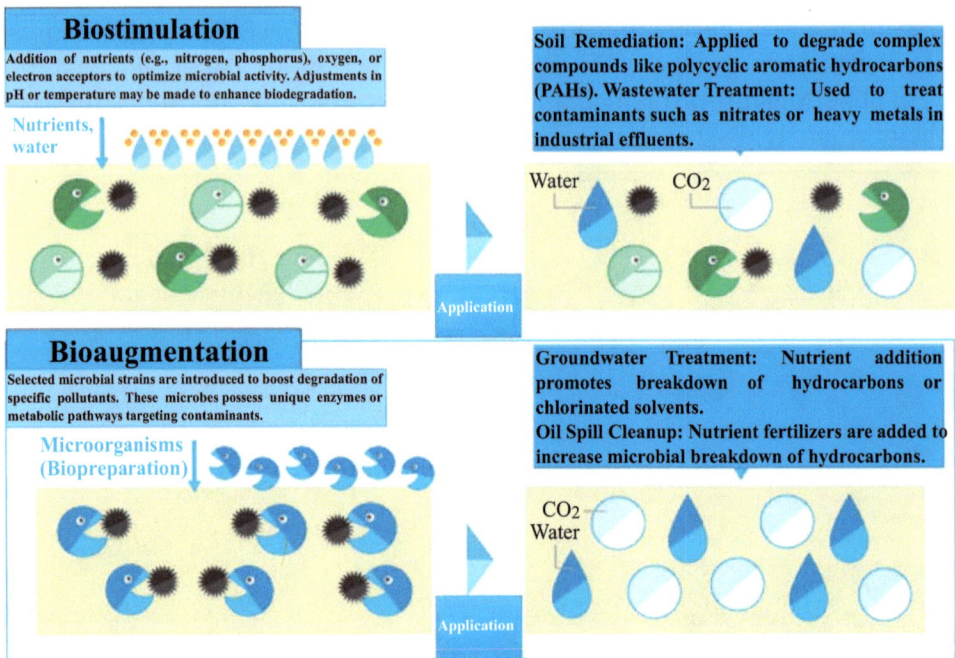

Biostimulation

Addition of nutrients (e.g., nitrogen, phosphorus), oxygen, or electron acceptors to optimize microbial activity. Adjustments in pH or temperature may be made to enhance biodegradation.

Nutrients, water

Soil Remediation: Applied to degrade complex compounds like polycyclic aromatic hydrocarbons (PAHs). **Wastewater Treatment:** Used to treat contaminants such as nitrates or heavy metals in industrial effluents.

Water　　　CO_2

Application

Bioaugmentation

Selected microbial strains are introduced to boost degradation of specific pollutants. These microbes possess unique enzymes or metabolic pathways targeting contaminants.

Microorganisms (Biopreparation)

Groundwater Treatment: Nutrient addition promotes breakdown of hydrocarbons or chlorinated solvents.
Oil Spill Cleanup: Nutrient fertilizers are added to increase microbial breakdown of hydrocarbons.

CO_2
Water

Application

Fig. (2). Mechanisms of bioaugmentation and biostimulation techniques in bioremediation [21].

The Use of Microbial Consortia for Enhanced Pollutant Degradation

The application of microbial consortia, assemblages of diverse microbial species collaborating synergistically, has garnered significant interest in pollution management (Fig. **3**). A consortium strategy leverages the metabolic diversity of microorganisms to digest a wider array of contaminants and improve the efficacy of bioremediation processes. The collaborative interactions among many microbial species facilitate the progressive breakdown of complex molecules, including persistent organic contaminants that may defy destruction by individual species [23]. Furthermore, microbial consortia may demonstrate synergistic effects, wherein one species generates compounds that enhance the growth or enzymatic activity of another, thereby elevating the total rate of pollutant breakdown. Consortia may be either naturally occurring or artificially designed to address certain pollution issues. Engineered microbial consortia have been employed in the treatment of industrial effluents and wastewater, as the synergistic capabilities of several microorganisms yield superior detoxification compared to monoculture methods. The utilisation of microbial consortia mitigates microbial competition, as various microorganisms may target disparate contaminants or function under unique environmental conditions [24].

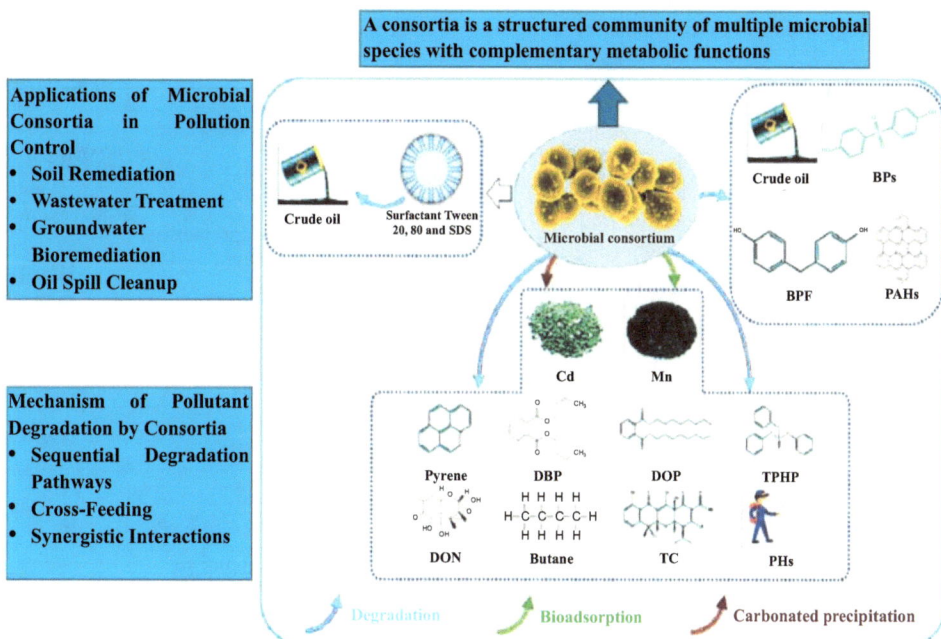

Fig. (3). Role of microbial consortia in pollutant degradation [23].

Advanced microbial technologies are progressively incorporated into environmental management methods owing to their efficiency, cost-effectiveness, and little environmental impact. Their ongoing advancement and implementation are poised to significantly contribute to alleviating the escalating global issues related to pollution and safeguarding ecosystem sustainability.

EMERGING MICROBIAL TECHNOLOGIES AND INNOVATIONS

In recent years, microbial technologies have progressed markedly, propelled by advancements in molecular biology, bioinformatics, and other leading-edge disciplines. These novel technologies improve the capacity to identify, assess, and mitigate environmental contaminants with more efficacy and efficiency. Progress in metagenomics, proteomics, biosensors, green nanotechnology, and the incorporation of Artificial Intelligence (AI) and Machine Learning (ML) into microbial ecology are transforming the field of environmental pollution management. These technologies enhance the efficacy of microbial-based remediation systems while facilitating real-time monitoring, accurate pollution identification, and sustainable solutions for environmental restoration (Table **2**).

Table 2. Emerging microbial technologies and innovations [25].

Technology/Innovation	Description	Applications	Advantages	Challenges
Advances in Metagenomics, Proteomics, and Microbiome Analysis	Metagenomics involves analyzing the genetic material from environmental samples to identify microbial communities, while proteomics explores microbial proteins and their functions.	Understanding microbial diversity and functions in polluted environments, identifying novel degradative pathways.	Allows comprehensive insights into microbial communities and their functional potential for pollution control.	Data complexity and high costs, interpretation of vast datasets, and need for specialized expertise.
Biosensors for Real-Time Pollution Monitoring	Biosensors are devices that use microorganisms or microbial components to detect pollutants in real time.	Monitoring environmental pollutants (*e.g.*, heavy metals, organic compounds, pathogens) in water, soil, and air.	Provides rapid, cost-effective, and accurate detection of pollutants at low concentrations.	Limited sensitivity for some pollutants, environmental interference, and need for calibration.
Green Nanotechnology and Microbial Remediation	Combines nanotechnology and microbiology for pollutant degradation. Nanoparticles (*e.g.*, metal nanoparticles) interact with microbes to enhance pollutant breakdown and removal.	Remediation of heavy metals, organic pollutants, and petroleum products in contaminated environments.	Environment-friendly, scalable, and enhances microbial degradation processes.	Potential toxicity of nanoparticles and the complexity of controlling their interaction with microbes.
Artificial Intelligence (AI) and Machine Learning (ML) in Microbial Ecology	AI and ML are employed to predict microbial behavior, optimize bioremediation processes, and identify novel biocatalysts.	Predicting microbial responses to pollutants, optimizing bioremediation strategies, and ecosystem modeling.	Facilitates data-driven decisions, improves efficiency, and personalizes remediation approaches.	Requires large datasets, training models for specific pollutants, and data quality issues.

Advances in Metagenomics, Proteomics, and Microbiome Analysis

Metagenomics and proteomics have transformed our comprehension of microbial populations and their function in environmental processes, such as pollution management. Metagenomics facilitates the sequencing and analysis of genetic material directly from environmental samples, enabling researchers to reveal the extensive diversity of microorganisms in diverse habitats, including those participating in pollutant degradation. This approach offers insights into the functional capabilities of microbial communities, encompassing the discovery of genes and pathways associated with the breakdown of particular pollutants. Metagenomics facilitates an in-depth examination of the microbiome, revealing the interactions and functional roles of bacteria in contaminated habitats [26].

Proteomics enhances metagenomics by concentrating on the protein expression profiles of microbes, thereby offering a comprehensive insight into microbial metabolic activities. This facilitates the discovery of enzymes and metabolic pathways crucial for pollution degradation, hence enhancing bioremediation procedures. Integrating metagenomic and proteome data enables researchers to obtain a comprehensive understanding of microbial reactions to contaminants, facilitating the discovery of innovative bioremediation technologies and biomarkers for environmental monitoring. These advancements are especially beneficial for creating customised microbial treatments targeting specific environmental pollutants, including heavy metals, hydrocarbons, and agrochemicals [26].

Biosensors for Real-Time Pollution Monitoring

Biosensors serve as a promising instrument for real-time pollution surveillance, facilitating continuous, on-site identification of contaminants in air, water, and soil. These devices operate on the premise of utilising microbes, enzymes, or other bio-receptors to identify specific pollutants. Microbial biosensors are specifically designed to detect environmental pollutants, including heavy metals, organic chemicals, or pathogens, by producing quantifiable signals (*e.g.*, fluorescence, colorimetric changes, or electrical responses). The incorporation of biosensors into pollution monitoring systems facilitates the swift identification of environmental threats, delivering prompt data for informed decision-making and mitigation approaches [27].

A primary advantage of microbial biosensors is their specificity and sensitivity. They can be engineered to identify minimal quantities of contaminants within intricate environmental matrices, yielding high-resolution data for monitoring pollutant levels over time. Biosensors provide benefits such as mobility, cost-effectiveness, and user-friendliness, rendering them suitable for field applications in environmental monitoring. Microbial biosensors provide the early identification of pollutants, thereby averting environmental degradation, assuring adherence to regulations, and improving the efficacy of pollution control measures.

Green Nanotechnology and Microbial Remediation

Green nanotechnology encompasses the utilisation of nanomaterials and nanostructures in eco-friendly processes, including the remediation of environmental contaminants. Microbial remediation, in conjunction with nanotechnology, has considerable potential for resolving intricate pollution challenges. Nanoscale materials, including nanoparticles, can engage with microbial systems to improve pollutant breakdown, increase the bioavailability of pollutants, and aid in the elimination of dangerous compounds from the environment. Nanoparticles can facilitate the degradation of enduring contaminants such as heavy metals or organic chemicals, or they might promote microbial proliferation by increasing the surface area available for microbial adhesion [28].

The amalgamation of green nanotechnology with microbial processes presents numerous benefits. Nanomaterials enhance bioremediation efficacy by augmenting the surface area for microbial interactions, promoting electron transfer during microbial metabolism, and allowing for the precise delivery of nutrients or microbial inoculants. Moreover, green nanomaterials, including biocompatible nanoparticles or bio-based nanomaterials, provide negligible environmental hazards relative to conventional nanomaterials, rendering them a compelling option for sustainable remediation approaches. The integration of microbial remediation with green nanotechnology is developing as an effective method for addressing environmental contamination, especially in intricate, polluted environments [23, 28].

Artificial Intelligence and Machine Learning in Microbial Ecology

The utilisation of AI and ML in microbial ecology is revolutionising the research and management of microbial communities in contaminated environments. AI and ML systems can scrutinise extensive datasets derived from metagenomics, proteomics, and environmental sensors to discern trends, forecast microbial behaviour, and enhance pollution management tactics. These technologies enable the creation of prediction models that can anticipate the behaviour of contaminants based on microbial community dynamics and environmental factors [29].

In pollution control, AI and ML are very effective in determining the optimal microbial species or consortia for specific toxins and environmental conditions.

Through the analysis of extensive ecological data, these technologies can propose optimised strategies for bioremediation and inform the development of microbial treatments specifically designed for distinct contaminants. Moreover, AI and ML can augment the efficacy of microbial biosensors by enhancing their sensitivity and specificity, resulting in more precise and dependable real-time pollution monitoring. The incorporation of AI and ML into microbial ecology has significant opportunities for enhancing pollution management tactics, offering intelligent, data-driven solutions for environmental sustainability [30 - 32].

Together, these emerging microbial technologies represent a powerful suite of tools for advancing pollution control and environmental sustainability. By integrating innovations in genomics, nanotechnology, biosensing, and data analysis, microbial-based solutions are becoming more efficient, scalable, and sustainable, offering promising alternatives to traditional environmental remediation methods.

FUTURE PERSPECTIVES AND CHALLENGES

The future of microbial technologies in pollution control holds immense potential, yet challenges remain in scaling these technologies, ensuring regulatory compliance, and integrating them into broader sustainability frameworks. As environmental pollution continues to escalate globally, the demand for effective, eco-friendly remediation methods is growing. Microbial technologies, particularly those focused on bioremediation and pollutant degradation, are emerging as promising alternatives to traditional chemical and physical methods. However, to achieve their full potential, certain hurdles need to be addressed, including scaling for industrial applications, navigating regulatory and safety concerns, and integrating these technologies into circular economy models. Furthermore, there are numerous opportunities for innovation and research, especially in developing more efficient, cost-effective, and sustainable microbial solutions for environmental management.

Scaling Microbial Technologies for Industrial Applications

One of the major challenges facing the widespread adoption of microbial technologies in pollution control is the scaling of laboratory-based processes to industrial and field levels. While microbial bioremediation has shown significant promise in laboratory settings, transferring these processes to large-scale operations involves several complexities. These include maintaining optimal environmental conditions for microbial activity, ensuring sufficient microbial population density, and dealing with the variability of contaminated sites [33].

In industrial applications, the efficiency of microbial systems can be hindered by factors such as nutrient limitations, environmental stresses (*e.g.*, temperature, pH, salinity), and competition from indigenous microbial communities. To address these challenges, bioreactor designs and bioprocess optimization strategies must be developed to ensure consistent performance in real-world conditions. Additionally, the integration of synthetic biology and genetic engineering holds promise for enhancing the resilience and activity of microbial strains, enabling them to function more effectively in challenging industrial environments. Moreover, developing strategies for long-term microbial survival and reproduction in contaminated sites is essential for the sustained success of microbial-based remediation efforts [34].

Addressing Regulatory and Safety Concerns

As microbial technologies advance, particularly in the context of Genetically Modified Organisms (GMOs) and engineered microbial strains, regulatory and safety concerns become increasingly significant. Regulatory bodies must evaluate the safety and environmental impact of releasing genetically modified or engineered microorganisms into the environment. Potential risks include unintended horizontal gene transfer, disruption of local ecosystems, and the persistence of engineered microbes in non-target environments [35].

To address these concerns, it is critical to establish clear regulatory frameworks that balance the potential benefits of microbial technologies with environmental and health safety standards. These regulations should ensure that any microbial technologies deployed for pollution control are thoroughly tested for ecological compatibility, biodegradability, and non-toxicity. Furthermore, comprehensive risk assessments and monitoring protocols should be implemented to track the fate of microbial agents in the environment. Public acceptance of microbial-based remediation strategies is also an important factor, as concerns about the environmental release of GMOs may hinder widespread adoption. Transparent communication, scientific research, and public engagement will play a key role in addressing safety and regulatory challenges [36].

Integrating Microbial Technologies with Circular Economy Models

The concept of a circular economy, which aims to minimize waste and promote the reuse and recycling of materials, aligns well with the principles of sustainable environmental management. Microbial technologies have the potential to play a pivotal role in the circular economy by facilitating the recycling of nutrients, the degradation of waste products, and the production of valuable by-products from environmental contaminants. For example, microbes can convert organic waste into biogas or other forms of bioenergy, thereby contributing to renewable energy production. Similarly, microbial processes can recover valuable metals from contaminated waste streams through bioremediation and biosorption [37].

Integrating microbial technologies into circular economy models requires the development of closed-loop systems where microbial processes are harnessed for waste valorization, resource recovery, and pollution mitigation. This may include utilizing microbial consortia to treat wastewater, recover nutrients (*e.g.*, nitrogen, phosphorus), and produce bioproducts such as biofuels, bioplastics, and biochemicals. By incorporating microbial technologies into circular economy frameworks, industries can reduce their environmental footprint, minimize waste generation, and contribute to the sustainable use of natural resources. The key to success in this integration lies in optimizing microbial processes for large-scale applications while maintaining ecological balance and cost-effectiveness [38].

Opportunities for Research and Innovation in Microbial Pollution Control

Research and innovation will continue to drive the evolution of microbial technologies in pollution control, opening new avenues for addressing environmental challenges. Opportunities for innovation are particularly abundant in the areas of synthetic biology, microbial ecology, and environmental monitoring. Advances in genetic engineering and synthetic biology offer the potential to design microbes with enhanced pollutant degradation capabilities tailored to specific environmental conditions and contaminants. This could lead to the development of highly specialized microbial strains capable of breaking down a wide range of pollutants, from heavy metals to complex organic compounds [39].

Additionally, microbial ecology offers valuable insights into the interactions between microbes and pollutants, as well as between different microbial species in contaminated environments. Understanding these interactions can help optimize the use of microbial consortia, which may enhance the degradation of pollutants through synergistic metabolic pathways. Furthermore, advances in microbial monitoring technologies, such as biosensors and metagenomics, will enable real-time tracking of microbial populations and pollutant degradation processes, facilitating more effective management of polluted sites [40].

Another exciting area of research is the exploration of microbial systems for waste-to-resource technologies, where microbes are used not only for pollutant degradation but also for the recovery of valuable resources such as metals, rare earth elements, and biofuels. The development of efficient, cost-effective microbial technologies for resource recovery can transform waste treatment processes into productive, sustainable systems. Finally, collaboration between academia, industry, and regulatory bodies will be crucial for accelerating the commercialization and widespread adoption of microbial pollution control technologies. By fostering interdisciplinary research, developing partnerships, and ensuring that research findings are translated into practical solutions, microbial technologies can become a cornerstone of sustainable environmental management [36, 37, 39].

CONCLUSION

The introduction to environmental pollution and microbial technologies highlights the urgent need for innovative and sustainable solutions to combat the growing environmental challenges posed by pollutants. As human activities continue to release a wide array of contaminants into the environment, traditional remediation methods often fall short in terms of cost-effectiveness, efficiency, and sustainability. Microbial technologies, particularly those focused on bioremediation and pollutant degradation, present a promising alternative, offering natural and eco-friendly solutions to detoxify polluted ecosystems. Microorganisms, with their metabolic versatility, are crucial in regulating environmental processes, including nutrient cycling, pollutant detoxification, and ecosystem stabilization. The potential of microbial-based technologies lies not only in their ability to break down contaminants but also in their role in enhancing the resilience of ecosystems and promoting sustainability. Emerging technologies, such as the use of microbial consortia, synthetic biology, and advanced monitoring tools, offer even greater potential for tackling complex pollution challenges. However, to fully realize the benefits of microbial technologies, several challenges must be addressed, including scaling these technologies for industrial applications, ensuring regulatory compliance, and integrating them into broader circular economy models. As research progresses and innovations emerge, microbial-based solutions will play an increasingly significant role in pollution control, offering cost-effective, sustainable, and safe alternatives to conventional methods. The integration of microbial technologies into environmental management systems, coupled with ongoing research and development, promises to drive a new era of ecological restoration and sustainable development, making microbial solutions a cornerstone of the global effort to protect the environment and achieve long-term sustainability goals.

REFERENCES

[1] Li X, Shen X, Jiang W, Xi Y, Li S. Comprehensive review of emerging contaminants: detection technologies, environmental impact, and management strategies. Ecotoxicol Environ Saf 2024; 278: 116420.
[http://dx.doi.org/10.1016/j.ecoenv.2024.116420]

[2] Goutam Mukherjee A, Ramesh Wanjari U, Eladl MA, *et al*. Mixed contaminants: Occurrence, interactions, toxicity, detection, and remediation. Molecules 2022; 27(8): 2577.
[http://dx.doi.org/10.3390/molecules27082577] [PMID: 35458775]

[3]   N. Ngweme a, Dhafer Mohammed M. Al Salah b c , Amandine Laffite b Georgette, *et al.* Occurrence of organic micropollutants and human health risk assessment based on consumption of Amaranthus viridis, Kinshasa in the Democratic Republic of the Congo. Sci Total Environ 2021; 754: 142175.
[http://dx.doi.org/10.1016/j.scitotenv.2020.142175]

[4]   Filote C, Roşca M, Hlihor R, *et al.* Sustainable application of biosorption and bioaccumulation of persistent pollutants in wastewater treatment: Current practice. Processes (Basel) 2021; 9(10): 1696.
[http://dx.doi.org/10.3390/pr9101696]

[5]   Fang W, Leilei X, Kelvin S-Y L , *et al.* Emerging contaminants: a one health perspective. Innov 2024; 5(4): 100612.
[http://dx.doi.org/10.1016/j.xinn.2024.100612]

[6]   Laxmi K B, Prangya R, Farhana R , *et al.* Status of Micro/Nano-Plastic (MNP) Pollution in the Aquatic System. In: Gaur N, Sharma E, Nguyen TA, Bilal M, Melkania NP, Eds. Societal and environmental ramifications of plastic pollution. Hershey, PA, USA: IGI Global 2025; pp. 43-62.

[7]   Jain H. From pollution to progress: groundbreaking advances in clean technology unveiled. Innov Green Dev 2024; 3(2): 100143.
[http://dx.doi.org/10.1016/j.igd.2024.100143]

[8]   Trends in biological processes in iWastewater treatment. M. P. Shah, ed.: IOP Publishing 2024.
[http://dx.doi.org/10.1088/978-0-7503-5678-7]

[9]   Laxmi K B, Prangya R, Shubhansh B , *et al.* COVID-19 and the interplay with antibacterial drug resistance. In: Grewal AS, Dhingra AK, Nepali K, Deswal G, Srivastav AL, Eds. Frontiers in combating antibacterial resistance: Current perspectives and future horizons. Hershey, PA, USA: IGI Global 2024; pp. 246-73.
[http://dx.doi.org/10.4018/979-8-3693-4139-1.ch010]

[10]   Meng S, Peng T, Liu X, *et al.* Ecological role of bacteria involved in the biogeochemical cycles of mangroves based on functional genes detected through geoChip 5.0. MSphere 2022; 7(1): e00936-21.
[http://dx.doi.org/10.1128/msphere.00936-21] [PMID: 35019668]

[11]   Bilal M, Iqbal H M N. Microbial bioremediation as a robust process to mitigate pollutants of environmental concern. Case Stud Chem Environ Eng 2020; 2: 100011.
[http://dx.doi.org/10.1016/j.cscee.2020.100011]

[12]   Shah MP. Emerging innovative trends in the application of biological processes for industrial wastewater treatment. Elsevier 2024.

[13]   Peng Q, Jiang S, Chen J, *et al.* Unique microbial diversity and metabolic pathway features of fermented vegetables from hainan, China. Front Microbiol 2018; 9: 399.
[http://dx.doi.org/10.3389/fmicb.2018.00399] [PMID: 29559966]

[14]   Mahmoud DAR. Mechanism of microbial biodegradation: secrets of biodegradation. In: Ali GAM, Makhlouf ASH, Eds. Handbook of biodegradable materials. Cham: Springer International Publishing 2022; pp. 1-15.
[http://dx.doi.org/10.1007/978-3-030-83783-9_6-1]

[15]   Braga R M, Dourado M N, Araújo W L. Microbial interactions: ecology in a molecular perspective. Braz J Microbiol 2016; 47: 86-98.
[http://dx.doi.org/10.1016/j.bjm.2016.10.005]

[16]   Freilich S, *et al.* Competitive and cooperative metabolic interactions in bacterial communities. Nat Commun 2011; 2(1): 589.
[http://dx.doi.org/10.1038/ncomms1597]

[17]   Tedesco P, Balzano S, Coppola D, Esposito F P, de Pascale D, Denaro R. Bioremediation for the recovery of oil polluted marine environment, opportunities and challenges approaching the blue growth. Mar Pollut Bull 2024; 200: 116157.
[http://dx.doi.org/10.1016/j.marpolbul.2024.116157]

[18] Aziz Z S, Jazza S H, Dageem H N, Banoon S R, Balboul B A, Abdelzaher M A. Bacterial biodegradation of oil-contaminated soil for pollutant abatement contributing to achieve sustainable development goals: a comprehensive review. Results Eng 2024; 22: 102083.
[http://dx.doi.org/10.1016/j.rineng.2024.102083]

[19] Kuppan N, Padman M, Mahadeva M, Srinivasan S, Devarajan R. A comprehensive review of sustainable bioremediation techniques: eco friendly solutions for waste and pollution management. Waste Manag Bull 2024; 2(3): 154-71.
[http://dx.doi.org/10.1016/j.wmb.2024.07.005]

[20] Das S, Dash HR. 1 - microbial bioremediation: A potential tool for restoration of contaminated areas. In: Das S, Ed. Microbial Biodegradation and Bioremediation. Oxford: Elsevier 2014; pp. 1-21.
[http://dx.doi.org/10.1016/B978-0-12-800021-2.00001-7]

[21] Tyagi M, da Fonseca MMR, de Carvalho CCCR. Bioaugmentation and biostimulation strategies to improve the effectiveness of bioremediation processes. Biodegradation 2011; 22(2): 231-41.
[http://dx.doi.org/10.1007/s10532-010-9394-4] [PMID: 20680666]

[22] Myazin VA, Korneykova MV, Chaporgina AA, Fokina NV, Vasilyeva GK. The effectiveness of biostimulation, bioaugmentation and sorption-biological treatment of soil contaminated with petroleum products in the russian subarctic. Microorganisms 2021; 9(8): 1722.
[http://dx.doi.org/10.3390/microorganisms9081722] [PMID: 34442801]

[23] Zhang T, Zhang H. Microbial consortia are needed to degrade soil pollutants. Microorganisms 2022; 10(2): 261.
[http://dx.doi.org/10.3390/microorganisms10020261] [PMID: 35208716]

[24] Cao Z, Yan W, Ding M, Yuan Y. Construction of microbial consortia for microbial degradation of complex compounds. Review. 2022; 10.
[http://dx.doi.org/10.3389/fbioe.2022.1051233]

[25] Zhang L, *et al.* Advances in metagenomics and its application in environmental microorganisms. Review. 2021; 12.
[http://dx.doi.org/10.3389/fmicb.2021.766364]

[26] Dweh TJ, Pattnaik S, Sahoo JP. Assessing the impact of meta-genomic tools on current cutting-edge genome engineering and technology. Int J Biochem Mol Biol 2023; 14(4): 62-75.
[PMID: 37736390]

[27] González-Plaza JJ, Furlan C, Rijavec T, *et al.* Advances in experimental and computational methodologies for the study of microbial-surface interactions at different omics levels. Front Microbiol 2022; 13: 1006946.
[http://dx.doi.org/10.3389/fmicb.2022.1006946] [PMID: 36519168]

[28] Rathod S, Preetam S, Pandey C, Bera S P. Exploring synthesis and applications of green nanoparticles and the role of nanotechnology in wastewater treatment. Biotechnol Rep 2024; 41: e00830.
[http://dx.doi.org/10.1016/j.btre.2024.e00830]

[29] Ghannam R B, Techtmann S M. Machine learning applications in microbial ecology, human microbiome studies, and environmental monitoring. Comput Struct Biotechnol J 2021; 19: 1092-107.
[http://dx.doi.org/10.1016/j.csbj.2021.01.028]

[30] Tripathi P, Srivastava A, Dubey CK, Mishra V, Dwivedi S, Madeshiya AK. Chapter Two - Implementation of Artificial Intelligence (AI) and Machine Learning (ML) in microbiology. In: Srivastava A, Mishra V, Eds. Academic Press. Methods in Microbiology 2024; 55: pp. 29-41.

[31] Jain H. Data analytics enabled by the Internet of Things and artificial intelligence for the management of Earth's resources. In: Kumar D, Tewary T, Shekhar S, Eds. Data analytics and artificial intelligence for earth resource management. Elsevier 2025; pp. 19-36.
[http://dx.doi.org/10.1016/B978-0-443-23595-5.00002-4]

[32] Jain H, Dhupper R, Shrivastava A, Kumar D, Kumari M. Leveraging machine learning algorithms for

improved disaster preparedness and response through accurate weather pattern and natural disaster prediction. Review. 2023; 11.
[http://dx.doi.org/10.3389/fenvs.2023.1194918]

[33]   Jadhav D A, *et al*. Scalability of microbial electrochemical technologies: applications and challenges. Bioresour Technol 2022; 345: 126498.
[http://dx.doi.org/10.1016/j.biortech.2021.126498]

[34]   Zhong C. Industrial-scale production and applications of bacterial cellulose. Review. 2020; 8.
[http://dx.doi.org/10.3389/fbioe.2020.605374]

[35]   Karalis DT, Karalis T, Karalis S, Kleisiari AS. Genetically modified products, perspectives and challenges. Cureus 2020; 12(3): e7306.
[PMID: 32313747]

[36]   Lensch A, *et al*. Safety aspects of microorganisms deliberately released into the environment. EFB Bioecon J 2024; 4: 100061.
[http://dx.doi.org/10.1016/j.bioeco.2023.100061]

[37]   Glockow T, Kaster A-K, Rabe K S, Niemeyer C M. Sustainable agriculture: leveraging microorganisms for a circular economy. Appl Microbiol Biotechnol 2024; 108(1): 452.
[http://dx.doi.org/10.1007/s00253-024-13294-0]

[38]   Krüger A, Schäfers C, Busch P, Antranikian G. Digitalization in microbiology – paving the path to sustainable circular bioeconomy. New Biotechnol 2020; 59: 88-96.
[http://dx.doi.org/10.1016/j.nbt.2020.06.004]

[39]   Sharma P, Bano A, Singh S P, Dubey N K, Chandra R, Iqbal H M N. Recent advancements in microbial-assisted remediation strategies for toxic contaminants. Clean Chem Eng 2022; 2: 100020.
[http://dx.doi.org/10.1016/j.clce.2022.100020]

[40]   Li J, Tao Y, Li G, Feng C, Chen R, Hua M. Biological processes for pollution control: current research and emerging technologies 2020. Archaea 2021; 2021(1): 9852531.
[http://dx.doi.org/10.1155/2021/9852531]

CHAPTER 2

Microbial Ecology in Polluted Environments

Niketa Bhati[1,*], Harshita Jain[1] and **Renu Dhupper[1]**

[1] *Amity Institute of Environmental Sciences, Amity University, Noida, Gautam Budh Nagar, Uttar Pradesh-201313, India*

Abstract: Microorganisms are key to the ecological dynamics of polluted environments with respect to contaminant persistence and remediation. This chapter explores the various dimensions of microbial ecology in polluted environments, their resiliency, adaptive strategies, and ecological functions. The ecology of rhizospheric, polluted soil, and the interaction of microbes therein and their degradation of pollutants is examined. Additionally, how biofilms fulfill their role in pollution control is discussed, with an emphasis on their unique structural and metabolic attributes. The microbial loop is further explained, and the impacts on nutrient cycling and availability are elucidated. As a niche of microbial activity, the phycosphere is examined, with emphasis on its role in pollutant mineralization. The notion of geomicrobiology as an interdisciplinary domain of microbial activity and geochemical processes in pollutant transformation is introduced. The chapter concludes by drawing our attention to the importance of microbial communities in supporting the resilience of ecosystems and potential applications in the sustainability of the environment.

Keywords: Biofilms, Ecosystem resilience, Geomicrobiology, Microbial ecology, Nutrient cycling.

INTRODUCTION

Microbial ecology studies everything concerning their relationships, the environment, and host organisms [1]. While these microbial communities are microscopic, they are important to ecological processes and to ensure environmental health. Microbial communities are, in most polluted environments, the first line of attack against contaminants and define their propensity to persist, transform, or be removed [2]. Given the escalating environmental pollution, the function of microorganisms in bioremediation and pollution management is of paramount significance, as it affects ecosystems, biodiversity, and human health. The study of microbial ecology in contaminated settings is an evolving discipline that offers critical insights into microbial responses to and mitigation of pollutant

* **Corresponding author Niketa Bhati:** Amity Institute of Environmental Sciences, Amity University, Noida, Gautam Budh Nagar, Uttar Pradesh- 201313, India; E-mail: hjain@amity.edu

threats. A variety of substances will be introduced into the environment *via* industrial activity, agricultural practices, urbanisation, and waste disposal, resulting in pollution. The effects of these pollutants on the natural balance can also cause a disturbance of ecosystems (both abiotic and biotic components). Microorganisms have proven the ability of higher organisms to adapt to and survive in contaminated environments, while these environments can severely pollute higher organisms. They are well adapted, due to their metabolic flexibility, high genetic diversity, and high rate of reproduction, to utilizing a variety of contaminants, which play important ecological roles in biogeochemical cycles and pollutant degradation [3]. The adaptability of microbes has led to their centrality in many environmental cleanup strategies, illuminating their importance in pollution control.

The resilience of microbial communities is one of the most important factors in microbial ecology in polluted environments. Variation in metabolic pathways and the formation of protective biofilms is just one group of mechanisms whereby microorganisms have adapted to survive in harsh conditions, including resistance to toxic substances [4]. They utilize these strategies not only to withstand polluted environments but also to help degrade contaminants. A better understanding of the ways that these microbial strategies have evolved will enable the development of sustainable, effective bioremediation strategies for addressing diverse pollution settings. In this chapter, some important aspects of microbial ecology in pollution environment are discussed, along with some special ecological niches and mechanisms of interaction of microorganisms with pollution pollutants and environmental remediation. Concepts, including rhizospheric ecology, biofilms, the microbial loop, phycosphere, geomicrobiology, and ecosystem resilience, which are essential for understanding the microbial roles in polluted environments, are explained.

RHIZOSPHERIC ECOLOGY OF CONTAMINATED ENVIRONMENTS

Microbial Interactions in Polluted Rhizospheres

The rhizosphere of polluted environments is a complex, dynamic environment where microbial interaction stimulates pollutant degradation and stabilization [5]. The rhizosphere, with its high microbial activity and diversity, represents a hotspot of ecological interactions structured by root exudates. These exudates consist of sugars, organic acids, amino acids, and phenolics, which serve as nutrient sources to attract and sustain microbial communities. These interactions are further altered by pollutants in the soil as selective pressures, increasing their capacity to promote the growth of pollutant-tolerant or degrading microbes. A complex network of interaction exists among key microbial players in polluted rhizospheres consisting of bacteria, fungi, and archaea [6].

As a result of their metabolic versatility and ability to degrade a wide variety of organic pollutants, such as hydrocarbons as well as pesticides, bacteria, such as Pseudomonas and *Bacillus*, are often dominant [7]. Mycorrhizal fungi modify the acquisition of nutrients by plants and help immobilize toxic metals in hyphal networks. Although less studied, Archaea are important degraders of complex organic compounds in extreme environments, including those with high salinity or heavy metals contamination [7]. These microorganisms do not work in isolation: Together, they make the rhizosphere a different place, chemically and biologically, and one that is pleasant to a plant and conducive to pollutant remediation. The effect of the presence of pollutants on the interaction between microbial species in the rhizosphere is to disrupt an ecosystem-level balance or to promote cooperative behaviours. For instance, microbial consortia often display synergistic interactions of the form that different species metabolize different components of complex pollutants [8]. Biofilms that can enable microbial communities to degrade pollutants together are also formed, and greater resilience to contaminants is enhanced [9]. An appreciation of these interactions is essential for the design of bioremediation strategies to utilize microbial potential in bioremediation.

Mechanisms of Pollutant Uptake and Degradation

Physical, chemical, and biological mechanisms mediated by plants and microbes are coupled to remove or transform pollutants in contaminated environments [10]. These are given synergistically, and plants and microbes use separate but connected ways to uptake, metabolize, or sequester contaminants (Fig. **1**). Phytoremediation relies primarily on planners to handle pollutants by strategies like phytoextraction, phytoremediation, and phytodegradation [11]. In phytoextraction, heavy metals are absorbed through the root, subsequently translocated to their aerial parts, and removed through harvesting [12]. Phytostabilization is the process by which plants stabilize polluting materials within the root zone, preventing them from migrating to groundwater or surrounding areas. In phytodegradation, organic pollutants are metabolized by some plants into other less toxic forms, thereby rendering them harmless [13]. Plant based mechanisms do not substitute microbial degradation in systems that metabolize organic pollutants through enzymatic pathways (Table **1**).

Specific oxygenases & dehydrogenases are produced by Pseudomonas, Sphingomonas, and Mycobacterium microorganisms to break down hydrocarbons, pesticides and other toxic compound [14]. But microbes are very versatile when it comes to metabolism, and can degrade a broad range of contaminants — from easy hydrocarbons, to those complex aromatic molecules. Detoxification of heavy metals is also facilitated by some microbes by biosorption, precipitation or the complexation with organic ligands [15]. The bioavailability of pollutants to plant uptake is increased by microbial activity. For example, some microbes produce biosurfactant or chelating agents to solubilize hydrophobic or insoluble contaminants (that became more readily absorbed and degraded) [16]. Even microbes produce phytohormones and stress alleviators such as ACC deaminase that help plants to withstand toxic effects of pollutants themselves; enabling the remediation process. Although not interacting, these interconnected mechanisms reinforce the role of plant microbe synergy in pollutant treatment [17].

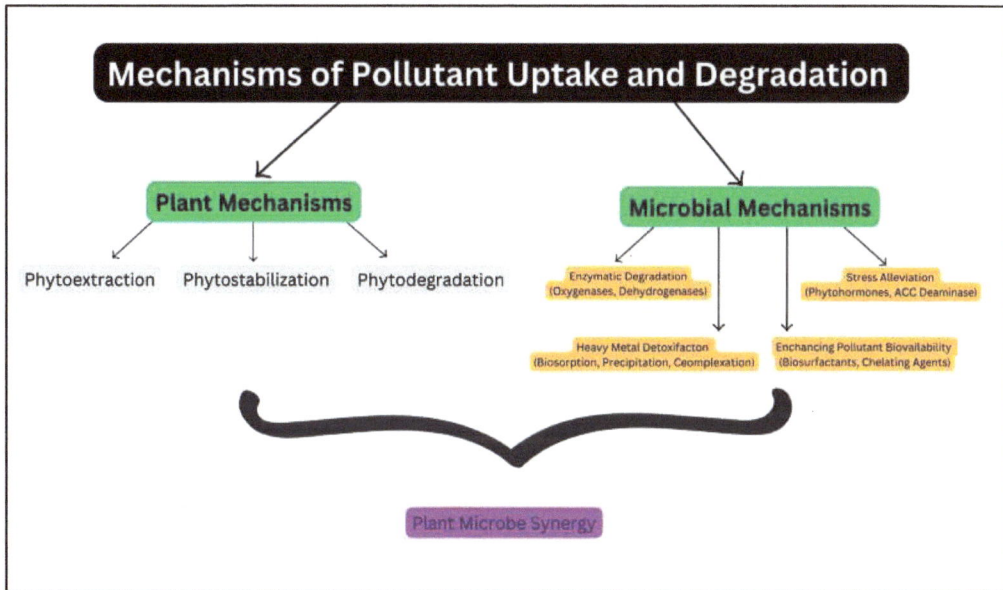

Fig. (1). Mechanisms of pollutant uptake and degradation.

Table 1. Microbial Degradation of Pollutants in Aquatic and Terrestrial Ecosystems [10 - 15].

Pollutant Type	Microbial Players	Degradation Process	End Products
Hydrocarbons (Oil Spills)	Hydrocarbon-degrading bacteria (*e.g., Alcanivorax, Marinobacter*)	Hydrocarbon oxidation	Carbon dioxide, water, microbial biomass
Heavy Metals	Metal-reducing bacteria (*e.g., Pseudomonas, Bacillus*)	Metal precipitation, biosorption	Insoluble metal sulfides, reduced bioavailability
Pesticides	Pesticide-degrading bacteria (*e.g., Pseudomonas, Bacillus*)	Biodegradation of pesticide chemicals	Non-toxic metabolites, reduced toxicity
Polychlorinated Biphenyls (PCBs)	PCB-degrading bacteria (*e.g., Burkholderia*)	Aerobic and anaerobic degradation of PCBs	Dechlorinated products, reduced environmental impact

Plant-microbe Partnerships in Bioremediation

The symbiosis between plants and microbes in the rhizosphere is a potent bioremediation tool which utilizes plants' physiological and biochemical capability combined with microbe's biology, to reverse contaminated environments [18]. For organic and many inorganic pollutants, plant microbiome partnerships are powerful, ecofriendly, sustainable methodologies as remediation alternatives to conventional methods. And among the key of these partnerships is the exchange of resources as well as benefits. Root exudates are energy sources of plant released to microbial communities, microbes improve plant growth by improving nutrient availability and protecting plants from stresses [19]. This symbiosis is even more critical in contaminated environments as plants provide a stable habitat and consistent nutrient supply for microbes, which degrade pollutant. For an example, under conditions of heavy metal exposure, Brassica juncea associated with metal tolerant bacteria demonstrates increased heavy metal uptake and efficiency of phytoextraction [20].

Plant microbe partnerships facilitated by mycorrhizal fungi involve expanding root network and help in pollutant immobilization [21]. Because their hyphal networks function like conduits for nutrient exchange and pollutant stabilization, especially in soils with heavy metals, they have significant ecological importance. As with the endophytic microorganisms that live within plant tissue, polluting microorganisms present within plant tissue also degrade pollutants within plant tissue, furthering the bioremediation [22]. Many case studies show that plant microbe partnerships are successful in bioremediation. When Populus species are combined with hydrocarbon degraders, such as Alcanivorax, petroleum hydrocarbon degradation is accelerated in hydrocarbon contaminated soil. Pseudomonas microbe is found to degrade organophosphate pesticides and prevents the rise of the residue tall while fostering crop growth in agricultural settings [23]. Plant-microbe systems demonstrated versatility and efficacy in these examples of contamination scenarios.

The potential of plant-microbe partnerships in bioremediation is also being further advanced by biotechnology. Genetic engineering of plants enables enhanced pollutant uptake and degradation, and synthetic microbe consortia can be engineered to treat specified contaminants [24]. Other techniques, such as bio stimulation, which increase the abundance of indigenous microorganisms, or bioaugmentation, which introduces specialized strains, are being explored to optimize bioremediation outcomes [25]. There are, however, technical challenges to plant-microbe partnerships in bioremediation, including toxicity of pollutants, variability in the environment, and the practical challenges of scaling bioremediation efforts. We further propose future research aimed at identifying a suite of robust plant and microbial species that can tolerate a wide range of pollutants, and to develop methods for monitoring and assessment of progress. The integration of plant and microorganisms into land reclamation and sustainable agricultural practices provides a means to deal with pollution and to reconstruct the ecological health of the soil.

ROLE OF BIOFILMS IN ENVIRONMENTAL POLLUTION CONTROL

Biofilms are structured communities comprised of microorganisms adherent to a surface within natural or engineered environments, or to industrial substrates, enclosed within a self-produced Extracellular Polymeric Substance (EPS) matrix [26]. Environmental pollution control relies on these microbial communities that act as a robust driver to degrade pollutants and stabilize contaminants [27]. Innovative and sustainable solutions for environmental pollution management, in particular, wastewater treatment and bioreactor technologies, can be offered by exploiting their structural complexity and metabolic diversity.

Biofilm Formation and Structural Dynamics

Biofilm formation is a highly organized multi stage process beginning with plaitctonic microorganisms adhering to a surface. Electrostatic interactions, hydrophobic forces and release of adhesive substances together support this attachment [28]. After becoming attached, the microbial cells flourish and begin to produce the EPS, a gelatinous matrix made up of polysaccharides, proteins, lipids, and nucleic acids. The structure of this matrix ensures the biofilm community structural stability, protects biofilm community from environmental stressors, and improved biofilm community resilience to toxic compounds and antibiotics. Within the biofilm, as the biofilm matures, it will form a three-dimensional architecture; microcolonies, water channels, gradients of nutrients and oxygen. But these structural features are critical for pollutant degradation by the biofilm [29].

The nutrient gradients provide not only niche spacing between microbes, but also create gradients of nutrients which promote specialized efficient uptake of nutrients and disposal of waste products by the water channels. In this spatial organization biofilms can degrade a wide variety of pollutants, ranging from simple organic compounds to complex industrial chemicals. Environmental factors affecting biofilm formation are pH, temperature, nutrient availability and presence of pollutants [30]. EPS production induced by heavy metals or hydrocarbons is also reported for many biofilms, resulting in increased EPS production to immobilize toxic compounds and protect the microbial community. Biofilms have favourable adaptability and resilience properties that make them an ideal choice for environmental applications where high pollutant loads and fluctuating environmental conditions are common [31].

Metabolic Pathways in Pollutant Degradation

Critical to the environmental pollutant degradation capacity of biofilm associated microorganisms is their metabolic versatility. Biofilms are microbial communities with the capacity for division of labor, in which different species or functional groups occupy equivalent but specialized metabolic pathways. This cooperation facilitates the decomposition of difficult to metabolize complex pollutants by each bacterium separately. When it comes to organic pollutants, biofilms use enzymatic pathways by which hydrocarbons, pesticides and other contaminants are degraded into simpler, less toxic material. Bacteria in biofilms produce hydrocarbon oxidizing enzymes (*e.g.* oxygenases; dehydrogenases), fungi secrete ligninolytic enzymes *e.g.* (laccases; peroxidases) to degrade recalcitrant pollutants such as Polycyclic Aromatic Hydrocarbons (PAHs) [32]. Degradation of chlorinated compounds and nitrates is also supported by anaerobic zones in biofilms by reductive dechlorination and denitrification processes.

Biofilm mediated mechanisms, either biosorption, bioaccumulation or biotransformation, are used to address heavy metal pollutants. The functional groups that bind metal ions to the EPS matrix of biofilms reduces metal ion mobility and bioavailability. Other biofilm associated microorganisms can change toxic metals into less toxic forms ($Cr3+$ from $Cr6+$) through enzymatic processes [33]. Moreover, biofilms are talented at bioleaching, *e.g.*, at degrading emerging contaminants, such as pharmaceuticals and personal care products, which (in an unintended manner) find their way into wastewater. Biofilms possess advanced metabolic pathways and microbial consortia that enable breakdown of these recalcitrant pollutants, becoming a promising remedy to face up to emerging environmental problems.

Applications in Wastewater Treatment and Bioreactors

Modern wastewater treatment systems depend heavily on biofilms to degrade organic and inorganic pollutants for environmental purification [33]. The most common application is in the form of biofilm bioreactors like trickling filters, rotating biological contactors, (but especially) Moving Bed Biofilm Reactors (MBBRs) [34]. The microbial communities can take advantage of these surfaces to retain high resistance to fluctuating conditions while efficiently treating wastewaters. Biofilms are integral to organic matter, nitrogen, phosphorus, and other contaminant removal in wastewater treatment. As an example, wastewater biofilms house nitrifying and denitrifying bacteria who convert ammonia to nitrate to nitrogen gas sequentially, removing the nitrogen load from wastewater

[35]. Like phosphorus, phosphorus removal is accomplished *via* uptake and storage of phosphorus compounds in biofilm associated polyphosphate accumulating organisms.

The biofilm reactors are also used in industrial effluent treatment where pollutants such as heavy metals, dyes and hydrocarbons are treated. For example, biofilm systems in the treatment of effluents from textile, petrochemical and mining industries have been demonstrated to be successful [36]. Since they are robust biofilms, it is because these systems are capable of tolerating high pollutant concentrations and toxic shocks, and hence, used for industrial applications due to the preference of these systems. Pollution control and resource recovery are based on emerging biofilm technologies including Microbial Fuel Cells (MFCs) and Anaerobic Membrane Bioreactors (AnMBRs) [37]. Biofilm based microbial communities are employed by MFCs to degrade organic matter in wastewater as they produce electricity as a by-product. In contrast to the conventional treatment systems, AnMBRs integrate biofilm processes with membrane filtration to provide high efficiency wastewater treatment and biogas production [38]. The innovative technologies presented here provide an illustration of how biofilms can foster sustainable environmental management.

MICROBIAL LOOP AND NUTRIENT AVAILABILITY

Microbial loop is a critical component of ecosystem dynamics that acts as a major participant in nutrient cycling and organic matter turnover (Fig. **2**) [39]. Microbial process, however, are defined as interactions among microorganisms and the environment, and the role that microbial processes play in determining the availability of nutrients in both aquatic and terrestrial ecosystems [40]. In polluted environments, the microbial loop becomes more important, contributing to the reduction of contaminant effects and to ecosystem recovery.

The Concept of the Microbial Loop

In specific, the microbial loop refers to the flow of energy and nutrients through microbial pathways first introduced in the context of aquatic ecosystems. The microbial loop differs from traditional food chains focused on energy transfer from primary producers to higher trophic levels and instead concentrates on the recycling of organic matter by microorganisms [41]. Dissolved Organic Matter (DOM), that is, organic matter released by phytoplankton, plants or decaying organisms, acts as a substrate of heterotrophic bacteria in that loop [42]. For example, DOM is metabolized by these bacteria, the POM produced by it is incorporated into it's biomass. The passively fed microbes then represent a food source for protozoa, ciliates and other micrograzers, who return the energy and nutrients back to the normal food web [42].

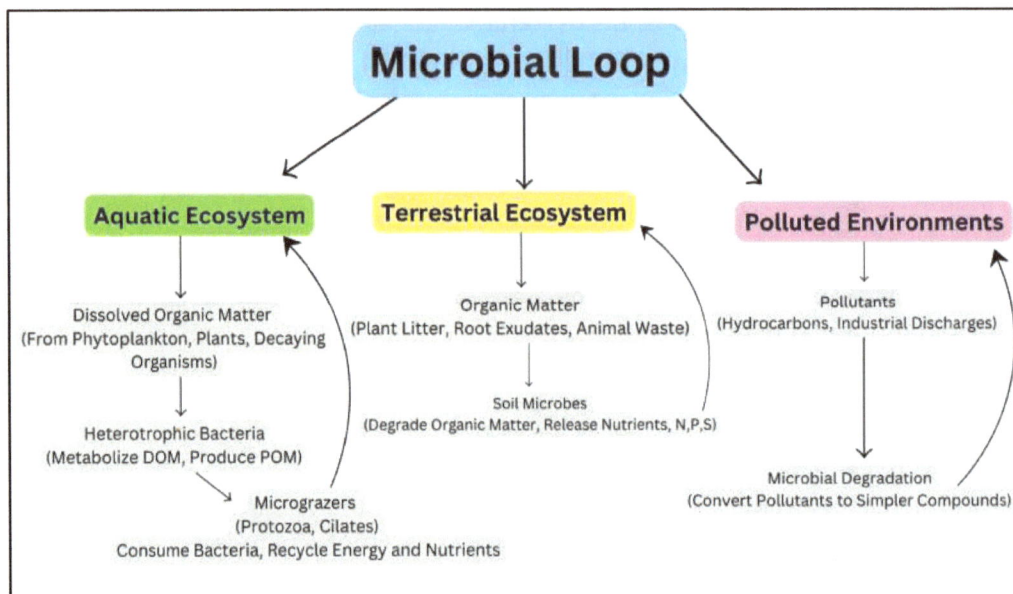

Fig. (2). Microbial loop as a critical component of a ecosystem.

The microbial loop operates in the soil matrix in terrestrial ecosystems and in which microbes degrade the organic matter leached from plant litter, root exudates and animal waste. They release nutrients — nitrogen, phosphorus, and sulphur — that are available to plants and other organisms in the soil. This microbial loop therefore provides a continuous source of nutrient, with a resultant ecosystem productivity and stability [43]. However, microbial loop is all the more important in polluted environments where it breaks down contaminants and recovers nutrient balance. As an example, microbial communities in areas that have been affected by oil spills or industrial discharges degrade hydrocarbons and other pollutants into simpler compounds which can be re-incorporated into the nutrient cycle [44] (Table **2**).

Table 2. Key Microbial Players in the Microbial Loop and Their Roles [41 - 50].

Microorganism Type	Key Functions	Pollutant Degradation Mechanisms
Heterotrophic Bacteria	Decompose organic matter, mineralize nutrients	Degrade hydrocarbons, heavy metals, Persistent Organic Pollutants (POPs)
Fungi (Saprotrophic)	Breakdown of lignin, cellulose, and other recalcitrant compounds	Mineralize organic carbon and release nutrients
Protozoa	Grazing on bacteria and small phytoplankton, nutrient cycling	Indirectly assist in contaminant breakdown by releasing nutrients back to food web

(Table 2) cont.....

Microorganism Type	Key Functions	Pollutant Degradation Mechanisms
Algae	Primary producers, oxygen production, nutrient cycling	Release of organic carbon, support microbe activity, aid pollutant degradation
Archaea	Methanogenesis, sulfur cycling	Break down complex organic materials, nitrogen cycling
Viruses	Control microbial populations, influence microbial dynamics	Influence pollutant breakdown *via* microbial lysis

Role in Organic Matter Turnover and Nutrient Recycling

Process which underpins ecosystem functioning: decomposition of organic matter and recycling of nutrients, the microbial loop is central to the process. The bulk of organic matter turnover is mediated by microorganisms: bacteria, fungi, archaea [45]. In aquatic systems, DOM following phytoplankton or terrestrial runoff is a substrate for heterotrophic bacteria. These bacteria degrade organic carbon into biomass while releasing inorganic nutrients such as ammonium, phosphate and nitrate through enzymatic degradation. Once these nutrients are picked up by primary producers and complete the nutrient cycle. In aquatic systems, the efficiency of the microbial loop can greatly influence whole ecosystem productivity, by determining the amount of available nutrients for phytoplankton and other primary producers [45].

Microbial communities live in the soil and decompose organic matter of plant and animal origin, basically releasing best nutrients for plant growth. Saprotrophic fungi have very important roles in breaking down recalcitrant compounds such as lignin and cellulose [46]. This process is supplemented by bacteria that metabolizes more simple organic compounds to release nutrients in bioavailable forms. Microbes also play a dominant role with nitrogen cycling, key processes of nutrient recycling like nitrogen fixation, nitrification and denitrification. While the microbial loop is well adapted to deal with contaminants in polluted environments, the process is enhanced towards organic matter turnover and nutrient recycling. Such microbial species may attenuate the toxicity of metals, which can happen through biosorption or precipitation of the metal in contaminated soils in which metals are themselves heavy metals, such as in soils contaminated with heavy metal ions. Like oil pollution waters, this line of studies addresses hydrocarbon degrading bacteria that metabolize oil compounds into carbon dioxide and biomass that reenfires the microbial loop [47].

Implications for Polluted Aquatic and Terrestrial Ecosystems

The implications of this microbial loop for recovery and resilience of polluted ecosystems are profound, both aquatic and terrestrial. Microbial communities both drive the breakdown of pollutants and positively impact nutrient availability and contribute to ecosystem remediation and restoration. However, the incidence of both organic matter accumulation and their pretreatment and nutrient accumulation through sewage discharge and agricultural runoff is often found in aquatic ecosystems from pollution [48]. Mitigating these effects, the microbial loop consists of bacteria and protozoa that degrade excess organic matter and recycle nutrients to prevent algal blooms and hypoxic conditions. The microbiomes can also degrade persistent organic pollutants such as polychlorinated biphenyls (PCBs), Polycyclic Aromatic Hydrocarbons (PAHs) and many other POP's, which are frequently present in contaminated water bodies [49]. In addition to helping to detoxify the environment these degradation processes also reintegrate carbon and nutrients into the ecosystem.

Soil structure and nutrient cycling can be disturbed by pollution from mining, chemical spills and excessive pesticide use, in terrestrial ecosystems. That's where the microbial loop comes in to counteract these effects and break down contaminants, helping to restore soil fertility. To point, hydrocarbon-degrading bacteria in soils contaminated with petroleum hydrocarbons metabolize these compounds to less toxic forms to enhance plant regrowth and recovery of soil. The restoration of nutrient cycling and plant growth in heavy metal contaminated soils relies on microbial species able to immobilize and detoxify metal. The microbial loop in polluted environments is very resilient and allows for a wide number of microbial communities, which are variable and tolerant to environmental pollutants [50]. Often these are communities of specialized species, which have evolved the ability to metabolize contaminants or are hard to live with. The retention of this functional diversity provides this buffer to the microbial loop in the face of stress, the results of which are used to preserve its function under environmental degradation. In addition, bioremediation strategies to enhance the microbial loop by deliberately introducing stimulations of the pollutant degrading microorganisms. Regular bioaugmentation treats pollutants with specific microbial strains that can degrade them, while regular bio stimulation adds electron donors or nutrients which stimulate existing microbes to degrade pollutants [51]. Instead, these approaches exploit the underlying natural processes of the microbial loop to speed up the remediation of polluted environments.

PHYCOSPHERE AND MINERALIZATION OF POLLUTANTS

Algal cells inhabit a phycosphere ecological niche, surrounded by a host of microbes with which algal cells interact in ways that profoundly alter pollutant breakdown and nutrient cycling in aquatic ecosystem [52]. These intercellular interactions are key in reducing environmental pollution as they mediate degradation organic and inorganic pollutants and disseminate available nutrients *via* nutrient cycling balancing ecosystems. The focus of this section is on microbial communities in the phycosphere, their attachment to algae in pollutant degradation and in nutrient cycling.

Microbial Communities in the Phycosphere

As a microenvironment supporting high concentrations of organic compounds released by algal cells, the phycosphere is compared with the rhizosphere of plants. Carbohydrate, amino acids, lipids and secondary metabolites of these exudates act as chemoattractant for the microorganisms [53]. It turns out that this nutrient rich zone is a hotbed for a high density, diverse group of microbes including bacteria, archaea, fungi, and even viruses. Heterotrophic bacteria dominate among the microbial inhabitants of the phycosphere because of their ability to utilise algal derived organic matter as an energy source [54]. Genera such as Alteromonas, Vibrio, and Pseudomonas are known because of their metabolic versatility and ability to degrade many pollutants are such bacteria [55]. Cyanobacteria, which co-exist with algae, also participate simultaneously in phycosphere as e competitors and c collaborators in the usage of nutrients [56].

Archaea are important phycosphere residents besides bacteria. For example, thermal degradation of algal material by methanogenic archaea allows the use of metabolic byproducts to produce methane, a process involved in carbon cycling [57]. Less studied in the phycosphere, but also as complex organic pollutant breaks down accomplices, fungi also work in synergistic ways with bacteria. Microbial population dynamics within the phycosphere are regulated by lysing of bacterial cells by viruses, especially bacteriophages. Lysis of this *Lysistrichia* releases organic matter back into the environment and is additional resources for microbial communities and influences nutrient cycling [58]. Many of the microbial communities of the phycosphere display dynamic interactions in which cooperation and competition impact the functional outcomes of pollutant degradation. The production of bioactive compounds and gene exchange, and the quorum sensing mediated relationships, result in a complex network of ecological relationships.

Algal-microbe Interactions in Pollutant Breakdown

Algal microbe's interactions in phycosphere are highly important for pollutant mineralization. These are mutualistic interactions that include algae contributing organic substrates and oxygen from photosynthesis and microbes providing pollutant degradation and nutrient cycling. The degradation of hydrocarbons is one of the most studied mechanism of microalgae-microbe interaction in pollutant breakdown. As a by-product of photosynthesis, algae release oxygen; it creates an aerobic environment that is good for the growth of bacteria that oxidizes hydrocarbons. These bacteria convert hydrocarbons to carbon dioxide, water or biomass. Say, in oil polluted waters, algae activate the activity of bacteria like *Alcanivorax* and *Marinobacter*, which are capable of degrading long chain hydrocarbon [59].

Degradation of heavy metals and Persistent Organic Pollutants (POPs) also depends on cooperation between algae and microbes. Extracellular Polymeric Substances (EPS) produced by algal cells can exorbit heavy metals and decrease the bioavailability [60]. At the same time, various microbes of the algae may precipitate metals as insoluble sulfides, detoxifying the environment. For instance, consortia of algae and bacteria have been proved to be effective in reducing cadmium, lead and mercury concentrations in contaminated waters. In pesticide pollution cases, algae and microbes break these chemicals down into less toxic metabolites together. Pesticides are used as carbon sources to microbes like *Pseudomonas* and *Bacillus*; and algae are the conditions which fuel microbial activity. The synergistic system has been exploited in bioremediation projects aiming to eliminate agricultural runoff [61]. Signalling molecules exchanged between algae and microbes are a critical factor in these interactions. Special compounds that algae release upregulate the expression of pollutant degrading enzymes in microbes. However, many microbial metabolites modulate algal growth and metabolic pathways, thereby enriching bioactive compounds that promote pollutant breakdown.

Contributions to Ecosystem Nutrient Cycling

Nutrient cycling occurs in the phycosphere, a hotspot for reactive element turnover along with carbon, nitrogen, and phosphorus by aquatic ecosystems. In this manner, the interactions between algae and microbes are important and serve to ensure the perpetual availability of the nutrients for number 1 manufacturing, as well as to maintain stability in surrounding conditions [62]. Algae contribute to the carbon cycle by liberating Dissolved Natural Carbon (DOC) into the phycosphere through solving carbon dioxide *via* photosynthesis. This DOC is metabolised by microbial groups to incorporate into microbial biomass or respired as carbon dioxide. Today the rampant respiration of this microbes additionally no longer only closes the carbon cycle it additionally helps to enhance the heterotrophic organisms in the atmosphere [63].

Nitrogen cycling within the phycosphere can be mechanisms including nitrogen fixations, nitrification, and denitrification. In the phycosphere inhabited and utilized by algae and various other organisms, atmospheric nitrogen is converted to extractable paperwork (such as ammonium) *via* the excretion of nitrogen solving bacteria. Ammonium is oxidized by nitrifying microorganisms to nitrite and often to nitrate, a key step of algal increases. Rather, Denitrifying microorganism will be located in nigh the loop, converting nitrates back into atmospheric nitrogen, thereby preventing eutrophication in aquatic environments [64]. The phycosphere also perturbs phosphorus cycling, which in turn can be a limiting factor in aquatic ecosystems. The organic phosphates released by algae become readopted by microbes, and they will be mineralized back to inorganic phosphate. Releasing this inorganic phosphate provides a good source for algae, primary producers and all other organisms maintaining the productivity of the ecosystem. In polluted environments, microbial immobilization of excess phosphorus in the phycosphere is possible, and can help alleviate eutrophication risk. The phycosphere is also an essential determinant of nutrient cycling and for detoxifying noxious compounds and recycling traces elements [65]. As an example, microbial communities of the phycosphere can sequester heavy metals in environments contaminated by these metals, reducing their bioavailability and toxicity. Algae participate in the cycle of incorporation of trace elements in their biomass, and the eventual reimmersion of the biomass into the nutrient cycle upon decomposition. Phycosphere contribution to nutrient cycling is especially important in polluted environments, helping to restore ecosystem [66]. Algal – micro beer consortia breakdown pollutants and recycle nutrients, to help restore balance to aquatic ecosystems favouring productivity and biodiversity.

GEOMICROBIOLOGY

Geomicrobiology is the study of microorganism-mineral interactions and their relevance in contaminated environments. Its role in environmental remediation, pollutant transformation and elements cycling *via* geochemical and microbial processes has made this discipline one of the hottest arenas of research. This has been used to understand the biogeochemical transformations that take place in polluted environments leading to a better understanding of how microbes contribute to mineralisation of pollutants [67]. The second section will discuss the role of microorganisms in contaminated sites, biogeochemical transformations of pollutants, and looks at relevant case studies of geomicrobiological applications.

Microbe-mineral Interactions in Contaminated Sites

Minerals can be altered, and pollutants can be transformed in contaminated environments by microbes. Microorganisms in contaminated sites interact with minerals in various ways that can affect the fate of pollutants, *e.g.* by transforming, immobilizing or degrading [68]. Because it is these interactions that can be highly specific, they include the breakdown of complex contaminant and mineral components by microbial enzymes and metabolic pathways in concert, sometimes leading to the precipitation or dissolution of mineral phases. In contaminated environments, Mineral surfaces are niche for microbial colonization. These mineral surfaces are the site for microorganism modification by organic acids and/or enzymes and other extracellular compounds excreted by microorganisms that result both in the solubility and mobility of pollutants [68]. For example, metal reducing bacteria, such as *Geobacter* and *Desulfovibrio*, can reduce metal oxides and mobilize toxic metals into ground water such as arsenic, cadmium and chromium. However, for some microbial processes, the reduction of bioavailability and toxicity of pollutants has beneficial consequences. One example of this is a process by which some bacteria such as *Pseudomonas* spp. immobilize metals by reducing them into insoluble forms which normally resulting bound to insoluble forms, preventing their mobility and their effects as poisons [69].

Ruhring found that organic pollutant degradation often requires these pollutants to attach to mineral surfaces. For example, microbes often adsorb hydrocarbons onto mineral surfaces before the hydrocarbons are degraded by microbes, and such adsorption may promote the interaction of the microbial enzymes with the pollutants [70]. Consequently, this interaction is particularly promising for remediation of petroleum or pesticide contaminated soils, where mineral associated microbial activity might accelerate the breakdown of complex organic pollutants. In addition to performing other forms of bioremediation, microorganisms are able to bind metals, reduce their toxicity through the formation of biofilm, and through the biofilm matrix efficiently sequester pollutants. Elemental cycling, for example, nitrogen, sulfur, and phosphorus, is also dependent upon microbe-mineral interactions [71]. Examples of sulfate reducing bacteria includes that it can use sulfate as an electron acceptor in the absent of oxygen and produce hydrogen sulfide. Results this can can precipitate metal sulfides and uptake metal in the environment such as copper and zinc could be reduced. As with other pollution problems, microorganisms serving as a nitrate reducing denitrification system can help in the control of excessive nitrate in polluted waters and soils. Microbe mineral interactions have important effects on pollutant dynamics and geochemical environment of contaminated sites through these routes [72].

Biogeochemical Transformations of Pollutants

The central process for environmental remediation involves the biogeochemical transformations of pollutants [73]. Some of these can be effectuated by microorganisms *via* reduction, oxidation, methylation, and hydrolysis metabolism. They usually remove the harm of pollution leading to detoxification or immobilization of the polluting elements. The particular microbe processes affected vary with the chemical nature of the pollutant, the constituting microbe community and their environmental conditions. Reduction of toxic metals and metalloids are one of the most important biogeochemical processes in geomicrobiology [74]. For example, contaminated water and soils are often polluted with arsenic (a common contaminant) and microorganisms are important players in its transformation. Mainly, microbially reductive AS(V) is reduced to more mobile and toxic AS(III), which occurs with microbes such as *Shewanella spp.* and *Geobacter spp.* However, arsenic may be immobilized by other microbial processes - for example, through the conversion of arsenic to less soluble forms (such as arsenic sulfides), using sulfur reducing bacteria [75] (Table 3).

Table 3. Key Biogeochemical Processes in Pollutant Transformation [70 - 77].

Biogeochemical Process	Microbe(s) Involved	Pollutants Processed	End Product
Reduction	*Shewanella spp., Geobacter spp.*	Arsenic, Chromium, Mercury	Less toxic, often insoluble forms
Oxidation	*Pseudomonas spp., Bacillus spp.*	Hydrocarbons, PAHs	Oxygenated or simpler compounds
Methylation	*Desulfovibrio spp.*	Mercury, Arsenic	Methylated, more toxic forms
Biodegradation	*Sphingomonas spp., Dehalococcoides spp.*	PAHs, Chlorinated solvents	Biodegradation to non-toxic forms

The microbial reduction of chromium (Cr(VI)) to chromium (Cr(III)), another critical pollutant transformation to a less toxic and often insoluble form, is another critical fact [76]. This transformation is due to the action of *Pseudomonas spp.* and other bacteria and is especially significant in areas of chromium contamination for which environmental protection is of major concern. Like mercury (Hg(0)), which is toxic when found in its elemental form, methylmercury (CH3Hg+), is also produced from mercury by some bacteria, a more bioaccumulative and toxic form [77]. But some microorganisms can also demethylate methylmercury, making it once again less toxic inorganic mercury forms. In addition to metals, diverse organic pollutants including pesticides, Polycyclic Aromatic Hydrocarbons (PAHs), and chlorinated solvents can be metabolized by different types of microbes. Biodegradation, by organic contaminants by microorganisms into simpler, oftentimes less toxic, compounds, is the process. *Sphingomonas spp.* are known to degrade PAHs while *Dehalococcoides spp.* can dechlorinate chlorinated solvents, such as trichloroethylene (TCE), a commonly used industrial solvent. In contaminated environments, such as groundwater and soils, however, these microbial transformations are critical for reducing the persistence of organic pollutants in the environment [78]. In addition, microorganisms used in bioremediation strategies are also able to degrade a large number of organic contaminants. Degradation can be *via* aerobes or anaerobes. The microbial breakdown of organic pollutants is carried out in aerobic as well as anaerobic processes, namely during the presence of oxygen, as well as in the absence of oxygen, but the latter, instead of oxygen, is using alternative electron acceptors, *e.g.* nitrate, sulfate, carbon dioxide [79]. The ability to use this flexibility of metabolic pathways to address a broad range of pollutants, from hydrocarbons to pesticides and solvents, is important.

Case Studies in Geomicrobiological Applications

Application of the field of geomicrobiology to numerous contaminated sites worldwide, and in particular the role of microbial minerals in transforming and removing pollutants, has been demonstrated. Human assessment and application of geomicrobiology in environmental cleanup and pollution mitigation are reflected in several case studies. The most well-known case of application is the remedy of arsenic contaminated groundwater in Bangladesh and other areas of South Asia [80]. Drinking water contamination by arsenic has long been an important health issue in these regions and microbial processes are key to arsenic transformation and removal. Sulfate reducing bacteria are also found to immobilize the arsenic and convert it to arsenic sulfide, a less mobile and toxic form of arsenic reported researchers [81]. Bioremediation processes for removal of arsenic from groundwater have already been exploited as a way to address the problem of a widespread environmental issue in a natural and cost-effective manner (Table **4**).

Table 4. Key Case Studies of Microbial Applications in Geomicrobiology [80 - 84].

Case Study	Pollution Type	Microbial Process	Outcome/Impact
Arsenic contamination in Bangladesh	Arsenic	Sulfate reduction, Biofilm formation	Arsenic immobilization in groundwater
Deepwater Horizon Oil Spill, Gulf of Mexico	Oil hydrocarbons	Hydrocarbon degradation by Alcanivorax spp., Marinobacter spp.	Natural oil degradation, ecosystem recovery
Heavy metal contamination in the USA	Lead, Cadmium	Precipitation of insoluble lead compounds	Soil decontamination, reduced bioavailability of lead
Wastewater treatment	Organic pollutants, Nitrates	Biodegradation, Denitrification	Removal of organic pollutants, nitrogen control

This second case study refers to the cleanup of oil contaminated sites in the Gulf of Mexico immediately after the period of the Deepwater Horizon oil spill in 2010 [81]. Extensive contamination of marine ecosystems occurred because millions of barrels of crude oil were spilled into the ocean. Biodegradation of the oil was dependent on microbial communities in the ocean, which acted by hydrocarbon degrading bacteria *Alcanivorax* and *Marinobacter* [82]. The microbes, however, break hydrocarbons down into lower compounds, which minimizes the hazardous effect of the spill on the surroundings. Furthermore, algae and phytoplankton in the area cooperatively worked together with bacteria to break down oil and specific bacteria speeded up the natural recovery of that area. Geomicrobiology has been applied in the industrial contexts to remediation of soils contaminated with the heavy metals and organic pollutants. A useful example is the bioremediation of soils contaminated with lead in the USA [83]. Bacteria have also been found that can immobilize lead by precipitating lead as insoluble forms, including lead sulfide. In order to reduce the risk of lead poisoning for the surrounding environment, organic amendments, which improve the growth of these bacteria, increase bioavailability of lead in the soil.

Geomicrobiological processes have also been used to treat wastewater. Bioreactors are used to use microorganisms to degrade organic pollutants, nitrogen compounds, and heavy metals in wastewater. These systems can be reduced by a combination of aerobic and anaerobic processes by the microbial communities in these systems. *Pseudomonas spp.* are frequently utilized to degrade organic matter in wastewater treatment plants, and *Desulfovibrio spp.* reduce sulphur compounds and metals. Cellular activity in wastewater treatment can be boosted through optimized conditions to increase nutrient recovery, decrease the environmental load from pollutants [84]. In these case studies, the application of geomicrobiology indicates that microbial mineral interactions may have application in environmental pollution control. By using microorganisms' power to transform pollutants we can develop sustainable, effective ways to address pollution on soils, water, and air. These applications are not only preventing the worst effects of the pollution but also to restore the ecological status of polluted environments.

MICROBIAL COMMUNITIES IN ECOSYSTEM RESILIENCE

Biofilm formation, which is a frequent microbial phenomenon, is important for ecosystem resilience due to its influence in the response of ecosystems to environmental stressors (*e.g.* pollution, climate change, habitat degradation). They are microorganisms located everywhere and very flexible that can live and generate in harsh environments and they are key for maintaining ecosystem functions and recovery after perturbation. Due to their ability to mediate key biogeochemical cycles, support plant growth, and degrading pollutants, they represent an unusually versatile food web activity when pooled together. In this section, adaptation of microorganisms to stress is investigated, its role in ecological restoration, and future perspectives in microbial pollution management.

Mechanisms of Microbial Adaptation to Stress

Microorganisms exploit a number of these mechanisms to increase their chances of survival at extreme conditions and to adapt to environmental stress. Such mechanisms permit microbial communities to persist in polluted or otherwise degraded environments in which other organisms would not. Metabolic plasticity, ability for protective structures and genetic adaptation are major primary strategies which the microorganisms depend on for their adaptation. Metabolic plasticity is one of the most important forms of microbial stress adaption [85]. What this means is that there are a ton of microorganisms who are able to turn on and off between metabolic pathways depending on whether or not there is sufficient resources around or in the presence of stressors. For example, some of the microbes shift from being aerobic to anaerobic respiration in polluted ecosystems where oxygen is limited thus using alternative electron acceptors: sulfate, carbon dioxide, nitrate. This metabolic shift allows them to live in environments that are oxygen deprived — which are rife with contaminated soils and aquifer water [86]. Some bacteria are even able to detoxify or degrade pollutants through certain enzymatic pathways, by which they can remain in high concentrations of toxic substances, like heavy metals, pesticides, or hydrocarbons.

And another essential adaptation mechanism is that microbes produce biofilm. That's basically an aggregate of microorganisms that attach to surfaces and over time a self-synthesized extracellular matrix. This matrix provides the microbes a protective shelter where the microbe is protected from environmental stress like desiccation, UV radiation, and toxic chemicals [87]. Microbes residing in biofilms in polluted environments are able to survive greater levels of contamination by reduced exposure to toxic chemicals and increased uptake of nutrients. Therefore, biofilm producing bacteria are important factors in wastewater treatment; biofilm producing bacteria metabolize organic matter and toxicants, and provide stable dwelling sites for various populations of microbes. In fact, some microorganisms also employ genetic mechanisms for the adaptation to the stress. Horizontal Gene Transfer (HGT)—that is, microbes acquiring new genes from their environment, including those that confer resistance to pollutants or allow them to metabolize new substrates—allows them to adapt both to a changing external environment and to compete with other bacteria [88]. Imagine bacteria gaining genes for antibiotic resistance, heavy metal resistance or break down of complex organic compounds. Ecological transformation of contaminated ecosystems is witnessed as microbial communities develop the ability to degrade pollutants more efficiently and thus gain greater resistance to pollutants over time. In addition, microbes can become mutated to be better able to survive in polluted environments.

In some cases, some microbes respond to exceptionally arduous environmental conditions, high salinity, temperature or acidity to go into a dormant state where their metabolic activity is reduced and energy to be conserved until conditions improve. Cellular dormancy is a strategy used by microbes to survive stress periods, from drought to chemical contamination, and begin activity again when conditions are more favorable. A few bacterial species that are known to produce spores which are tolerant of extreme environmental stress; high temp, UV radiation or harmful agent. Bacterial species such as *Bacillus* and *Clostridium* [89]. The ability to develop these mechanisms is critical to microbial survival in polluted ecosystems and for promoting resilience to environmental stress in microbial communities. These adaptations help toward restoration and the long-term health of ecosystems damaged by pollution by allowing microbes to degrade pollutants, cycle nutrients, and maintain ecosystem functioning.

Contributions to Ecological Restoration

Restoration of ecosystems polluted, degraded, or subjected to other forms of environmental stress is supported to a great extent by microbial communities. Strong agents for ecological recovery are their ability to adapt to and change their environment. Integrated within ecosystems restoration are microbial driven processes such as bioremediation, nutrient cycling, and plant growth promotion. Bioremediation, the utilization of microorganisms to degrade, transform, or immobilize a pollutant in the environment, constitutes one of the most well-known microbial community contributions to ecological restoration. Microbes can process hazardous chemicals in contaminated soils, sediments and groundwater including petroleum hydrocarbons, pesticides, heavy metals, organic solvents and many others [90]. This also reduces pollution concentration and cleaner the natural ecosystem balance by lowering the environment's toxicity. Many microbial processes take place through bioremediation allowing for the process of biodegradation, biotransformation, and biosorption.

Microbial communities are key Gregory 1996) in the degradation of hydrocarbons in cases of oil spills. As carbon source *Alcanivorax spp* and *Marinobacter spp* microbes are able to metabolize hydrocarbons as their sole carbon source, thereby breaking down the oil and decreasing its impact on the environment. The microbial communities that inhabit more marine environments are generally the first responders to natural elements that get contaminated with oil and their activity can totally wipe out oil presence in the ecosystem. In cases of organic pollutant spills or oil spills, the use of bioaugmentation (adding to the contaminated site-specific microbial strains) has been used for enhanced natural bioremediation. Sustaining ecosystem health and stability is reliant not only on nutrient cycling by inorganic cycles such as water and soil, but also on microbial communities [91]. Cycling of key nutrients, such as nitrogen, sulfur and phosphorus, are essential nutrients for plant growth and ecosystem function, many of which are also mediated by microorganisms. For example, in nitrogen cycling nitrifying bacteria such as bacteria that convert ammonium to nitrate that can be taken up by plants. By contrast however, denitrifying bacteria convert excess nitrate into nitrogen gas, keeping harmful levels of nitrogen from accumulating in the ecosystem [92]. Sulfur reducing bacteria also transform sulfate to hydrogen sulfide, which is part of the sulfur cycle, and is of importance to the recovery of ecosystems where acidification or heavy metal contamination occurs [93].

They're also a huge part of ecosystem restoration in that they facilitate plant growth. A common example of symbiotic relationships in which a lot of microbes assist plants are mycorrhizal fungi on one end, or nitrogen fixing bacteria on the other end, which help plants by providing much needed nutrients, and also aid plants in coping with stress. A well-known example of mycorrhizal fungus form beneficial associations with plant roots, increasing the plant's ability to utilize nutrients, primarily phosphorus, that reside in the soil. These fungal associations are vital to plant establishment and growth in nutrient poor soils or polluted environments [94]. Similarly nitrogen fixing bacteria, like Rhizobium form symbiotic relationships with leguminous plants and converts the atmospheric nitrogen into form which plants can use to increase soil fertility as well as plant growth. In addition, microbial communities may be important to soil health, which is crucial for ecosystem restoration [95]. Microbes break down organic matter spitting out more nutritious nutrients, which are available for plants to use. Because organic matter has been depleted from many ecosystems due to pollution or degradation this decomposition process is critical for maintaining soil structure and fertility. As a key component of ecological recovery in polluted environments, the restoration of soil microbial communities is thus at stake. Microorganisms also affect habitat restoration beside bioremediation and nutrient cycling. For example, in the restoration of wetlands or other aquatic ecosystems, microbial communities provide maintenance of water quality by degrading organic matter to prevent high organic loads and concentrations of waste degrading pollutants. Also microorganisms can sometimes help to revive the physical structure of degraded habitats, for example by building up biofilms that cement the sediments in rivers or lakes [96].

Future Perspectives in Microbial-Based Pollution Management

The future of microbial pollution management holds promise and ongoing research and technology advancement continual promise more efficient and more effective microbial powered environmental remediation strategy development. Novel bioremediation strategies, the use of advanced molecular tools, and the incorporation of microbial communities into ecosystem restoration efforts will be important for overcoming pollution management challenges in the next decades. The use of synthetic biology to engineer microorganisms with boosted abilities of polymonitors degradation represents one exciting area for future research [97]. Researchers can make such strains by manipulating the genetic makeup of the microorganisms so that the pathways through which they metabolize certain pollutants will work better. Genetically engineered bacteria could be engineered to degrade complex industrial chemicals or to more efficiently detoxify heavy metals, for example [98]. This approach has the potential to completely transform the quantification capabilities and bioremediation strategies in place.

The other intriguing twist is the use of metagenomics, in which the full range of microbial community is studied rather than separate species. Science learns which genes and pathways are responsible for breaking down pollutants by sequencing DNA in microbial communities polluted by those products, and which species are previously unknown microbes with their own pollutant degraders [99]. This approach can be used to decipher natural microbial processes in polluted ecology, which will further help to foresee the new bioremediation strategies.

Further development of microbial bioremediation technologies, along with application of microbial communities with other environmental management strategies, including phytoremediation and nanotechnology, is of increasing interest. Integration of microbes with plants or nanoparticles might synergistically increase the overall efficiency of pollution management by combining the biological and physical processes. *Ex*: We can have plants and microbes working together to use a toxic pollutant in soils or sediments or we can use nanoparticles to deliver specific microbial strains to polluted areas [3]. As pollution management continues to grow in complexity, interdisciplinary approaches uniquely that use science, microbiology, and engineering will be critical for creating sustainable and effective solutions. These strategies will continue to rely heavily on microbial communities as they will help to ensure ecosystem resilience and facilitate pollutant restoration. Microbial based pollution management is not only promising but imperative to the future of generation environmental challenges of 21st century.

CONCLUSION

In this chapter the role of microbial ecology in remediation of environmental pollution as a system- with its tremendous resilience, adaptability and functional implications is elucidated. The primary agents in pollutant degradation and transformation are microbial communities, as microbial metabolic flexibility and genetic diversity impact biogeochemical cycles and microbial ecosystem health. Ecological niches identified as key, such as the rhizosphere, biofilms, microbial loop, phycosphere, and geomicrobiological interfaces, were explored to demonstrate their particular roles in pollutant breakdown and nutrient cycling. For example, rhizospheric zone plant-microbe partnerships enhance bioremediation, biofilms improve pollution control and structural as well as metabolic advantages in wastewater and bioreactors [100]. It is important in nutrient recycling and pollutant mineralization and is integral to the maintenance of aquatic and terrestrial ecosystems in the microbial loop; the phycosphere. Geomicrobiology completes the picture by linking microbial activity to geochemical processes in innovative ways to transform pollutants. The critical role of microbial adaptation mechanisms, such as biofilm formation and metabolic pathway variation, in ecosystem resilience and restoration is underscored by their importance in understanding. By developing these insights, sustainable and effective microbial based remediation strategies developed for diverse pollution scenarios are possible. Microbial ecology holds the promise of microbial ecological solutions to environmental sustainability. Microbial dynamics and ecological niches serve as a basis for promoting robust remediation solutions and enhancing ecosystem resilience in the setting of escalating environmental problems.

REFERENCES

[1]   Mony C, Vandenkoornhuyse P, Bohannan BJM, Peay K, Leibold MA. A landscape of opportunities for microbial ecology research. Front Microbiol 2020; 11: 561427. Available from: https://www.frontiersin.org/journals/microbiology/articles/10.3389/fmicb.2020.561427/full [http://dx.doi.org/10.3389/fmicb.2020.561427] [PMID: 33329422]

[2]   Kugarajah V, Nisha KN, Jayakumar R, Sahabudeen S, Ramakrishnan P, Mohamed SB. Significance of microbial genome in environmental remediation. Microbiol Res 2023; 271: 127360. Available from: https://www.sciencedirect.com/science/article/pii/S0944501323000629 [http://dx.doi.org/10.1016/j.micres.2023.127360] [PMID: 36931127]

[3]   Kochhar N, K I K, Shrivastava S, Ghosh A, Rawat V S, Sodhi K K, *et al.* Perspectives on the microorganism of extreme environments and their applications. Curr Res Microb Sci 2022; 3: 100134-4. Available from: https://www.sciencedirect.com/science/article/pii/S2666517422000311

[4]   Aryal M. Rhizomicrobiome dynamics: A promising path towards environmental contaminant mitigation through bioremediation. J Environ Chem Eng 2024; 12(2): 112221. Available from: https://www.sciencedirect.com/science/article/abs/pii/S2213343724003518 [http://dx.doi.org/10.1016/j.jece.2024.112221]

[5]   Wu L, Weston LA, Zhu S, Zhou X. Editorial: Rhizosphere interactions: root exudates and the rhizosphere microbiome. Front Plant Sci 2023; 14: 1281010. Available from: https://pmc.ncbi.nlm.nih.gov/articles/PMC10509041/

[http://dx.doi.org/10.3389/fpls.2023.1281010] [PMID: 37736613]

[6]   Kebede G, Tafese T, Abda EM, Kamaraj M, Assefa F. Factors influencing the bacterial bioremediation of hydrocarbon contaminants in the soil: Mechanisms and impacts. J Chem 2021; 2021: 1-17. Available from: https://onlinelibrary.wiley.com/doi/10.1155/2021/9823362 [http://dx.doi.org/10.1155/2021/9823362]

[7]   Martínez-Espinosa RM. Halophilic archaea as tools for bioremediation technologies. Appl Microbiol Biotechnol 2024; 108(1): 401. Available from: https://pmc.ncbi.nlm.nih.gov/articles/PMC11217053/ [http://dx.doi.org/10.1007/s00253-024-13241-z] [PMID: 38951176]

[8]   Cao Z, Yan W, Ding M, Yuan Y. Construction of microbial consortia for microbial degradation of complex compounds. Front Bioeng Biotechnol 2022; 10: 1051233. Available from: https://pmc.ncbi.nlm.nih.gov/articles/PMC9763274/ [http://dx.doi.org/10.3389/fbioe.2022.1051233] [PMID: 36561050]

[9]   Saini S, Tewari S, Dwivedi J, Sharma V. Saini S, Tewari S, Dwivedi J, Sharma V. Biofilm-mediated wastewater treatment: a comprehensive review. Mater Adv 2023; 4(6): 1415-43. Available from: https://pubs.rsc.org/en/content/articlehtml/2023/ma/d2ma00945e [http://dx.doi.org/10.1039/D2MA00945E]

[10]   Ayilara MS, Babalola OO. Ayilara M S, Babalola O O. Bioremediation of environmental wastes: the role of microorganisms. Front Agron 2023; 5 Available from: https://www.frontiersin.org/journals/agronomy/articles/10.3389/fagro.2023.1183691/full [http://dx.doi.org/10.3389/fagro.2023.1183691]

[11]   Agarwal S, Jain H. Investigating the potential of floral waste as a vermicompost and dual-functional biosorbent for sustainable environmental management. Water Air Soil Pollut 2024; 235(5): 322-2. [http://dx.doi.org/10.1007/s11270-024-07144-y]

[12]   Zhakypbek Y, Kossalbayev BD, Belkozhayev AM, *et al.* Reducing heavy metal contamination in soil and water using phytoremediation. Plants 2024; 13(11): 1534. Available from: https://www.mdpi.com/2223-7747/13/11/1534 [http://dx.doi.org/10.3390/plants13111534] [PMID: 38891342]

[13]   Islam Md M, Saxena N, Sharma D. Phytoremediation as a green and sustainable prospective method for heavy metal contamination: a review. RSC Sustain 2024; 2(5): 1269-88. Available from: https://pubs.rsc.org/en/content/articlehtml/2024/su/d3su00440f

[14]   Bhandari S, Poudel DK, Marahatha R, *et al.* Microbial enzymes used in bioremediation. J Chem 2021; 2021: 1-17. Available from: https://onlinelibrary.wiley.com/doi/10.1155/2021/8849512 [http://dx.doi.org/10.1155/2021/8849512]

[15]   Igiri BE, Okoduwa SIR, Idoko GO, Akabuogu EP, Adeyi AO, Ejiogu IK. Toxicity and bioremediation of heavy metals contaminated ecosystem from tannery wastewater: A review. J Toxicol 2018; 2018: 1-16. Available from: https://pmc.ncbi.nlm.nih.gov/articles/PMC6180975/ [http://dx.doi.org/10.1155/2018/2568038] [PMID: 30363677]

[16]   Eras-Muñoz E, Farré A, Sánchez A, Font X, Gea T. Microbial biosurfactants: a review of recent environmental applications. Bioengineered 2022; 13(5): 12365-91. Available from: https://pmc.ncbi.nlm.nih.gov/articles/PMC9275870/ [http://dx.doi.org/10.1080/21655979.2022.2074621] [PMID: 35674010]

[17]   Trends in biological processes in industrial wastewater treatment. M. P. Shah, ed.: IOP Publishing, 2024. [http://dx.doi.org/10.1088/978-0-7503-5678-7]

[18]   Raklami A, Meddich A, Oufdou K, Baslam M. Plants—microorganisms-based bioremediation for heavy metal cleanup: Recent developments, phytoremediation techniques, regulation mechanisms, and molecular responses. Int J Mol Sci 2022; 23(9): 5031. [http://dx.doi.org/10.3390/ijms23095031] [PMID: 35563429]

[19]   Upadhyay SK, Srivastava AK, Rajput VD, *et al.* Root exudates: Mechanistic insight of plant growth promoting rhizobacteria for sustainable crop production. Front Microbiol 2022; 13: 916488.
[http://dx.doi.org/10.3389/fmicb.2022.916488] [PMID: 35910633]

[20]   Bortoloti GA, Baron D. Phytoremediation of toxic heavy metals by Brassica plants: A biochemical and physiological approach. Environ Adv 2022; 8: 100204.
[http://dx.doi.org/10.1016/j.envadv.2022.100204]

[21]   Wahab A, Muhammad M, Munir A, *et al.* Role of arbuscular mycorrhizal fungi in regulating growth, enhancing productivity, and potentially influencing ecosystems under abiotic and biotic stresses. Plants 2023; 12(17): 3102.
[http://dx.doi.org/10.3390/plants12173102] [PMID: 37687353]

[22]   Al-Enazi NM, AlTami MS, Alhomaidi E. Unraveling the potential of pesticide-tolerant *Pseudomonas* sp. augmenting biological and physiological attributes of *Vigna radiata* (L.) under pesticide stress. RSC Advances 2022; 12(28): 17765-83.
[http://dx.doi.org/10.1039/D2RA01570F] [PMID: 35765317]

[23]   Jain H. From pollution to progress: Groundbreaking advances in clean technology unveiled. Innov Green Dev 2024; 3(2): 100143-3.
[http://dx.doi.org/10.1016/j.igd.2024.100143]

[24]   Medaura MC, Guivernau M, Moreno-Ventas X, Prenafeta-Boldú FX, Viñas M. Bioaugmentation of native fungi, an efficient strategy for the bioremediation of an aged industrially polluted soil with heavy hydrocarbons. Front Microbiol 2021; 12: 626436.
[http://dx.doi.org/10.3389/fmicb.2021.626436] [PMID: 33868189]

[25]   Donlan RM. Biofilms: microbial life on surfaces. Emerg Infect Dis 2002; 8(9): 881-90.
[http://dx.doi.org/10.3201/eid0809.020063] [PMID: 12194761]

[26]   Liu L, Bilal M, Duan X, Iqbal HMN. Mitigation of environmental pollution by genetically engineered bacteria — Current challenges and future perspectives. Sci Total Environ 2019; 667: 444-54.
[http://dx.doi.org/10.1016/j.scitotenv.2019.02.390] [PMID: 30833243]

[27]   Jain H. Exploring the emergence of sustainable practices in healthcare research and application as a path to a healthier future. In: Prabhakar PK, Leal Filho W, Eds. In preserving health, preserving earth: The path to sustainable healthcare. Cham: Springer Nature Switzerland 2024; pp. 121-37.
[http://dx.doi.org/10.1007/978-3-031-60545-1_7]

[28]   Sauer K, Stoodley P, Goeres DM, *et al.* The biofilm life cycle: expanding the conceptual model of biofilm formation. Nat Rev Microbiol 2022; 20(10): 608-20.
[http://dx.doi.org/10.1038/s41579-022-00767-0] [PMID: 35922483]

[29]   Muhammad MH, Idris AL, Fan X, *et al.* Beyond risk: Bacterial biofilms and their regulating approaches. Front Microbiol 2020; 11: 928.
[http://dx.doi.org/10.3389/fmicb.2020.00928] [PMID: 32508772]

[30]   Lago A, Rocha V, Barros O, Silva B, Tavares T. Bacterial biofilm attachment to sustainable carriers as a clean-up strategy for wastewater treatment: A review. J Water Process Eng 2024; 63: 105368.
[http://dx.doi.org/10.1016/j.jwpe.2024.105368]

[31]   Pozdnyakova N, Dubrovskaya E, Schlosser D, *et al.* Widespread ability of ligninolytic fungi to degrade hazardous organic pollutants as the basis for the self-purification ability of natural ecosystems and for mycoremediation technologies. Appl Sci (Basel) 2022; 12(4): 2164.
[http://dx.doi.org/10.3390/app12042164]

[32]   Jain H. Groundwater vulnerability and risk mitigation: A comprehensive review of the techniques and applications. Groundw Sustain Dev 2023; 22: 100968-8.
[http://dx.doi.org/10.1016/j.gsd.2023.100968]

[33]   Eshamuddin M, Zuccaro G, Nourrit G, Albasi C. The influence of process operating conditions on the microbial community structure in the moving bed biofilm reactor at phylum and class level: A review.

J Environ Chem Eng 2024; 12(4): 113266.
[http://dx.doi.org/10.1016/j.jece.2024.113266]

[34]   Shah M P. Emerging innovative trends in the application of biological processes for industrial wastewater treatment. Elsevier 2024.

[35]   Aznaw A, Birlie B, Teshome B, Jemberie M. Textile effluent treatment methods and eco-friendly resolution of textile wastewater. Case Stud Chem Environ Eng 2022; 6: 100230-0.
[http://dx.doi.org/10.1016/j.cscee.2022.100230]

[36]   Malik S, Kishore S, Dhasmana A, *et al.* A perspective review on microbial fuel cells in treatment and product recovery from wastewater. Water 2023; 15(2): 316.
[http://dx.doi.org/10.3390/w15020316]

[37]   Elmoutez S, Abushaban A, Necibi MC, Sillanpää M, Liu J, Dhiba D, *et al.* Design and operational aspects of anaerobic membrane bioreactor for efficient wastewater treatment and biogas production. Environ Chall 2022; 10: 100671-1. Available from: https://www.sciencedirect.com/science/article/pii/S266701002200227X

[38]   Tariq M, Liu Y, Rizwan A, *et al.* Impact of elevated CO_2 on soil microbiota: A meta-analytical review of carbon and nitrogen metabolism. Sci Total Environ 2024; 950: 175354.
[http://dx.doi.org/10.1016/j.scitotenv.2024.175354] [PMID: 39117202]

[39]   Weiland-Bräuer N. Friends or foes—microbial interactions in nature. Biology (Basel) 2021; 10(6): 496.
[http://dx.doi.org/10.3390/biology10060496] [PMID: 34199553]

[40]   Evans C, Brandsma J, Meredith MP, *et al.* Shift from carbon flow through the microbial loop to the viral shunt in coastal antarctic waters during austral summer. Microorganisms 2021; 9(2): 460.
[http://dx.doi.org/10.3390/microorganisms9020460] [PMID: 33672195]

[41]   Philippe Cherabier, Régis Ferrière, Eco-evolutionary responses of the microbial loop to surface ocean warming and consequences for primary production The ISME Journa 2022; 16(4): 1130-9.
[http://dx.doi.org/10.1038/s41396-021-01166-8]

[42]   Aziz ZS, Jazza SH, Dageem HN, Banoon SR, Balboul BA, Abdelzaher MA. Bacterial biodegradation of oil-contaminated soil for pollutant abatement contributing to achieve sustainable development goals: a comprehensive review. Results Eng 2024; 22: 102083-3.
[http://dx.doi.org/10.1016/j.rineng.2024.102083]

[43]   Naylor D, McClure R, Jansson J. Trends in microbial community composition and function by soil depth. Microorganisms 2022; 10(3): 540.
[http://dx.doi.org/10.3390/microorganisms10030540] [PMID: 35336115]

[44]   Kang J, Luo Z, Mohamed HF, *et al.* Environmental regulation of photosynthetically produced dissolved organic carbon by phytoplankton along a subtropical estuarine bay. Front Mar Sci 2022; 9: 813401.
[http://dx.doi.org/10.3389/fmars.2022.813401]

[45]   Janusz G, Pawlik A, Sulej J, Świderska-Burek U, Jarosz-Wilkołazka A, Paszczyński A. Lignin degradation: microorganisms, enzymes involved, genomes analysis and evolution. FEMS Microbiol Rev 2017; 41(6): 941-62.
[http://dx.doi.org/10.1093/femsre/fux049] [PMID: 29088355]

[46]   Alabssawy AN, Hashem AH. Bioremediation of hazardous heavy metals by marine microorganisms: A recent review. Arch Microbiol 2024; 206(3): 103.
[http://dx.doi.org/10.1007/s00203-023-03793-5] [PMID: 38358529]

[47]   Akinnawo SO. Akinnawo S O. Eutrophication: causes, consequences, physical, chemical and biological techniques for mitigation strategies. Environ Chall 2023; 12: 100733-3. Available from: https://www.sciencedirect.com/science/article/pii/S2667010023000574

[48]   Popli S, Badgujar PC, Agarwal T, Bhushan B, Mishra V. Persistent organic pollutants in foods, their

interplay with gut microbiota and resultant toxicity. Sci Total Environ 2022; 832: 155084.
[http://dx.doi.org/10.1016/j.scitotenv.2022.155084] [PMID: 35395291]

[49]   Hellal J, Barthelmebs L, Bérard A, *et al.* Unlocking secrets of microbial ecotoxicology: recent achievements and future challenges. FEMS Microbiol Ecol 2023; 99(10): fiad102.
[http://dx.doi.org/10.1093/femsec/fiad102] [PMID: 37669892]

[50]   Bala S, Garg D, Thirumalesh BV, *et al.* Recent strategies for bioremediation of emerging pollutants: A review for a green and sustainable environment. Toxics 2022; 10(8): 484.
[http://dx.doi.org/10.3390/toxics10080484] [PMID: 36006163]

[51]   Kim H, Kimbrel JA, Vaiana CA, Wollard JR, Mayali X, Buie CR. Bacterial response to spatial gradients of algal-derived nutrients in a porous microplate. ISME J 2022; 16(4): 1036-45.
[http://dx.doi.org/10.1038/s41396-021-01147-x] [PMID: 34789844]

[52]   Fracchia F, Guinet F, Engle NL, Tschaplinski TJ, Veneault-Fourrey C, Deveau A. Microbial colonisation rewires the composition and content of poplar root exudates, root and shoot metabolomes. Microbiome 2024; 12(1): 173.
[http://dx.doi.org/10.1186/s40168-024-01888-9] [PMID: 39267187]

[53]   Ramanan R, Kim BH, Cho DH, Oh HM, Kim HS. Algae–bacteria interactions: Evolution, ecology and emerging applications. Biotechnol Adv 2016; 34(1): 14-29.
[http://dx.doi.org/10.1016/j.biotechadv.2015.12.003] [PMID: 26657897]

[54]   Bowman JP. Bioactive compound synthetic capacity and ecological significance of marine bacterial genus pseudoalteromonas. Mar Drugs 2007; 5(4): 220-41.
[http://dx.doi.org/10.3390/md504220] [PMID: 18463726]

[55]   Mehdizadeh Allaf M, Peerhossaini H. Cyanobacteria: Model microorganisms and beyond. Microorganisms 2022; 10(4): 696.
[http://dx.doi.org/10.3390/microorganisms10040696] [PMID: 35456747]

[56]   Volmer JG, McRae H, Morrison M. The evolving role of methanogenic archaea in mammalian microbiomes. Front Microbiol 2023; 14: 1268451.
[http://dx.doi.org/10.3389/fmicb.2023.1268451] [PMID: 37727289]

[57]   Tong D, Wang Y, Yu H, Shen H, Dahlgren RA, Xu J. Viral lysing can alleviate microbial nutrient limitations and accumulate recalcitrant dissolved organic matter components in soil. ISME J 2023; 17(8): 1247-56.
[http://dx.doi.org/10.1038/s41396-023-01438-5] [PMID: 37248401]

[58]   Fuentes J, Garbayo I, Cuaresma M, Montero Z, González-del-Valle M, Vílchez C. Impact of microalgae-bacteria interactions on the production of algal biomass and associated compounds. Mar Drugs 2016; 14(5): 100.
[http://dx.doi.org/10.3390/md14050100] [PMID: 27213407]

[59]   Touliabah HES, El-Sheekh MM, Ismail MM, El-Kassas H. A review of microalgae- and cyanobacteria-based biodegradation of organic pollutants. Molecules 2022; 27(3): 1141.
[http://dx.doi.org/10.3390/molecules27031141] [PMID: 35164405]

[60]   Raffa CM, Chiampo F. Bioremediation of agricultural soils polluted with pesticides: A review. Bioengineering (Basel) 2021; 8(7): 92.
[http://dx.doi.org/10.3390/bioengineering8070092] [PMID: 34356199]

[61]   Rothman DH. Slow closure of Earth's carbon cycle. Proc Natl Acad Sci USA 2024; 121(4): e2310998121.
[http://dx.doi.org/10.1073/pnas.2310998121] [PMID: 38241442]

[62]   Ni G, Leung PM, Daebeler A, *et al.* Nitrification in acidic and alkaline environments. Essays Biochem 2023; 67(4): 753-68.
[http://dx.doi.org/10.1042/EBC20220194] [PMID: 37449414]

[63]   Christie-Oleza JA, Sousoni D, Lloyd M, Armengaud J, Scanlan DJ. Nutrient recycling facilitates long-

term stability of marine microbial phototroph–heterotroph interactions. Nat Microbiol 2017; 2(9): 17100.
[http://dx.doi.org/10.1038/nmicrobiol.2017.100] [PMID: 28650444]

[64] Mushinski RM, Payne ZC, Raff JD, *et al.* Nitrogen cycling microbiomes are structured by plant mycorrhizal associations with consequences for nitrogen oxide fluxes in forests. Glob Change Biol 2021; 27: 1068-82.
[http://dx.doi.org/10.1111/gcb.15439]

[65] Cockell CS. Geomicrobiology beyond Earth: microbe–mineral interactions in space exploration and settlement. Trends Microbiol 2010; 18(7): 308-14.
[http://dx.doi.org/10.1016/j.tim.2010.03.005] [PMID: 20381355]

[66] Dong H, Huang L, Zhao L, *et al.* A critical review of mineral–microbe interaction and co-evolution: mechanisms and applications. Natl Sci Rev 2022; 9(10): nwac128.
[http://dx.doi.org/10.1093/nsr/nwac128] [PMID: 36196117]

[67] Wang C, Kuzyakov Y. Mechanisms and implications of bacterial–fungal competition for soil resources. ISME J 2024; 18(1): wrae073.
[http://dx.doi.org/10.1093/ismejo/wrae073] [PMID: 38691428]

[68] Wróbel M, Śliwakowski W, Kowalczyk P, Kramkowski K, Dobrzyński J. Bioremediation of heavy metals by the genus *bacillus*. Int J Environ Res Public Health 2023; 20(6): 4964.
[http://dx.doi.org/10.3390/ijerph20064964] [PMID: 36981874]

[69] Hussain B, Zhu H, Xiang C, *et al.* Evaluation of the immobilized enzymes function in soil remediation following polycyclic aromatic hydrocarbon contamination. Environ Int 2024; 194: 109106.
[http://dx.doi.org/10.1016/j.envint.2024.109106] [PMID: 39571295]

[70] Luo G, Xue C, Jiang Q, *et al.* Soil carbon, nitrogen, and phosphorus cycling microbial populations and their resistance to global change depend on soil C:N:P Sstoichiometry. mSystems 2020; 5(3)
[http://dx.doi.org/10.1128/msystems.00162-20] [PMID: 32606023]

[71] Radhakrishnan, A.; Balaganesh, P.; Vasudevan, M.; *et al.* Bioremediation of hydrocarbon pollutants: Recent promising sustainable approaches, scope, and challenges. Sustainability 2023; 15: 5847.
[http://dx.doi.org/10.3390/su15075847]

[72] Zeng S, Dai Z, Ma B, Dahlgren RA, Xu J. Environmental interactions and remediation strategies for co-occurring pollutants in soil. Deleted J 2024; 1(1): 100002-2.
[http://dx.doi.org/10.1016/j.ecz.2024.100002]

[73] Jing R, Kjellerup BV. Biogeochemical cycling of metals impacting by microbial mobilization and immobilization. J Environ Sci (China) 2018; 66: 146-54.
[http://dx.doi.org/10.1016/j.jes.2017.04.035] [PMID: 29628081]

[74] Drewniak L, Sklodowska A. Arsenic-transforming microbes and their role in biomining processes. Environ Sci Pollut Res Int 2013; 20(11): 7728-39.
[http://dx.doi.org/10.1007/s11356-012-1449-0] [PMID: 23299972]

[75] Plestenjak E, Kraigher B, Leskovec S, *et al.* Reduction of hexavalent chromium using bacterial isolates and a microbial community enriched from tannery effluent. Sci Rep 2022; 12(1): 20197.
[http://dx.doi.org/10.1038/s41598-022-24797-z] [PMID: 36418532]

[76] Wu YS, Osman AI, Hosny M, *et al.* The toxicity of mercury and its chemical compounds: molecular mechanisms and environmental and human health implications: A comprehensive review. ACS Omega 2024; 9(5): 5100-26.
[http://dx.doi.org/10.1021/acsomega.3c07047] [PMID: 38343989]

[77] Nikel PI, Pérez-Pantoja D, de Lorenzo V. Why are chlorinated pollutants so difficult to degrade aerobically? Redox stress limits 1,3-dichloprop-1-ene metabolism by *Pseudomonas pavonaceae*. Philos Trans R Soc Lond B Biol Sci 2013; 368(1616): 20120377.
[http://dx.doi.org/10.1098/rstb.2012.0377] [PMID: 23479756]

[78]   Elyamine AM, Kan J, Meng S, Tao P, Wang H, Hu Z. Aerobic and anaerobic bacterial and fungal degradation of pyrene: Mechanism pathway including biochemical reaction and catabolic genes. Int J Mol Sci 2021; 22(15): 8202.
[http://dx.doi.org/10.3390/ijms22158202] [PMID: 34360967]

[79]   Sarkar A, Paul B, Darbha GK. The groundwater arsenic contamination in the Bengal Basin-A review in brief. Chemosphere 2022; 299: 134369.
[http://dx.doi.org/10.1016/j.chemosphere.2022.134369] [PMID: 35318018]

[80]   2024. Deepwater Horizon – BP gulf of mexico oil spill | US EPA. US EPA. 2024. Available from: https://www.epa.gov/enforcement/deepwater-horizon-bp-gulf-mexico-oil-spill

[81]   Murphy SMC, Bautista MA, Cramm MA, Hubert CRJ. Diesel and crude oil biodegradation by cold-adapted microbial communities in the labrador sea. Appl Environ Microbiol 2021; 87(20): e00800-21.
[http://dx.doi.org/10.1128/AEM.00800-21] [PMID: 34378990]

[82]   Harun F A, Yusuf M R, Usman S, Shehu D, Babagana K, Sufyanu A J, *et al.* Bioremediation of lead contaminated environment by *Bacillus* cereus strain BUK_BCH_BTE2: isolation and characterization of the bacterium. Case Stud Chem Environ Eng 2023; 8: 100540-0. Available from: https://www.sciencedirect.com/science/article/pii/S2666016423002451

[83]   Sharma M, Agarwal S, Agarwal Malik R, *et al.* Recent advances in microbial engineering approaches for wastewater treatment: a review. Bioengineered 2023; 14(1): 2184518.
[http://dx.doi.org/10.1080/21655979.2023.2184518] [PMID: 37498651]

[84]   Pacciani-Mori L, Giometto A, Suweis S, Maritan A. Dynamic metabolic adaptation can promote species coexistence in competitive microbial communities. PLOS Comput Biol 2020; 16(5): e1007896.
[http://dx.doi.org/10.1371/journal.pcbi.1007896] [PMID: 32379752]

[85]   Lu Z, Imlay JA. When anaerobes encounter oxygen: mechanisms of oxygen toxicity, tolerance and defence. Nat Rev Microbiol 2021; 19(12): 774-85.
[http://dx.doi.org/10.1038/s41579-021-00583-y] [PMID: 34183820]

[86]   Sharma S, Mohler J, Mahajan SD, Schwartz SA, Bruggemann L, Aalinkeel R. Microbial biofilm: A review on formation, infection, antibiotic resistance, control measures, and innovative treatment. Microorganisms 2023; 11(6): 1614.
[http://dx.doi.org/10.3390/microorganisms11061614] [PMID: 37375116]

[87]   Burmeister AR. Horizontal Gene Transfer: Figure 1. Evol Med Public Health 2015; 2015(1): 193-4.
[http://dx.doi.org/10.1093/emph/eov018] [PMID: 26224621]

[88]   Dittmann C, Han HM, Grabenbauer M, Laue M. Dormant *Bacillus* spores protect their DNA in crystalline nucleoids against environmental stress. J Struct Biol 2015; 191(2): 156-64.
[http://dx.doi.org/10.1016/j.jsb.2015.06.019] [PMID: 26094877]

[89]   Elshafei AM, Mansour R. Elshafei A M, Mansour R. Microbial bioremediation of soils contaminated with petroleum hydrocarbons. Discover Soil 2024; 1(1)
[http://dx.doi.org/10.1007/s44378-024-00004-5]

[90]   Sayed K, Baloo L, Sharma NK. Bioremediation of Total Petroleum Hydrocarbons (TPH) by bioaugmentation and biostimulation in water with floating oil spill containment booms as bioreactor basin. Int J Environ Res Public Health 2021; 18(5): 2226.
[http://dx.doi.org/10.3390/ijerph18052226] [PMID: 33668225]

[91]   Holmes DE, Dang Y, Smith JA. Nitrogen cycling during wastewater treatment. Adv Appl Microbiol 2019; 106: 113-92.
[http://dx.doi.org/10.1016/bs.aambs.2018.10.003] [PMID: 30798802]

[92]   Govil T, Rathinam NK, Salem DR, Sani RK. Taxonomical diversity of extremophiles in the deep biosphere. Elsevier eBooks 2018; 631-56. Available from: https://www.sciencedirect.com/topics/agricultural-and-biological-sciences/sulfur-reducing-bacteria

[93] Etesami H, Jeong BR, Glick BR. Contribution of arbuscular mycorrhizal fungi, phosphate–solubilizing bacteria, and silicon to P uptake by plant. Front Plant Sci 2021; 12: 699618.
[http://dx.doi.org/10.3389/fpls.2021.699618] [PMID: 34276750]

[94] Zahran HH. Rhizobium-legume symbiosis and nitrogen fixation under severe conditions and in an arid climate. Microbiol Mol Biol Rev 1999; 63(4): 968-89.
[http://dx.doi.org/10.1128/MMBR.63.4.968-989.1999] [PMID: 10585971]

[95] Wang J, Long Y, Yu G, *et al.* A review on microorganisms in constructed wetlands for typical pollutant removal: species, function, and diversity. Front Microbiol 2022; 13: 845725.
[http://dx.doi.org/10.3389/fmicb.2022.845725] [PMID: 35450286]

[96] Nhgri. Synthetic Biology. Genome.gov. 2019. Available from: https://www.genome.gov/about-genomics/policy-issues/Synthetic-Biology

[97] Pande V, Pandey SC, Sati D, Bhatt P, Samant M. Microbial interventions in bioremediation of heavy metal contaminants in agroecosystem. Front Microbiol 2022; 13: 824084.
[http://dx.doi.org/10.3389/fmicb.2022.824084] [PMID: 35602036]

[98] Inglis GD, Enkerli J, Goettel MS. Laboratory techniques used for entomopathogenic fungi. Elsevier eBooks 2012; 189-253.
[http://dx.doi.org/10.1016/B978-0-12-386899-2.00007-5]

[99] Yamini V, Shanmugam V, Rameshpathy M, *et al.* Environmental effects and interaction of nanoparticles on beneficial soil and aquatic microorganisms. Environ Res 2023; 236(Pt 1): 116776.
[http://dx.doi.org/10.1016/j.envres.2023.116776] [PMID: 37517486]

[100] Rhizospheric microbiota-plant interactions: a bioremediation strategy for inorganic pollutants. Front Res Topic 2022. Available from: https://www.frontiersin.org/research-topics/35930/rhizospheri--microbiota-plant-interactions-a-bioremediation-strategy-for-inorganic-pollutants/magazine

Microbes-assisted Sequestration: A Sustainable Solution for Environmental Pollution

Paurabi Das[1] and Nilanjan Chakraborty[2,*]

[1] *Crop Production and Protection Division, CSIR- Central Institute of Medicinal and Aromatic Plants, Lucknow-226015, India*

[2] *Scottish Church College, Kolkata-700006, India*

Abstract: In recent years, environmental contamination by toxic pollutants has become a major concern due to irreversible ecological damage. The major hazardous substances like petrochemicals, agrochemicals, pharmaceuticals, nanomaterials, pesticides, and herbicides are generated by industrialization and urbanization. They are either consciously or inadvertently discharged into the water and soil system, endangering human health, animal health, and biodiversity. Numerous physicochemical techniques have been used for this. However, they have a lot of drawbacks, including high costs, labour costs, alteration to the soil properties, perturbation of the natural soil microflora, and the production of hazardous byproducts. To address this complex issue, namely, the removal, immobilization, and detoxification of these pollutants, microbe-assisted sequestration bioremediation techniques are gaining interest from researchers worldwide. Microorganisms have contributed reasonably to restoring the natural state of degraded environments with long-term environmental benefits by becoming resistant to intoxicants and developing the ability to remediate various pollutants. Microbes have a wide range of sequestration capabilities, making them suitable for biosorption interactions with pollutants. This chapter discusses how various microorganisms sequester and degrade different pollutants. A brief overview of molecular techniques like systemic biology, gene editing, and omics is also provided. These techniques have improved the bioremediation process enormously.

Keywords: Bioremediation, Detoxification, Immobilization, Microbes, Sequestration.

INTRODUCTION

In recent decades, various organic pollutants like petroleum fuels, solvents, microplastics, heavy metals, and pesticides have been synthesized and discharged into the environment in numerous ways [1]. These pollutants are highly persistent and low-degradable and can become trapped in soil and water bodies, leading to

---

* **Corresponding author Nilanjan Chakraborty:** Scottish Church College, Kolkata-700006, India; E-mail: nilanjan.chakraborty@scottishchurch.ac.in

Harshita Jain & Maulin P. Shah (Eds.)

bioaccumulation [2]. Their potential adverse effects on human health and multiple environments, including aquatic, terrestrial, and atmospheric, have raised significant environmental concerns, necessitating the restoration of contaminated sites [1]. Various ineffective and expensive methods for cleaning pollutants, such as pyrolysis, oxidative thermal treatment, air-sparging, and incineration, have led to the generation of toxic and recalcitrant by-products. With global environmental awareness increasing, microbes capable of degrading pollutants and their derivatives have gained attention. Utilizing native microbial strains for the elimination of aromatic pollutants offers benefits such as environmental safety, cost-effectiveness, efficiency, and sustainability, whether applied individually or in mixed cultures [3]. Bioremediation, a sustainable, eco-friendly, and cost-effective method for removing organic pollutants from contaminated sites, has been extensively studied for its ability to degrade pollutants [4]. However, bioremediation often fails due to various factors, including low degradability, viability, pollutant availability, nutrient depletion, pH, oxygen, temperature, and moisture at the site. Therefore, the use of microorganisms responsible for pollutant degradation at contaminated sites is crucial for successful bioremediation [5]. Enrichment cultures mimicking contaminated environments are effective approaches to identify and characterize key players responsible for pollutant degradation. This approach helps identify and isolate microorganisms capable of actively degrading organic pollutants at contaminated sites. Contaminated environments lack essential nutrients for microbial growth and metabolism. Bioremediation strategies, such as biostimulation and bioaugmentation, are popular due to the knowledge of participating microbial communities and their nutrient requirements. Biostimulation is supplementing microorganisms with nutrients to accelerate their growth and metabolic abilities. This can be achieved through water-soluble inorganic nutrients, slow-release fertilizers, and oxygenation. Bioaugmentation involves externally supplementing microbes, either as a single pure culture or a mixed culture, to enhance their overall metabolic activity for complete degradation. Pure cultures of *Pseudomonas*, *Flavobacterium*, *Sphingomonas*, *Achromobacter*, *Bacillus*, and *Rhodococcus* are promising bioaugmentation agents. Mixed bacterial cultures are more advantageous due to synergistic interactions among microbial species [3]. This chapter aims to comprehensively describe the pollutant removal mechanism and emphasizes the microbe-assisted sequestration and bioremediation of different contaminants.

MICROBES IN BIOREMEDIATION OF PESTICIDES

Modern agriculture has led to significant use of pesticides to meet crop production demands and their economic importance. However, prolonged use has made pests resistant to these synthetic chemicals, resulting in increased usage. This can lead to pollution and toxicity, as excessive pesticides can leach into groundwater and contaminate surface water bodies like lakes, ponds, and rivers. Excessive use of pesticides can lead to pollution and toxicity, as they can leach into adjacent water bodies and contaminate groundwater. Spray dispersion and runoff from agricultural areas are the primary sources of pollution [6, 7]. The chemical composition of the pesticide determines its degradation pathway [8]. According to their chemical composition, the pesticides can be divided into chlorinated and non-chlorinated groups. Organochlorine pesticides, like eldrin, dieldrin, and endosulfan, are persistent and easily absorbed by biotic components, leading to food contamination. They are biologically stable, persistent due to their lipophilic nature, and slow in natural degradation [9]. Their toxic properties vary based on the position of the chlorine molecule, with a decrease in toxicity when the chlorine atom is substituted with a methoxide group. Excessive usage of pesticides causes soil and water contamination since prohibited pesticides such as dichlorodiphenyltrichloroethane are detectable over 20 years after application [10]. Since they dissolve in organic solvents, Organophosphorus Pesticides (OPP) may penetrate the soil and contaminate the groundwater. They are used as insecticides on various fruits, vegetables, and ornamentals. Methyl parathion (O, O-dimethyl-O-p-nitro phenylphosphorothioate) is used regularly on various agricultural plants like rice, onion, spinach, peach, and strawberries [10]. A mixture of two or more pesticides is used to increase insecticide efficiency. Carbamate pesticides, primarily carbamic acid derivatives, are low-persistence pesticides used as insecticides, herbicides, or fungicides due to their versatility [11]. Carbamates are water soluble and thermally unstable, making them highly toxic to vertebrates. Pyrethroids are derived from pyrethrin, another widely used pesticide obtained from chrysanthemum flowers. The pyrethrins usually have two functional groups: acid and alcohol. Two kinds of pyrethrins can be distinguished by their chemical composition and pest-repelling mechanism. Permethrin is the type I pyrethrin, which lacks the cyano group. On the other hand, deltamethrin is the type II pyrethrin which contains the cyano group at the phenyl benzyl alcohol position. Pyrethroids are less persistent and prone to photodegradation. In insects, they impact the neurological system and cause muscle paralysis and death by delaying the opening of sodium channels [12]. The microbial degradation of various pesticides is represented in Tables **1** and **2**.

Lindane is a broad-spectrum organochlorine pesticide synthesized after WWII. It has insecticidal properties due to its excitatory action on the nervous system. Over the past seven decades, it has been widely utilized globally, targeting various crops, animals, and animal premises [13]. The production of hexachloro-cyclohexane produces four major isomers, with only the γ isomers possessing insecticidal properties [14]. Numerous lindane-degrading bacterial strains have been screened and implemented in lindane-contaminated sites, including

Microbacterium, Paracoccus, Burkholderia, Rhodococcus, Chromohalobacter, Staphylococcus, Streptomyces, Arthrobacter, Azotobacter, Sphingomonas, Xanthomonas, Pseudomonas, Klebsiella, Pleurotus, Fusarium, and *Actinobacteria* [15]. Bacteria are important in the biodegradation of lindane because of their chemical and physical interactions, which result in structural alterations or full disintegration. Different bacteria degrade lindane at different rates, with some being able to mineralize it completely. According to a prior study, *Microbacterium* spp. strain P27 obtained from *Phragmites karka* destroyed 82.7% lindane in 15 days while fostering plant development [16]. According to a different study, *Paracoccus sp.* NITDBR1 from Manipur used lindane as a carbon source and broke it down 90% in 8 days [17]. In eight days, the *Burkholderia spp.* strain IPL04 broke down 98% of the lindane [18]. Fungal biodegradation is an eco-friendly method for detoxifying POPs due to its strong activity of non-specific enzymes. However, only a few bacteria are known to mineralize lindane completely.

Table 1. The microbial degradation of different pesticides by various bacteria.

Degrading pesticide	Species (Bacteria)	Degradation rate (%)	References
Aldrin	*Pseudomonas fluorescens*	94.8	[19]
	Pseudomonas fluorescens	43.2	[20]
	Bacillus polymyxa	48.2	[19]
	Flavobacteria	27	
	Micrococcus	24.2	
Endosulfan	*Pseudomonas fluorescens*	100	[21]
Alpha endosulfan	*Klebsiella*	95	[22]
	Acinetobacter	90	[23]
	Flavobacterium	75	
	Bacillus	75	
	Bordetella sp.	80	
	Pandoraea sp.	95–100	
	Achromobacter xylosoxidans	100	
Beta endosulfan	*Klebsiella*	85	[23]
	Acinetobacter	90	
	Flavobacterium	95	
	Bacillus	80	
	Bordetella	86	
	Pandoraea sp.	95-100	

(Table 1) cont.....

Degrading pesticide	Species (Bacteria)	Degradation rate (%)	References
Lindane	*Streptomyces sp. A5*	-	[24]
	Microbacterium spp. strain P27	82.7	[16]
	Paracoccus sp.NITDBR1	90	[17]
	Achromobacter sp.strain	88.7	[25]
	Burkholderia sp. IPL04	98	[18]
	Rhodococcus wratislaviensis Ch628	32.3	[26]
	Chromohalobactersp.LD2	89.6	[27]
	Kocuriasp.DAB-1Y	94	[28]
	Staphylococcus sp. DAB-1W	98	[28]
	Streptomyces sp.M7	45	[29]
Dichlorodiphenyltrichloroethane	*Stenotrophomonas sp*	55.0	[30]
Dichlorodiphenyldichloroethylene	*Stenotrophomonas sp*	36.7	[30]
Carbofuran	*Enterobactercloacae TA7*	-	[31]
-	*Cupriavidussp. ISTL7*	-	[32]
-	*Enterobacter sp.*	-	[33]
-	*Sphingbiumsp.CFD-1*	-	[34]
Carbaryl	*Enterobactercloaceae TA7*	-	[31]
	Stenotrophomonas sp.YC-1	-	[35]
	Rhodopseudomonas capsulata	-	[36]
	Rhodopseudomonas sphaeroides	-	[37]
Methomyl	*Escherichia coli*	-	[38]
	Bacillus cereus and B. safensis	-	[39]
	Aminobacter sp.MDW-2 and Afipia sp. MDW-3	-	[40]
Oxamyl	*Pseudomonas sp. OXA20*	-	[41]
	Micrococcus luteus	-	[42]
Aldicarb	*Enterobacter cloaceae TA7*	-	[31]
	Trametes versicolor	-	[43]

(Table 1) cont.....

Degrading pesticide	Species (Bacteria)	Degradation rate (%)	References
Chlorpyrifos	*Shewanella sp. BT05*	94.3	[44]
	Pseudomonas nitroreducens AR-3	97	[45]
	Bacillus licheniformis BHUJP-P3	49	[46]
	Bacillus cereus BHUJP- P4	87	
	Pseudomonas sp.CB2	74.6	[47]
	Sphingomonas sp. HJY	96	[48]
	Pseudomonas kilonensis SRK1	86.6	[49]
	Ochrobactrum sp.FCp1 Achromobacter xylosoxidans JCp4	78.6	[50]
	Agrobacterium tumefaciens strain ECO1, Cellulosimicrobium funkei strain ECO2, Shinella zoogloeoides strain ECO3 and Bacillus aryabhattai strain ECO4	100	[51]

Table 2. The microbial degradation of different pesticides by various fungi.

Degrading pesticide	Species (Fungi)	Degradation rate (%)	References
Heptachlor	*Phlebia tremellosa*	71	[52]
	Phlebia Brevispora	74	[9]
	Phlebia acanthocystis	90	
Heptachlor epoxide	*Phlebia acanthocystis*	16	[52]
	Phlebia Brevispora	16	
	Phlebia lindtneri	22	
	Phlebia aurea	25	
Hexachlorocyclohexane	*Phanerochaete chrysosporium*	-	[53]
Lindane	*Phanerochaete chrysosporium*	> 10	[54]
	Pleurotus sajorcaju	> 10	[55]
	Pleurotus eryngii		[56]
	Ganoderma lucidum GL-2	75.5	[55]
Dichlorodiphenyltrichloroethane	*Cladosporium sp*	-	[30]
	Trichoderma hamatum	-	[57]
Dieldrin	*Phlebia aurea*	54	[19]

Chlorpyrifos (CP) is an extensively used organophosphorus insecticide with a half-life of 60-120 days in soil due to its hydrophobic nature, which varies depending on various factors. Its effectiveness ranges from 2 weeks to over 1 year. A consortium of bacterial strains, including *Agrobacterium tumefaciens*, *Cellulosimicrobium funkei*, *Shinella zoogloeoides*, and *Bacillus aryabhattai*, degraded 100% of CP within 6 days. GC-MS analysis confirmed the emergence and degradation of metabolites [58].

Carbofuran is a toxic, broad-spectrum, systemic pesticide used in agriculture to control soil-dwelling and leaf-feeding insects. It belongs to class I N-methyl carbamate pesticides with ester and amide linkages. Carbofuran-related research primarily involves bacterial degradation from *Flavobacterium*, *Pseudomonas*, *Sphingomonas*, *Archomobacter*, *Bacillus*, and *Enterococcus* [59]. Tondon *et al.* (2018) identified four bacterial strains—*Arthrobacter globiformis*, *Bhargavaea indica*, *Bacillus beijingensis*, and *Streptomyces* sp.—from contaminated soil, with degradation efficiencies of 43%, 55%, 35%, and 44%, respectively [60]. "Gupta *et al.* (2019) found that *Cupriavidus sp.* ISTL7 achieved 98% carbofuran degradation in 96 hours, a process aided by its exo-polysaccharide production [32]. *Stenotrophomonas maltophilia* OG2 degraded 81.53% of α-endosulfan in 10 days *via* hydrolysis, influenced by concentration, temperature, and pH [61].

Zaffar *et al.* (2018) found Stenotrophomonas maltophilia EN-1 effective in degrading endosulfan [62]. *Pseudomonas mendocina* ZAM1 degraded over 64.5% of endosulfan in 12 days [63]. Odukkathil and Vasudevan (2015) reported *Bordetella petrii* I and II degraded up to 89% of endosulfan with biosurfactant production [64]. Jimenez-Torres *et al.* (2016) found that *Enterobacter cloacae* PMM16 completely degraded β-endosulfan and 71.32% of α-endosulfan within 24 days through non-oxidative pathways [65].

Mechanism

Pesticide degradation occurs through hydrolysis, redox reactions, and expansion, forming new metabolites. Hydrolysis is influenced by hydrophilic environmental factors, followed by oxidation or reduction, which modifies the redox state. The chemical group determines whether oxidation or reduction occurs. Lastly, expansion leads to the formation of a new reaction product or metabolite. Microbial enzymes drive these processes, influencing biodegradation based on environmental factors, bioavailability, and metabolic potential. Key reactions include oxidation, dehalogenation, nitro group modification, ring cleavage, and hydroxylation [66].

Pesticide metabolism involves two systems: the first stage involves hydrolases, esterases, and mixed-function oxidases, whereas the second stage involves glutathione S transferases. Hydrolases degrade pesticides without a redox cofactor, aided by various hydrolyzing enzymes. Phosphotriesterases (PTEs) are proteins produced by microorganisms with the *opd* gene, which reduces the ability of organophosphates to inactivate Acetylcholinesterase (AchE). They detoxify and hydrolyze pesticides containing organophosphates by activating water molecules, producing an inverted configuration by attacking the phosphorus at the centre of the pesticides, and polarizing the active site. Esterases are essential hydrolases for the metabolism of organic pollutants, with two major types: Estrases A and Estrases B. Estrase A hydrolyzes pollutants with the –SH functional group, while Estrase B hydrolyzes endogenous compounds and pollutants with ester, amide, thioester, phosphate esters, and acid anhydrides in mammals. The organophosphate interacts with the Ser-OH bond, forming a P=O bond that is hydrolyzed by water. Mixed Function Oxidases (MFO) are enzyme systems that use membrane proteins, cytochrome P450 and NADPH-cytochrome P450 reductase, to hydrolyze and oxidize substrates. Glutathione-S-Transferase (GST) catalyzes GSH conjugation to an electrophilic site, which is crucial for pesticide detoxification by introducing xenobiotics. Oxidoreductases are enzymes that catalyze oxidation/reduction reactions, crucial in pesticide detoxification. The process involves three phases: Phase I, Phase II, and Phase III. Phase I produces water-soluble, less toxic products through oxidation, reduction, and hydrolysis. Phase II involves conjugating pesticides with sugar or amino acids to increase solubility and reduce toxicity. Phase II metabolites are converted into non-toxic secondary conjugates in the final phase [67, 68].

MICROBES IN MICROPOLLUTANTS (INCLUDING MICROPLASTICS) SEQUESTRATION

Researchers are exploring the prospects of function-based microbial species for the remediation of various contaminants. These organisms are intentionally optimized and organized for experimentation. The consortium is based on cellular communication *via* molecular signaling by metabolites, dividing labor to output specific functionality. Using multiple substrates in parallel fulfills the Maximum Power Principle (MPP), making it superior to single species for designing microbial interactions for bioremediation in environmental management. This approach aims to optimize microbial interactions for effective environmental management. The study found that *Bacillus subtilis* MZA-75 and *Pseudomonas aeruginosa* MZA-85 degraded polyester polyurethane [69]. Coculture of these microbes achieved maximum weight reduction in 20 days with higher CO_2 evolution. Thermophilic consortia from sewage and landfills enhanced PP, LDPE, and HDPE degradation, showing the highest weight reduction at various temperatures. Mixed floras (*Exiguobacterium* sp., *Halomonas* sp., *Ochrobactrum* sp.) outperform single bacteria in PET and PE degradation due to synergistic effects, improving adaptability, efficiency, and bioremediation potential [70]. PET Hydrolytic Enzymes (PHEs) enable plastic waste bioremediation, with *Ideonella sakaiensis* and a microbial consortium using PET and MHET hydrolases for degradation [71].

Xu *et al.* (2020) found tht *Paracoccus kondratievae* effectively degrades DEHP into MEHP, PA, benzoic acid, 2-ethylhexanol, and acetic acid within 150 hours [72]. Wright *et al.* (2020) found that *Halomonas* sp. and *Mycobacterium* sp. degrade PAEs, metabolizing DEHP into MEHP and PA, with benzoic acid as a key intermediate, likely *via* esterase, lipase, or cytochrome P450 [73]. Various bacteria, including *Pseudomonas* sp. HB01 and *Bacillus* sp. HBCD-sjtu, degrade HBCDs. Under anaerobic conditions, hexabromocyclododecane biotransforms into dibromocyclododecadiene, while tribromoneopentyl alcohol requires oxygen for degradation [74]. *Sphingopyxis sp. YC-JH3* degraded 96% of triphenyl phosphate in 7 days [75]. *Fusarium culmorum* completely degraded DEHP in 27 hours, producing butanol, hexanal, catechol, and acetic acid. These were further converted or mineralized, with eight esterase isoforms acting synergistically in degradation [76]. Table **3** lists the types of phthalates degraded by various microorganisms.

Awasthi *et al.* found that *Klebsiella pneumoniae* degrades HDPE after thermal treatment, increasing biofilm thickness and causing weight loss, tensile weakening, and surface corrosion visible in SEM and AFM images [77]. *P. chrysogenum* NS10 and *P. oxalicum* NS4 were tested for HDPE and LDPE degradation using response surface methodology. SEM and AFM confirmed biofilm formation and surface changes, including cracks, pits, and undulations [78]. The superworm *Zophobas atratus* can consume and degrade polystyrene, with respirometry tests showing 36.7% of Styrofoam carbon converted to CO_2 in 16 days [78]. *Pseudomonas lini* JNU01 and *Acinetobacter johnsonii* JNU01 degrade polystyrene (PS), with alkane-1-monooxygenase AlkB as the key enzyme, highlighting their potential as PS decomposers [79]. Alphaproteobacteria and Gammaproteobacteria dominated plastic-associated assemblages, enhancing PS degradation and reducing film weight [80]. Polypropylene films and bioriented polypropylene showed cracks and erosion after 11 months, indicating microbial degradation [81].

Mechanism

Factors like molecular weight, chemical structure, presence of microorganisms, and environmental factors influence Microplastic (MP) degradation. Other factors include mobility, crystallinity, molecular weight, functional groups, substituents, and additives like plasticizers or additives [99]. The processes linked to MP degradation are comparable to those of other polymers, as MPs are polymers. Both must first be broken down into monomers during the degradation process to become mineralized. MPs have a comparatively high particle size in relation to the pore size of the bacterium cell membrane. Thus, MPs must undergo depolymerization into monomers for microbial absorption and biodegradation [100, 101]. Microbial enzymes play an important role in the depolymerization of MP, which often involves the process of hydrolysis. Both hydrolyzable and non-hydrolyzable polymers can be changed *via* MP hydrolysis. The enzyme attaches to the MPs and then catalyzes hydrolysis. Different bacteria have extracellular or intracellular depolymerase that may break down MPs [81]. Intracellular degradation occurs when the accumulating microorganisms hydrolyze an endogenous carbon store. Conversely, MPs are broken down by extracellular enzymes of the microorganisms into short chains or smaller molecules (oligomers, dimers, and monomers) that are small enough to move through semipermeable membranes. This process, known as extracellular degradation, requires using an external carbon source, which is not always carried out by the collecting microorganisms. Depolymerization is the term used to describe this process. These short-chain molecules undergo a process known as mineralization, which turns them into final products (such as CO_2, H_2O, or CH_4), used as carbon and energy sources. *Ideonella sakaiensis* 201-F6 uses PET as its primary energy source, producing enzymes ISF6-4831 and ISF6-0224 to hydrolyze PET into terephthalic acid and ethylene glycol [102].

Microorganisms, including bacteria and fungi, can adapt to new habitats, excrete specific enzymes, and produce natural surfactants. They can degrade organo-pollutants like plastic additives through hydrolysis and oxidation enzymatic processes. Hydrolytic systems break down lateral structures, while oxidative systems degrade complex structures. The complete biodegradation of organic pollutants occurs inside cells through conjugation and reduction-oxidation reactions, with the cytochrome P450 enzyme family participating. Biosurfactants, amphiphilic molecules, aid in the biodegradation of hydrophobic organic environmental contaminants [103]. Surfactants are produced by microorganisms to attach to hydrophobic substrates, improve mobility, and enhance bioavailability. Some microorganisms produce surfactants exclusively on hydrophobic substrates, while others produce them on water-soluble and hydrophobic ones. The biodegradation of plastic additives occurs through microbial cell colonization of organopollutant molecules, resulting in the production of specific enzymes for pollutant degradation. Different pathways and intermediates are observed depending on the organisms and phthalate substrates involved.

Table 3. The microbial degradation of Phthalate by various bacteria and fungi.

	Organism	Phthalate metabolized	Phthalate concentration	References
Bacteria	*Paracoccus kondratievae*	DBP DEHP	400 400	[72]
	Bacillus velezensis SYBC H47	DMP, DEP, DPrP and DBP	DMP, 194; DEP, 222; DPrP, 250; DBP, 278	[82]
	Halomonas sp. Mycobacterium sp.	DBP	1	[73]
	Microbacterium sp. and Pandoraea	DBP	139	[83]
	Pseudomonas juntendi (Pj), P. putida (Pp) and Pseudomonas nitritireducens (Pn)	DEP	500, 400, 500 (for Pj, Pp and Pn, respectively)	[84]
	Gordonia sp. strain 5F	DEHP	1000	[85]
	Pseudomonas sp. QDF12	DMP 1000	DMP 1000	[86]
	Sphingobium yanoikuyae SHJ	DBP 50	DBP 50	[87]
	Gordonia sp. YC-JH1	DEHP 100	DEHP 100	[88]
-	*Halotolerant bacteria consortium (major species: Gordonia sp., Rhodococcus sp., Achromobacter sp., Microbacterium sp. and Cellumonas sp.*	DEHP 2000	DEHP 2000	[89]
	Pseudomonas sp	DEHP 500	DEHP 500	[90]
	Agromyces sp. MT-O	DEHP 1000	DEHP 1000	[91]
	Rhizobium sp	DBP 50,100,150 and 200	DBP 50,100,150 and 200	[92]
	Achromobacter denitrificans SP1	DEHP 4	DEHP 4	[93]
Fungi	*Ascocoryne sp, Acephala sp.*	DBP, DEP	174 (for DBP) and 139 (for DEP)	[94]
	Fusarium culmorum Fc1-CICBUAT	DEHP 1000	DEHP 1000	[95]
	F. culmorum Fc1-CICBUAT	DBP 1000	DBP 1000	[96]
	Pleurotus ostreatus COLPOS Po50	DEHP 1000	DEHP 1000	[97]
	P. ostreatus COLPOS Po50	DBP 500 and 1000	DBP 500 and 1000	[98]

PETROLEUM-BASED POLLUTANTS BIOREMEDIATION

The microbe-assisted bioremediation is the most prevalent technique for treating pollution caused by petroleum hydrocarbon. Microorganisms are essential for preserving the biosphere and ecosystem to create a sustainable environment. They aid in minimizing the negative impacts of pollutants, biomonitoring of ecosystems, reducing adverse effects of the pollutants, and bioremediation. Microorganisms inhabit diverse environments such as freshwater, saltwater, soil, and air, frequently isolated from hydrocarbon-contaminated soil, sediments, or water. Since hydrocarbons are naturally prevalent, these naturally occurring microorganisms have hydrocarbon-utilizing, sequestration, and degrading abilities [104]. Biodegradation of petroleum hydrocarbons might be considered energetically beneficial since microorganisms obtain "Carbon", a necessary component of cellular processes. The energy gained during hydrocarbon pollutant degradation can be used for various cellular metabolism. Some microorganisms can degrade alkanes and aromatics. These organisms adapt to the polluted areas, leading to genetic alterations in subsequent generations, function as hydrocarbon sequesters or degraders. Polluted areas have higher hydrocarbon-degrading microbial diversity than unpolluted sites, which are crucial for hydrocarbon bioremediation due to their indigenous hydrocarbon-sequestering microorganisms [3]. Bacteria, fungi, and algae degrade hydrocarbons, with bacteria as the primary and most active degraders. Some microorganisms can sequester and degrade aliphatics, monoaromatics, polyaromatics, and resins. Table **4** lists the types of hydrocarbons degraded by various microorganisms.

Table 4. The microbial degradation of different petroleum and polyaromatic compounds by various microorganisms.

Organism	Petroleum and polyaromatic hydrocarbon compounds	References
Paraburkholderia aromaticivorans BN5	benzene, toluene, ethylbenzene, and *o*-, *m*-, and *p*-xylene.	
Rhodotorula taiwanensis KKUY-0162	Octane	[105]
Rhodotorula ingeniosa KKUY 170	Pyrene	[105]
Rhodococcus sp. OBD-3	Petroleum hydrocarbon	[109]
Bacillus spp.	aliphatic and aromatic hydrocarbons	[116]
Ochrobactrum sp. (PWTJD)	Phenanthrene	[117]
Raoultella ornithinolytica, Serratia marcescens, Bacillus megaterium and *Aeromonas hydrophila*	acenaphthene and fluorene	[118]
Achromobacter xylosoxidans PY4	High molecular weight 4-ring Pyrene	[119]
Aeromonas hydrophila	2-ring Acenaphthene and Fluorene 2-ring	[118]
Cellulosimicrobium cellulans CWS2	5-ring B(a)P	[120]
Stenotrophomonas sp.	2-ring naphthalene 3-ring fluorene 3-ring phenanthrene 4-ring pyrene 5-ring B(a)P	[121]
Sphingobium sp.	Anthraquinone (ANTQ) and 3-ring anthracene	[122]
Thermoactinomycetaceae	2-ring naphthalene, 3-ring phenanthrene	[123]

(Table 4) cont.....

Organism	Petroleum and polyaromatic hydrocarbon compounds	References
Aspergillus sp. RFC-1	Crude oil, 2-ring naphthalene, 3-ring phenanthrene, and 4-ring pyrene	[124]
Aspergillus fellis	Crude petroleum and hydrocarbons	[125]
Aspergillus fumigatus	3-ring fluorene, 3-ring phenanthrene and 4-ring pyrene	[126]
Aspergillus favus	3-ring fluorene, 3-ring phenanthrene and 4-ring pyrene	[126]
Aspergillus ficuum	3-ring fluorene, 3-ring phenanthrene and 4-ring pyrene	[126]
Aspergillus luteonubrus	crude petroleum and hydrocarbons	[125]
Cladosporium sp. CBMAI 1237	3-ring anthracene	[127]
Coriolopsis byrsina APC5	4-ring pyrene	[128]
Paecilomyces variotii	Crude oil contamination	[129]
Penicillium citrinum	Crude oil contamination	[129]
Penicillium eryngii F032	5-ring B(a)P	[130]

Paraburkholderia aromaticivorans BN5 degrades naphthalene and BTEX but not n-hexadecane, with its genome encoding 29 monooxygenases for short-chain hydrocarbon degradation [3]. Several yeast strains degraded aliphatic and aromatic hydrocarbons such as *Yamadazyma mexicana* KKUY-0160, *Rhodotorula taiwanensis* KKUY-0162, *Pichia kluyveri* KKUY-0163, *Rhodotorula ingeniosa* KKUY-0170, *Candida pseudointermedia* KKUY-0192, with *Rhodotorula taiwanensis* KKUY-0162 specializing in octane degradation and *Rhodotorula ingeniosa* KKUY-0170 targeting pyrene [105]. Xing *et al.*, 2022 have reported that direct aerobic oxidation bacteria such as *Methylobacterium* and *Proteobacterium* were found to have tolerance against high concentrations of chlorinated aliphatic hydrocarbons and efficiently degrade Trans-1,--dichloroethylene, 1,2-dichloroethane, dichloromethane, vinyl chloride and cis-1,2-dichloroethylene [106]. Indigenous strains, namely *Pseudomonas boreopolis* IITR108, *Microbacterium schleiferi* IITR109, *Pseudomonas aeruginosa* IITR110, and *Bacillus velezensis* IITR111 degraded 80–89% of aliphatic and 71–78% of aromatic hydrocarbons, while their consortium achieved 93.2% and 85.5% degradation, respectively [107]. Isolates of *Pseudomonas sp., P. oryzihabitans*, *Roseomonas aestuarii, Pantoea agglomerans*, and *Arthrobacter sp.* exhibited hydrocarbon degradation, removing over 85% of Total Petroleum Hydrocarbons (TPH) [108]. *Rhodococcus* sp. OBD-3, isolated from petroleum-contaminated soil extracted, degraded petroleum hydrocarbon by 60.6-86.6% [109]. The consortium containing *Bacillus subtilis, Pseudomonas stutzeri*, and *Acinetobacter baumannii* revealed degradation of crude oil, diesel, dotriacontane, and tetracontane [110]. The strains of *Rhodopseudomonas* sp. DD4, DQ41, and FO2 efficiently degraded over 84.2% of polycyclic aromatic hydrocarbons [111]. Ortega-Gonzalez *et al.* (2015) reported *Amycolatopsis* sp. achieved 100% naphthalene degradation and reduced anthracene (37.87%), pyrene (25.10%), and fluoranthene (18.18%) in 45 days [112]. Biodegradation of pollutants involves enzyme-based steps, with hydrocarbons selectively metabolized by individual or consortium microbial strains. Consortiums have shown greater potential for metabolizing complete hydrocarbon pollutants than individual cultures. Microorganisms prefer biodegrading n-alkanes over PAHs due to petroleum hydrocarbon availability, environmental adaptation, and specialized enzymes. *Pseudomonas aeruginosa* revealed a strong degradation capability towards PAHs such as naphthalene, pyrene, and phenanthrene. Besides, *P. aeruginosa* W10 and DQ8 strains degrade phenanthrene, fluoranthene, and pyrene, making them crucial in removing these chemicals from petroleum-polluted environments [113]. Extremophiles degrade hydrocarbons and can sustain different environmental conditions, making them suitable for extreme environments contaminated by hydrocarbons. *Rhodococcus* strains have three structural genes, *narAa*, *narAb*, and *narB*, which encode Naphthalene dioxygenase, unlike *Pseudomonas* [114]. *Sphingomonas* effectively degrade PAHs and various hydrocarbons, including biphenyl, carbazole, fluorene, pyrene, and polyethylene glycol. *Sphingomonas*, like most Gram-negative bacteria, degrade aromatic compounds like anthracene, naphthalene, and phenanthrene. Large plasmids in sphingomonas contribute to their catabolic versatility, but their degradative genes are scattered or not regulated [115]. A flexible gene architecture, including a mix of conserved gene clusters, may facilitate the effective and efficient aromatic compounds and xenobiotics degradation.

MECHANISM OF PETROLEUM DEGRADATION

Microorganisms can utilize petroleum hydrocarbon pollutants to produce energy or directly assimilate into the cellular biomass. The catabolism of petroleum hydrocarbon pollutants can be either through aerobic or anaerobic degradation. The process of catabolism is mostly aerobic; however, some microorganisms follow anaerobic degradation. This degradation process involves reactions such as oxidation, reduction, hydroxylation, and dehydrogenation that increase polar chains with a gradual decrease of saturated and aromatic hydrocarbons [131]. During aerobic degradation, pathways such as terminal oxidation, sub-terminal oxidation, ω-oxidation, and β-oxidation [104]. Alkane degradation begins with methyl group oxidation to alcohol, which is dehydrogenated to carboxylic acid and further metabolized *via* β-oxidation [104]. The di-terminal pathway involves ω-hydroxylation of fatty acids, converting them into dicarboxylic acids and processing through β-oxidation. Aromatic hydrocarbons are less biodegradable than saturated hydrocarbons and have more harmful effects on the environment and life forms. Degrading aromatic pollutants involves initial oxidative attack, benzene ring cleavage, diol formation by bacteria, ring cleavage, and dicarboxylic acid formation by fungi and eukaryotes [131]. Microorganisms cleave the benzene ring, forming intermediates like protocatechuates and catechols. These are converted to Tricarboxylic Acid (TCA) cycle intermediates. Cyclic alkanes are converted to cyclic alcohols and ketones through an oxidase system, followed by a mono-oxygenase system and lactone hydrolase. Alkenes can be attacked by oxygenase, oxidating to epoxides and diols [104]. Anaerobic metabolism converts aromatic compounds into phenols, organic acids, fatty acids, CH_4, and CO_2 using electron acceptors like nitrate, ferrous iron, manganese, or sulfate ions [131].

XENOBIOTICS BIODEGRADATION

Xenobiotic compounds are environmental chemicals that threaten the biosphere due to their resistance to biological degradation. Despite their resistance, numerous microorganisms have been found capable of degrading these compounds. Key pollutants with xenobiotic structures include nitroaromatics, aromatic hydrocarbons, halogenated aliphatics, azo compounds, s-triazines, and organic sulfonic acids [132]. Organic sulfonic acids are used in detergents, dyes, personal care, food, dispersants, and pharmaceuticals. Common sulfonates include p-toluenesulfonic acid, naphthalenesulfonates, and alkylbenzene sulfonates, produced by sulfonation or as industrial by-products [132]. Microbial degradation of sulfonates varies by structure; benzene and mono-substituted naphthalene sulfonates are biodegradable, while complex substituted sulfonates are harder to degrade [133]. Sulfonates resist bacterial biodegradation due to specific transport enzymes. Their degradation primarily involves desulfonation by anaerobic and aerobic microorganisms. Aerobic degradation of sulfonated compounds is achieved using bacterial consortia and pure cultures. These organisms use sulfonates as carbon and energy sources. Various pure and consortial bacterial strains, including *Alcaligenes* sp, *Hydrogenophaga palleroni*, *Agrobacterium radiobacter*, *Pseudomonas pseudoalcaligenes*, *Comamonas testosteroni* A3, *Sphingomonas xenophaga* BN6, *Pseudomonas* sp. BN9, *Pseudaminobacter salicylatoxidans* BN12, and *Methylosulfonomonas methylovora* [132]. Coasta *et al*. 2020 demonstrated the high efficiency of *Penicillium chrysogenum* in the biodegradation of sodium dodecylbenzene sulfonate [134].

S-Triazine is a six-membered organic compound used as a herbicide to control grassy weeds and broadleaf in crops like macadamia nuts, sugarcane, pineapple, sorghum, and corn, as well as on golf courses and residential lawns [135]. Atrazine and Simazine are widely used s-Triazine-based herbicides for controlling annual grasses and weeds in crop fields. However, they are persistent in soil due to their prolonged degradation rate, damaging the immune and reproductive systems of reptiles, crustaceans, mammals, and amphibians. The study of s-triazine degradation by various bacterial species, including *Agrobacterium, Pseudomonas, Pseudominobacter, Stenptrophomonas, Rhodococcus,* and *Arthrobacter,* has been extensively discussed and reported in the literature [136]. Atrazine degradation in bacteria occurs through various routes, including chlorohydrolase-catalyzed dechlorination, cyanuric acid metabolism, biuret metabolization, allophanate, and allophanate hydrolase-catalyzed conversion. The synthesis of desisopropyl atrazine or deethylatrazine, catalyzed by atrazine monooxygenase or N-dealkylase, is further converted into ammonia and carbon dioxide through cyanuric acid metabolism [136]. Different species or microbial consortia prevent atrazine's effective degradation. A previous study proposed 16 different microorganism combinations for effective atrazine degradation [137]. The microbial degradation of xenobiotic compounds is listed in Table **5**.

Table 5. The microbial degradation of xenobiotic compounds by various microorganisms.

Organism	Xenobiotic compounds	References
Microalgae	TNT	[142]
Phanerochaete velutina, Gymnopilus Luteofolius, and Kuehneromyces mutabilis	TNT	[143]
Bacillus badius	3-Nitro benzoic acid, 4-nitrotoluene, 3- nitrotoluene, 1-chloro-2- nitro benzene	[144]
Shewanella oneidensis MR-1	2,6-Dinitrotoluene	[145]
Aeromonas hydrophila NIU01, A. salmonicida 741760, A. tecta 644, Enterobacter cancerogenus BYm30, Exiguobacterium acetylicum NIU–K2, Ex. indicum NIU–K4	Nitroaromatic compounds	[146]
Pseudomonas sp. X5	Dinitrotoluene sulfonates	[137]
Bacillus megaterium, Micrococcus luteus and Bacillus pumilus	RB dye	[147]
Aeromonas hydrophila	RB5	[148]
Pseudomonas extremorientalis BU118	CR	[149]
Laccase SilA produced by Streptomyces ipomoeae CECT 3341	IC, OII, Cresol Red	[150]
Mesophilic and thermophilic lactic acid bacteria (LAB)	Dorasyn Red azo dye	[151]
Bacterial consortium of Pantoea ananatis (KM502538), Alcaligenes faecalis (KM502541), Bre- vibacillus parabrevis strain GRG (KM502542) and Bordetella trematum (KP751929) organisms	AB193/ AB194	[152]

Nitroaromatic compounds, which contain at least one or more nitro groups (NO_2), are highly useful in the chemical industry and are crucial in the synthesis of explosives, including nitrobenzene, nitrotoluenes, and nitrophenols [132]. The aerobic degradation of nitroaromatic compounds in bacteria involves a three-step process. The first step involves changes in the substrate's substituent group by monooxygenases or dioxygenases enzymes, releasing nitro groups in the form of nitrate or nitro groups. Monooxygenase helps to add one oxygen to the benzene ring while releasing the nitro group. The dioxygenase enzyme catalyzes the addition of two hydroxyl groups, releasing two nitro groups in the form of nitrate [138]. Dioxygenase enzymes disrupt the carbon-carbon bond in the ring structure, producing unsaturated aliphatic acid. There are two types of dioxygenases enzymes: intradiol (or ortho) dioxygenases produce cis, cismuconic acid, or derivative, and extradiol (or meta) dioxygenases produce 2-hydroxymuconic semialdehyde or derivative. The derivatives generated after the cleavage of the ring structure are converted into small aliphatic compounds [138]. Anaerobic degradation of nitroaromatic compounds involves electron transport, facilitated by the nitroreductase enzyme, converting nitro groups into nitroso derivatives, hydroxylamines, and amines, often occurring in symbiont with aerobic microorganisms [138]. Chlorophenols are formed by hydroxylation at the ortho-position or para-position of the benzene ring structure, resulting in chlorocatechols and chlorohydroquinones [132]. The intermediate is catalyzed through ortho-cleavage, meta-cleavage, hydroxylation, or dehalogenation, leading to benzene ring structure disruption. Anaerobic conditions remove chlorine, while aerobic conditions allow dichlorination before or after ring disruption [139].

Microbial organisms use enzymes like azoreductase, laccases, lignin peroxidase, manganese peroxidase, and hydroxylases for the degradation of azo dyes [140]. Aerobic bacterial strains like *E. coli* can break the azo bond through the reduction process by Azoreductases in oxygen-rich environments [141]. The azoreductase enzyme in bacteria produces toxic intermediate aromatic amines, which are mineralized by the membrane-bound enzyme. Bacteria use azo dye as a carbon or nitrogen source and form a glucose product. The laccases enzyme, with multicopper atoms, can degrade various aromatic compounds. Bacteria like *Pleurotus ostreatus*, *Schizophyllum commune*, *Sclerotium rolfsii*, and *Neurospora crassa* can be used for azo dye degradation [132].

BIOTRANSFORMATION AND CO-METABOLISM OF POLLUTANTS

Co-metabolism is a microbial metabolism phenomenon where microorganisms metabolize primary metabolic substrates for growth while transforming non-essential co-metabolic substrates. The co-metabolized substrate is not typically used as energy or carbon but can be converted by the enzyme system, allowing the microorganism to utilize it for growth [153]. The microbial transformation process, or biotransformation, is based on mineralization and cometabolism. Mineralization converts organic substances into inorganic end products, typically non-toxic or less toxic than the substrate. Consortium microbes, which combine beneficial bacteria with biodegradative potential, perform better in biotransformation processes. Cometabolism is a partial transformation that forms an intermediate product, serving as an energy or carbon source. The biotransformation process can be aerobic or anaerobic, depending on the pollutant and microbe type. Aerobic biotransformation uses oxygen, while some microorganisms, like chemoautotrophic or lithotrophic bacteria, oxidize inorganic contaminants instead of using organic substrates [154]. Aerobic biotransformation occurs when oxygen levels deplete, causing anaerobic conditions. Anaerobic bacteria use terminal electron acceptors like nitrate, sulfate, or iron salts as an alternative to oxygen. This multi-step process begins with the hydrolysis of polymeric contaminants into monomeric substances. The ultimate step is carried out by anaerobic bacteria such as methanogens, denitrifying bacteria, or sulfate-reducing bacteria, resulting in alcohol, acids, hydrogen, and carbon dioxide [154].

Xylose acts as a potential co-substrate, enhancing the degradation of Reactive Black 5 (RB5) dye by *Klebsiella* sp. KL-1 through increased key enzyme activities. In this process, enzymes such as azoreductase, laccase, and manganese/lignin peroxidase are involved in the degradation process. Proteomics analysis showed differential expression of metabolic proteins like NADH-quinone oxidoreductase, pyruvate dehydrogenase, and NADH dehydrogenase [155]. *Aeromonas caviae* GLB-10 degrades a commonly found antibiotic sulfamethoxazole (SMX) into aniline and 3-amino-5-methylisoxazole. A mixed bacterial consortium, including *Vibrio diabolicus*, *Zobellella taiwanensis,* and *Microbacterium testaceum* achieves ultrahigh degradation efficiency, degrading 250 mg/L SMX in 3 days, producing additional metabolites like acetanilide and hydroquinone *via* co-metabolism [156].

Pseudomonas aeruginosa degraded 50% of TBBPA in 7 days using glucose as a co-substrate. Optimal conditions, including pH and glucose concentration, were identified, with pyocyanin generating H_2O_2 and $OH\cdot$. Five metabolites, including debromination products, were detected. Co-metabolism boosted early bacterial growth, reducing total organic carbon by 78%, potentially causing pyocyanin auto-poisoning [157].

The study found that BDE-47 was completely degraded within 7 days using methanol as a co-substrate, with a 3.26 times higher efficiency. However, excessive methanol combined with BDE-47 caused significant stress on microbial cells. Methanol enhanced debromination, hydroxylation, and bond breakage but also induced microbial stress, including increased levels of reactive oxygen species, superoxide dismutase activity, catalase activity, and malondialdehyde content. Despite increased oxidative stress markers due to the addition of methanol, *Methylobacterium* helped to maintain the stability of the QY2 community [153]. The study reveals that co-metabolism is a promising method for optimizing the biodegradation of p-Chloroaniline (PCA). *Pseudomonas* sp. CA-1 reduced 76.57% of PCA, with an increase of 12.50% degradation efficiency, when aniline was added as a co-substrate. The study also explains the response and co-metabolism mechanism of CA-1 to PCA, which causes deformation and damage on its surface, followed by adsorption on protein molecules. After the entry of PCA inside the cell, it was degraded by the action of various oxidoreductases. Aniline enhances the antioxidant capacity of CA-1, stimulates catechol 2,3-dioxygenase expression, and promotes PCA meta-cleavage efficiency [158].

The study explores the co-metabolism of Sulfamethoxazole (SMX) by *Chlorella pyrenoidosa* when glucose and lysine are used as co-substrates. Results showed that lysine has a 1.5-fold higher SMX's microalgae-mediated co-metabolism compared to glucose, with the highest removal rate observed in the 9-mM-Lys co-metabolic system. The study also reveals the catalytic reactions included bond breaking, ring cleavage, hydroxylation, acylation, and glutamyl conjugation facilitated by Cytochrome P450, serine hydrolase, and peroxidase [159]. In a sodium acetate system, *C. pyrenoidosa* facilitated SMX dissipation (14%) after 11 days of cultivation. This system increased SMX dissipation from 6.05% to 99.3% in 5 days [160]. Pyridine wastewater poses environmental challenges due to its recalcitrance and toxicity. Co-metabolic degradation, using glucose and phenol as substrates, has emerged as a promising solution. The study found that all reactors achieved high removal efficiencies, surpassing 98.5%, and better pyridine-N removal performance. Glucose supplementation promoted nitrogen assimilation, resulting in the highest removal rate and efficiency. The high abundance of *Saccharibacteria* and enrichment of GLU and glnA supported this finding. However, phenol delayed pyridine oxidation due to its higher affinity for phenol hydroxylase [161].

CONCLUSION

In summary, a sustainable and efficient method of reducing environmental contamination brought on by a variety of dangerous materials, including petrochemicals, agrochemicals, and medications, is microbe-assisted sequestration. With no negative impact on the environment, this bioremediation method restores ecological balance by using microorganisms' innate ability to break down, detoxify, and immobilize contaminants. Microbial bioremediation, in contrast to traditional physicochemical techniques, is economical, ecologically benign, and able to maintain soil health and biodiversity. By combining cutting-edge molecular approaches like system biology, gene editing, and omics, microbial sequestration technologies have become even more effective and versatile, providing encouraging opportunities for extensive environmental restoration projects. Using microbial systems to their full potential is a major step in the direction of long-term environmental resilience and sustainability.

REFERENCES

[1]   Naidu R. Recent advances in contaminated site remediation. Water Air Soil Pollut 2013; 224: 1-11.

[2]   Biswas B, Sarkar B, Rusmin R, Naidu R. Bioremediation of PAHs and VOCs: Advances in clay mineral–microbial interaction. Environ Int 2015; 85: 168-81.
[http://dx.doi.org/10.1016/j.envint.2015.09.017] [PMID: 26408945]

[3]   Lee Y, Lee Y, Jeon CO. Biodegradation of naphthalene, BTEX, and aliphatic hydrocarbons by Paraburkholderia aromaticivorans BN5 isolated from petroleum-contaminated soil. Sci Rep 2019; 9(1): 860.
[http://dx.doi.org/10.1038/s41598-018-36165-x] [PMID: 30696831]

[4]   Fuentes S, Méndez V, Aguila P, Seeger M. Bioremediation of petroleum hydrocarbons: catabolic genes, microbial communities, and applications. Appl Microbiol Biotechnol 2014; 98(11): 4781-94.
[http://dx.doi.org/10.1007/s00253-014-5684-9] [PMID: 24691868]

[5]   Rayu S, Karpouzas DG, Singh BK. Emerging technologies in bioremediation: constraints and opportunities. Biodegradation 2012; 23(6): 917-26.
[http://dx.doi.org/10.1007/s10532-012-9576-3] [PMID: 22836784]

[6]   Agarwal S, Jain H. Investigating the potential of floral waste as a vermicompost and dual-functional biosorbent for sustainable environmental management. Water Air Soil Pollut 2024; 235(5): 322.
[http://dx.doi.org/10.1007/s11270-024-07144-y]

[7]   Jain H. From pollution to progress: groundbreaking advances in clean technology unveiled. Innov Green Dev 2024; 3(2): 100143.
[http://dx.doi.org/10.1016/j.igd.2024.100143]

[8]   Bose S, Kumar PS, Vo DVN, Rajamohan N, Saravanan R. Microbial degradation of recalcitrant pesticides: a review. Environ Chem Lett 2021; 19(4): 3209-28.
[http://dx.doi.org/10.1007/s10311-021-01236-5]

[9]   Satish GP, Ashokrao DM, Arun SK. Microbial degradation of pesticide: A review. Afr J Microbiol Res 2017; 11(24): 992-1012.
[http://dx.doi.org/10.5897/AJMR2016.8402]

[10]   Valenzuela EF, Menezes HC, Cardeal ZL. Passive and grab sampling methods to assess pesticide residues in water. A review. Environ Chem Lett 2020; 18(4): 1019-48.
[http://dx.doi.org/10.1007/s10311-020-00998-8]

[11]   Mustapha MU, Halimoon N, Johar WLW, Abd Shukor MY. An overview on biodegradation of carbamate pesticides by soil bacteria. Pertanika J Sci Technol 2019; 27(2)

[12]   Ullah S, Li Z, Zuberi A, Arifeen MZU, Baig MMFA. Biomarkers of pyrethroid toxicity in fish. Environ Chem Lett 2019; 17(2): 945-73.
[http://dx.doi.org/10.1007/s10311-018-00852-y]

[13]   Dominguez CM, Oturan N, Romero A, Santos A, Oturan MA. Lindane degradation by electrooxidation process: Effect of electrode materials on oxidation and mineralization kinetics. Water Res 2018; 135: 220-30.

[http://dx.doi.org/10.1016/j.watres.2018.02.037] [PMID: 29477060]

[14]   Vijgen J, Weber R, Lichtensteiger W, Schlumpf M. The legacy of pesticides and POPs stockpiles—a threat to health and the environment. Environ Sci Pollut Res Int 2018; 25(32): 31793-8.
[http://dx.doi.org/10.1007/s11356-018-3188-3] [PMID: 30280348]

[15]   Zhang W, Lin Z, Pang S, Bhatt P, Chen S. Insights into the biodegradation of lindane (γ-hexachlorocyclohexane) using a microbial system. Front Microbiol 2020; 11: 522.
[http://dx.doi.org/10.3389/fmicb.2020.00522] [PMID: 32292398]

[16]   Singh T, Singh DK. *Rhizospheric Microbacterium sp.* P27 showing potential of lindane degradation and plant growth promoting traits. Curr Microbiol 2019; 76(7): 888-95.
[http://dx.doi.org/10.1007/s00284-019-01703-x] [PMID: 31093691]

[17]   Sahoo B, Ningthoujam R, Chaudhuri S. Isolation and characterization of a lindane degrading bacteria *Paracoccus sp.* NITDBR1 and evaluation of its plant growth promoting traits. Int Microbiol 2019; 22(1): 155-67.
[http://dx.doi.org/10.1007/s10123-018-00037-1] [PMID: 30810939]

[18]   Kumar D. Biodegradation of γ-hexachlorocyclohexane by *Burkholderia sp.* IPL04. Biocatal Agric Biotechnol 2018; 16: 331-9.
[http://dx.doi.org/10.1016/j.bcab.2018.09.001]

[19]   Purnomo AS. Microbe-assisted degradation of aldrin and dieldrin. Microbe-induced degradation of pesticides. 1-22. 2017; pp.
[http://dx.doi.org/10.1007/978-3-319-45156-5_1]

[20]   Doolotkeldieva T, Konurbaeva M, Bobusheva S. Microbial communities in pesticide-contaminated soils in Kyrgyzstan and bioremediation possibilities. Environ Sci Pollut Res Int 2018; 25(32): 31848-62.
[http://dx.doi.org/10.1007/s11356-017-0048-5] [PMID: 28884389]

[21]   Jesitha K, Nimisha KM, Manjusha CM, Harikumar PS. Biodegradation of Endosulfan by *Pseudomonas fluorescens.* Environ Process 2015; 2(1): 225-40.
[http://dx.doi.org/10.1007/s40710-015-0059-5]

[22]   Ozdal M, Ozdal OG, Algur OF. Isolation and characterization of α-endosulfan degrading bacteria from the microflora of cockroaches. Pol J Microbiol 2016; 65(1): 63-8.
[http://dx.doi.org/10.5604/17331331.1197325] [PMID: 27281995]

[23]   Kafilzadeh F, Ebrahimnezhad M, Tahery Y. Isolation and identification of endosulfan-degrading bacteria and evaluation of their bioremediation in kor river, iran. Osong Public Health Res Perspect 2015; 6(1): 39-46.
[http://dx.doi.org/10.1016/j.phrp.2014.12.003] [PMID: 25737830]

[24]   Kumar D, Pannu R. Perspectives of lindane (γ-hexachlorocyclohexane) biodegradation from the environment: a review. Bioresour Bioprocess 2018; 5(1): 29.
[http://dx.doi.org/10.1186/s40643-018-0213-9]

[25]   Singh T, Singh DK. Lindane degradation by root epiphytic bacterium *Achromobacter* sp. strain A3 from *Acorus calamus* and characterization of associated proteins. Int J Phytoremediation 2019; 21(5): 419-24.
[http://dx.doi.org/10.1080/15226514.2018.1524835] [PMID: 30648424]

[26]   Egorova DO, Buzmakov SA, Nazarova EA, Andreev DN, Demakov VA, Plotnikova eg. Bioremediation of hexachlorocyclohexane-contaminated soil by the new *Rhodococcus wratislaviensis* strain Ch628. Water Air Soil Pollut 2017; 228(5): 183.
[http://dx.doi.org/10.1007/s11270-017-3344-2]

[27]   Bajaj S, Sagar S, Khare S, Singh DK. Biodegradation of γ-hexachlorocyclohexane (lindane) by halophilic bacterium *Chromohalobacter sp.* LD2 isolated from HCH dumpsite. Int Biodeterior Biodegradation 2017; 122: 23-8.

[http://dx.doi.org/10.1016/j.ibiod.2017.04.014]

[28] Kumar D, Kumar A, Sharma J. Degradation study of lindane by novel strains *Kocuria* sp. DAB-1Y and *Staphylococcus* sp. DAB-1W. Bioresour Bioprocess 2016; 3(1): 53.
[http://dx.doi.org/10.1186/s40643-016-0130-8] [PMID: 28090433]

[29] Sineli PE, Tortella G, Dávila Costa JS, Benimeli CS, Cuozzo SA. Evidence of α-, β- and γ-HCH mixture aerobic degradation by the native *actinobacteria Streptomyces sp.* M7. World J Microbiol Biotechnol 2016; 32(5): 81.
[http://dx.doi.org/10.1007/s11274-016-2037-0] [PMID: 27038951]

[30] Xie H, Zhu L, Wang J, Jiang J, Wang J. Biodegradation of DDE and DDT by bacterial strain *Stenotrophomonas sp.* DXZ9. J Environ Anal Toxicol 2017; 7(496): 2161-0525.1000496.
[http://dx.doi.org/10.4172/2161-0525.1000496]

[31] Fareed A, Zaffar H, Rashid A, Maroof Shah M, Naqvi TA. Biodegradation of *N-methylated* carbamates by free and immobilized cells of newly isolated strain *Enterobacter cloacae* strain TA7. Bioremediat J 2017; 21(3-4): 119-27.
[http://dx.doi.org/10.1080/10889868.2017.1404964]

[32] Gupta J, Rathour R, Singh R, Thakur IS. Production and characterization of extracellular polymeric substances (EPS) generated by a carbofuran degrading strain *Cupriavidus sp.* ISTL7. Bioresour Technol 2019; 282: 417-24.
[http://dx.doi.org/10.1016/j.biortech.2019.03.054] [PMID: 30884462]

[33] Ekram MAE, Sarker I, Rahi MS, Rahman MA, Saha AK, Reza MA. Efficacy of soil-borne *Enterobacter* sp. for carbofuran degradation: HPLC quantitation of degradation rate. J Basic Microbiol 2020; 60(5): 390-9.
[http://dx.doi.org/10.1002/jobm.201900570] [PMID: 32115726]

[34] Jiang W, Gao Q, Zhang L, *et al.* Identification of the key amino acid sites of the carbofuran hydrolase CehA from a newly isolated carbofuran-degrading strain *Sphingbium sp.* CFD-1. Ecotoxicol Environ Saf 2020; 189: 109938.
[http://dx.doi.org/10.1016/j.ecoenv.2019.109938] [PMID: 31759739]

[35] Yang C, Xu X, Liu Y, *et al.* Simultaneous hydrolysis of carbaryl and chlorpyrifos by *Stenotrophomonas sp.* strain YC-1 with surface-displayed carbaryl hydrolase. Sci Rep 2017; 7(1): 13391.
[http://dx.doi.org/10.1038/s41598-017-13788-0] [PMID: 29042673]

[36] Wu P, Xie L, Mo W, *et al.* The biodegradation of carbaryl in soil with *Rhodopseudomonas capsulata* in wastewater treatment effluent. J Environ Manage 2019; 249: 109226.
[http://dx.doi.org/10.1016/j.jenvman.2019.06.127] [PMID: 31442909]

[37] Wu Pan W P, *et al.* Carbaryl waste-water treatment by Rhodopseudomonas sphaeroides 2019.

[38] Kulkarni AG, Kaliwal BB. Bioremediation of methomyl by *Escherichia coli*. Toxicity and Biodegradation Testing. 2018; pp. 75-86.

[39] Roy T, Das N. Isolation, characterization, and identification of two methomyl-degrading bacteria from a pesticide-treated crop field in West Bengal, India. Microbiology 2017; 86(6): 753-64.
[http://dx.doi.org/10.1134/S0026261717060145]

[40] Zhang C, Yang Z, Jin W, *et al.* Degradation of methomyl by the combination of *Aminobacter* sp. MDW-2 and *Afipia* sp. MDW-3. Lett Appl Microbiol 2017; 64(4): 289-96.
[http://dx.doi.org/10.1111/lam.12715] [PMID: 28083911]

[41] Rousidou C, Karaiskos D, Myti D, *et al.* Distribution and function of carbamate hydrolase genes *cehA* and *mcd* in soils: the distinct role of soil pH. FEMS Microbiol Ecol 2017; 93(1): fiw219.
[http://dx.doi.org/10.1093/femsec/fiw219] [PMID: 27797966]

[42] Mohamed E. Oxamyl utilization by micrococcus luteus OX, isolated from Egyptian soil. International Journal of Applied Environmental Sciences 2017; 12(5): 999-1008.

[43]   Rodríguez-Rodríguez CE, Madrigal-León K, Masís-Mora M, Pérez-Villanueva M, Chin-Pampillo JS. Removal of carbamates and detoxification potential in a biomixture: Fungal bioaugmentation versus traditional use. Ecotoxicol Environ Saf 2017; 135: 252-8.
[http://dx.doi.org/10.1016/j.ecoenv.2016.10.011] [PMID: 27750092]

[44]   Govarthanan M, Ameen F, Kamala-Kannan S, *et al.* Rapid biodegradation of chlorpyrifos by plant growth-promoting psychrophilic *Shewanella sp.* BT05: An eco-friendly approach to clean up pesticide-contaminated environment. Chemosphere 2020; 247: 125948.
[http://dx.doi.org/10.1016/j.chemosphere.2020.125948] [PMID: 32069723]

[45]   Aswathi A, Pandey A, Sukumaran RK. Rapid degradation of the organophosphate pesticide – Chlorpyrifos by a novel strain of *Pseudomonas nitroreducens* AR-3. Bioresour Technol 2019; 292: 122025.
[http://dx.doi.org/10.1016/j.biortech.2019.122025] [PMID: 31466023]

[46]   Jaiswal DK, Verma JP, Krishna R, Gaurav AK, Yadav J. Molecular characterization of monocrotophos and chlorpyrifos tolerant bacterial strain for enhancing seed germination of vegetable crops. Chemosphere 2019; 223: 636-50.
[http://dx.doi.org/10.1016/j.chemosphere.2019.02.053] [PMID: 30798059]

[47]   Zhang Q, Li S, Ma C, Wu N, Li C, Yang X. Simultaneous biodegradation of bifenthrin and chlorpyrifos by *Pseudomonas* sp. CB2. J Environ Sci Health B 2018; 53(5): 304-12.
[http://dx.doi.org/10.1080/03601234.2018.1431458] [PMID: 29431579]

[48]   Feng F, Li Y, Ge J, *et al.* Degradation of chlorpyrifos by an endophytic bacterium of the *Sphingomonas* genus (strain HJY) isolated from Chinese chives (*Allium tuberosum*). J Environ Sci Health B 2017; 52(10): 736-44.
[http://dx.doi.org/10.1080/03601234.2017.1356675] [PMID: 28937878]

[49]   Khalid S, Hashmi I, Khan SJ. Bacterial assisted degradation of chlorpyrifos: The key role of environmental conditions, trace metals and organic solvents. J Environ Manage 2016; 168: 1-9.
[http://dx.doi.org/10.1016/j.jenvman.2015.11.030] [PMID: 26692411]

[50]   Akbar S, Sultan S. Soil bacteria showing a potential of chlorpyrifos degradation and plant growth enhancement. Braz J Microbiol. 2016; 47: pp. (3)563-70.
[http://dx.doi.org/10.1016/j.bjm.2016.04.009]

[51]   Uniyal S, Sharma RK. Technological advancement in electrochemical biosensor based detection of Organophosphate pesticide chlorpyrifos in the environment: A review of status and prospects. Biosens Bioelectron 2018; 116: 37-50.
[http://dx.doi.org/10.1016/j.bios.2018.05.039] [PMID: 29857260]

[52]   Qiu L, Wang H, Wang X. Conversion mechanism of heptachlor by a novel bacterial strain. RSC Advances 2018; 8(11): 5828-39.
[http://dx.doi.org/10.1039/C7RA10097C] [PMID: 35539625]

[53]   Chen H, Gao B, Wang S, Fang J. Microbial degradation of Hexachlorocyclohexane (HCH) pesticides. Adv Biodegrad Bioremediat Ind Waste 2015; 10: 181-210.

[54]   Asemoloye MD, Ahmad R, Jonathan SG. Synergistic rhizosphere degradation of γ-hexachlorocyclohexane (lindane) through the combinatorial plant-fungal action. PLoS One 2017; 12(8): e0183373.
[http://dx.doi.org/10.1371/journal.pone.0183373] [PMID: 28859100]

[55]   Kaur H, Kapoor S, Kaur G. Application of ligninolytic potentials of a white-rot fungus *Ganoderma lucidum* for degradation of lindane. Environ Monit Assess 2016; 188(10): 588.
[http://dx.doi.org/10.1007/s10661-016-5606-7] [PMID: 27670886]

[56]   Xiao P, Kondo R. Potency of Phlebia species of white rot fungi for the aerobic degradation, transformation and mineralization of lindane. J Microbiol. 2020; 58: pp. 395-404.
[http://dx.doi.org/10.1007/s12275-020-9492-x]

[57] Russo F, Ceci A, Pinzari F, *et al.* Bioremediation of Dichlorodiphenyltrichloroethane (DDT)-contaminated agricultural soils: potential of two autochthonous saprotrophic fungal strains. Appl Environ Microbiol 2019; 85(21): e01720-19.
[http://dx.doi.org/10.1128/AEM.01720-19] [PMID: 31444208]

[58] Uniyal S, Sharma RK, Kondakal V. New insights into the biodegradation of chlorpyrifos by a novel bacterial consortium: Process optimization using general factorial experimental design. Ecotoxicol Environ Saf 2021; 209: 111799.
[http://dx.doi.org/10.1016/j.ecoenv.2020.111799] [PMID: 33360782]

[59] Mishra S, Zhang W, Lin Z, *et al.* Carbofuran toxicity and its microbial degradation in contaminated environments. Chemosphere 2020; 259: 127419.
[http://dx.doi.org/10.1016/j.chemosphere.2020.127419] [PMID: 32593003]

[60] Tondon S, Deore R, Parab A. Isolation, identification and the use of carbofuran degrading microorganisms for the removal of carbofuran pesticide from contaminated waters. Global Journal of Bio-Science and Biotechnology 2018; 6: 89-95.

[61] Ozdal M, Ozdal OG, Algur OF, Kurbanoglu EB. Biodegradation of α-endosulfan via hydrolysis pathway by *Stenotrophomonas maltophilia* OG2. 3 Biotech. 2017; 7: pp. 1-7.
[http://dx.doi.org/10.1007/s13205-017-0765-y]

[62] Zaffar H, Sabir SR, Pervez A, Naqvi TA. Kinetics of endosulfan biodegradation by *Stenotrophomonas maltophilia* EN-1 isolated from pesticide-contaminated soil. Soil Sediment Contam 2018; 27(4): 267-79.
[http://dx.doi.org/10.1080/15320383.2018.1470605]

[63] Mir ZA, *et al.* Degradation and conversion of endosulfan by newly isolated *Pseudomonas mendocina* ZAM1 strain. 3 Biotech. 2017; 7: pp. 1-12.
[http://dx.doi.org/10.1007/s13205-017-0823-5]

[64] Odukkathil G, Vasudevan N. Biodegradation of endosulfan isomers and its metabolite endosulfate by two biosurfactant producing bacterial strains of *Bordetella petrii*. J Environ Sci Health B 2015; 50(2): 81-9.
[http://dx.doi.org/10.1080/03601234.2015.975596] [PMID: 25587777]

[65] Jimenez-Torres C, Ortiz I, San-Martin P, Hernandez-Herrera RI. Biodegradation of malathion, α- and β-endosulfan by bacterial strains isolated from agricultural soil in Veracruz, Mexico. J Environ Sci Health B 2016; 51(12): 853-9.
[http://dx.doi.org/10.1080/03601234.2016.1211906] [PMID: 27715499]

[66] Trends in biological processes in industrial wastewater treatment. M. P. Shah, ed.: IOP Publishing 2024.
[http://dx.doi.org/10.1088/978-0-7503-5678-7]

[67] Jain H. Data analytics enabled by the Internet of Things and artificial intelligence for the management of Earth's resources. In: Kumar D, Tewary T, Shekhar S, Eds. Data Anal Artif Intell Earth Resour Manag. Elsevier 2025; pp. 19-36.
[http://dx.doi.org/10.1016/B978-0-443-23595-5.00002-4]

[68] Jain H, Dhupper R, Shrivastava A, Kumari M. Enhancing groundwater remediation efficiency through advanced membrane and nano-enabled processes: a comparative study. Groundw Sustain Dev 2023; 23: 100975.
[http://dx.doi.org/10.1016/j.gsd.2023.100975]

[69] Shah Z, Gulzar M, Hasan F, Shah AA. Degradation of polyester polyurethane by an indigenously developed consortium of Pseudomonas and Bacillus species isolated from soil. Polym Degrad Stabil 2016; 134: 349-56.
[http://dx.doi.org/10.1016/j.polymdegradstab.2016.11.003]

[70] Vivek K, Sandhia GS, Subramaniyan S. Extremophilic lipases for industrial applications: A general

review. Biotechnol Adv 2022; 60: 108002.
[http://dx.doi.org/10.1016/j.biotechadv.2022.108002] [PMID: 35688350]

[71]   Taniguchi I, Yoshida S, Hiraga K, Miyamoto K, Kimura Y, Oda K. Biodegradation of PET: current status and application aspects. ACS Catal 2019; 9(5): 4089-105.
[http://dx.doi.org/10.1021/acscatal.8b05171]

[72]   Xu Y, Minhazul KAHM, Wang X, *et al.* Biodegradation of phthalate esters by *Paracoccus kondratievae* BJQ0001 isolated from Jiuqu (Baijiu fermentation starter) and identification of the ester bond hydrolysis enzyme. Environ Pollut 2020; 263(Pt B): 114506.
[http://dx.doi.org/10.1016/j.envpol.2020.114506] [PMID: 32268225]

[73]   Wright RJ, Bosch R, Gibson MI, Christie-Oleza JA. Plasticizer degradation by marine bacterial isolates: a proteogenomic and metabolomic characterization. Environ Sci Technol 2020; 54(4): 2244-56.
[http://dx.doi.org/10.1021/acs.est.9b05228] [PMID: 31894974]

[74]   Peng X, Wei D, Huang Q, Jia X. Debromination of hexabromocyclododecane by anaerobic consortium and characterization of functional bacteria. Front Microbiol 2018; 9: 1515.
[http://dx.doi.org/10.3389/fmicb.2018.01515] [PMID: 30042751]

[75]   Wang J, Khokhar I, Ren C, *et al.* Characterization and 16S metagenomic analysis of organophosphorus flame retardants degrading consortia. J Hazard Mater 2019; 380: 120881.
[http://dx.doi.org/10.1016/j.jhazmat.2019.120881] [PMID: 31307001]

[76]   González-Márquez A, Loera-Corral O, Santacruz-Juárez E, *et al.* Biodegradation patterns of the endocrine disrupting pollutant di(2-ethyl hexyl) phthalate by *Fusarium culmorum*. Ecotoxicol Environ Saf 2019; 170: 293-9.
[http://dx.doi.org/10.1016/j.ecoenv.2018.11.140] [PMID: 30530181]

[77]   Awasthi S, Srivastava P, Singh P, Tiwary D, Mishra P K. Biodegradation of thermally treated high-density polyethylene (HDPE) by *Klebsiella pneumoniae* CH001. 3 Biotech 2017; 7(5): 332.
[http://dx.doi.org/10.1007/s13205-017-0959-3]

[78]   Ojha N, Pradhan N, Singh S, *et al.* Evaluation of HDPE and LDPE degradation by fungus, implemented by statistical optimization. Sci Rep 2017; 7(1): 39515.
[http://dx.doi.org/10.1038/srep39515] [PMID: 28051105]

[79]   Kim HW, Jo JH, Kim YB, *et al.* Biodegradation of polystyrene by bacteria from the soil in common environments. J Hazard Mater 2021; 416: 126239.
[http://dx.doi.org/10.1016/j.jhazmat.2021.126239] [PMID: 34492990]

[80]   Xu X, Wang S, Gao F, *et al.* Marine microplastic-associated bacterial community succession in response to geography, exposure time, and plastic type in China's coastal seawaters. Mar Pollut Bull 2019; 145: 278-86.
[http://dx.doi.org/10.1016/j.marpolbul.2019.05.036] [PMID: 31590788]

[81]   Yuan J, Ma J, Sun Y, Zhou T, Zhao Y, Yu F. Microbial degradation and other environmental aspects of microplastics/plastics. Sci Total Environ 2020; 715: 136968.
[http://dx.doi.org/10.1016/j.scitotenv.2020.136968] [PMID: 32014782]

[82]   Huang L, Meng D, Tian Q, *et al.* Characterization of a novel carboxylesterase from *Bacillus velezensis* SYBC H47 and its application in degradation of phthalate esters. J Biosci Bioeng 2020; 129(5): 588-94.
[http://dx.doi.org/10.1016/j.jbiosc.2019.11.002] [PMID: 31761671]

[83]   Lu M, Jiang W, Gao Q, Zhang M, Hong Q. Degradation of Dibutyl Phthalate (DBP) by a bacterial consortium and characterization of two novel esterases capable of hydrolyzing PAEs sequentially. Ecotoxicol Environ Saf 2020; 195: 110517.
[http://dx.doi.org/10.1016/j.ecoenv.2020.110517] [PMID: 32220793]

[84]   Khadka S, Nshimiyimana JB, Zou P, Koirala N, Xiong L. Biodegradation kinetics of diethyl phthalate

by three newly isolated strains of Pseudomonas. Sci Am 2020; 8: e00380.

[85]   Huang H, Zhang XY, Chen TL, Zhao YL, Xu DS, Bai YP. Biodegradation of structurally diverse phthalate esters by a newly identified esterase with catalytic activity toward di (2-ethylhexyl) phthalate. J Agric Food Chem 2019; 67(31): 8548-58.
[http://dx.doi.org/10.1021/acs.jafc.9b02655] [PMID: 31266305]

[86]   Mo J, Zhang Z, Wang Z, *et al.* Isolation and identification of a psychrotolerant dimethyl phthalate-degrading bacterium from selected frozen soil of high-latitude areas in China and optimization of its fermentation conditions using response surface methodology. Biotechnol Biotechnol Equip 2019; 33(1): 1706-20.
[http://dx.doi.org/10.1080/13102818.2019.1696703]

[87]   Feng L, Liu H, Cheng D, *et al.* Characterization and genome analysis of a phthalate esters-degrading strain *Sphingobium yanoikuyae* SHJ. BioMed Res Int 2018; 2018(1): 3917054.
[PMID: 30065937]

[88]   Fan S, Wang J, Yan Y, Wang J, Jia Y. Excellent degradation performance of a versatile phthalic acid esters-degrading bacterium and catalytic mechanism of monoalkyl phthalate hydrolase. Int J Mol Sci 2018; 19(9): 2803.
[http://dx.doi.org/10.3390/ijms19092803] [PMID: 30231475]

[89]   Li F, Liu Y, Wang D, *et al.* Biodegradation of di-(2-ethylhexyl) phthalate by a halotolerant consortium LF. PLoS One 2018; 13(10): e0204324.
[http://dx.doi.org/10.1371/journal.pone.0204324] [PMID: 30321184]

[90]   Singh N, Dalal V, Mahto JK, Kumar P. Biodegradation of Phthalic Acid Esters (PAEs) and *in silico* structural characterization of Mono-2-Ethylhexyl Phthalate (MEHP) hydrolase on the basis of close structural homolog. J Hazard Mater 2017; 338: 11-22.
[http://dx.doi.org/10.1016/j.jhazmat.2017.04.055] [PMID: 28531656]

[91]   Zhao HM, Du H, Lin J, *et al.* Complete degradation of the endocrine disruptor di-(2-ethylhexyl) phthalate by a novel Agromyces sp. MT-O strain and its application to bioremediation of contaminated soil. Sci Total Environ 2016; 562: 170-8.
[http://dx.doi.org/10.1016/j.scitotenv.2016.03.171] [PMID: 27099998]

[92]   Tang WJ, Zhang LS, Fang Y, Zhou Y, Ye BC. Biodegradation of phthalate esters by newly isolated *Rhizobium* sp. LMB-1 and its biochemical pathway of di- *n* -butyl phthalate. J Appl Microbiol 2016; 121(1): 177-86.
[http://dx.doi.org/10.1111/jam.13123] [PMID: 26970545]

[93]   Pradeep S, Sarath Josh MK, Binod P, *et al. Achromobacter denitrificans* strain SP1 efficiently remediates di(2-ethylhexyl)phthalate. Ecotoxicol Environ Saf 2015; 112: 114-21.
[http://dx.doi.org/10.1016/j.ecoenv.2014.10.035] [PMID: 25463861]

[94]   Carstens L, Cowan AR, Seiwert B, Schlosser D. Biotransformation of phthalate plasticizers and bisphenol A by marine-derived, freshwater, and terrestrial fungi. Front Microbiol 2020; 11: 317.
[http://dx.doi.org/10.3389/fmicb.2020.00317] [PMID: 32180766]

[95]   González-Márquez A, *et al.* Ability of Fusarium culmorum to degrade the endocrine disruptor di (2-ethyl hexyl) phthalate: enzymes production and pathway of biodegradation. New Biotechnol. Elsevier Radarweg 29, 1043 Nx Amsterdam, Netherlands 2018; 44: pp. S49-50.
[http://dx.doi.org/10.1016/j.nbt.2018.05.081]

[96]   Ahuactzin-Pérez M, *et al.* Mineralization of high concentrations of the endocrine disruptor dibutyl phthalate by *Fusarium culmorum*. 3 Biotech. 2018; 8: pp. 1-10.
[http://dx.doi.org/10.1007/s13205-017-1065-2]

[97]   Ahuactzin-Pérez M, Tlecuitl-Beristain S, García-Dávila J, *et al.* A novel biodegradation pathway of the endocrine-disruptor di(2-ethyl hexyl) phthalate by *Pleurotus ostreatus* based on quantum chemical investigation. Ecotoxicol Environ Saf 2018; 147: 494-9.
[http://dx.doi.org/10.1016/j.ecoenv.2017.09.004] [PMID: 28915396]

[98] Ahuactzin-Pérez M, Tlécuitl-Beristain S, García-Dávila J, *et al.* Kinetics and pathway of biodegradation of dibutyl phthalate by *Pleurotus ostreatus*. Fungal Biol 2018; 122(10): 991-7.
[http://dx.doi.org/10.1016/j.funbio.2018.07.001] [PMID: 30227934]

[99] Velez JFM, Shashoua Y, Syberg K, Khan FR. Considerations on the use of equilibrium models for the characterisation of HOC-microplastic interactions in vector studies. Chemosphere 2018; 210: 359-65.
[http://dx.doi.org/10.1016/j.chemosphere.2018.07.020] [PMID: 30007190]

[100] Shah MP. Application of Sewage Sludge in Industrial Wastewater Treatment. John Wiley & Sons 2024.
[http://dx.doi.org/10.1002/9781119857396]

[101] Shah MP, Shah N. Environmental Approach to Remediate Refractory Pollutants from Industrial Wastewater Treatment Plant. Elsevier 2024.

[102] Yoshida S, Hiraga K, Takehana T, *et al.* A bacterium that degrades and assimilates poly(ethylene terephthalate). Science 2016; 351(6278): 1196-9.
[http://dx.doi.org/10.1126/science.aad6359] [PMID: 26965627]

[103] Selva Filho AAP, Converti A, Soares da Silva RCF, Sarubbo LA. Biosurfactants as multifunctional remediation agents of environmental pollutants generated by the petroleum industry. Energies 2023; 16(3): 1209.
[http://dx.doi.org/10.3390/en16031209]

[104] Abbasian F, Lockington R, Mallavarapu M, Naidu R. A comprehensive review of aliphatic hydrocarbon biodegradation by bacteria. Appl Biochem Biotechnol 2015; 176(3): 670-99.
[http://dx.doi.org/10.1007/s12010-015-1603-5] [PMID: 25935219]

[105] Hashem M, Alamri SA, Al-Zomyh SSAA, Alrumman SA. Biodegradation and detoxification of aliphatic and aromatic hydrocarbons by new yeast strains. Ecotoxicol Environ Saf 2018; 151: 28-34.
[http://dx.doi.org/10.1016/j.ecoenv.2017.12.064] [PMID: 29304415]

[106] Xing Z, Su X, Zhang X, Zhang L, Zhao T. Direct Aerobic Oxidation (DAO) of chlorinated aliphatic hydrocarbons: A review of key DAO bacteria, biometabolic pathways and in-situ bioremediation potential. Environ Int 2022; 162: 107165.
[http://dx.doi.org/10.1016/j.envint.2022.107165] [PMID: 35278801]

[107] Tripathi V, Gaur VK, Thakur RS, Patel DK, Manickam N. Assessing the half-life and degradation kinetics of aliphatic and aromatic hydrocarbons by bacteria isolated from crude oil contaminated soil. Chemosphere 2023; 337: 139264.
[http://dx.doi.org/10.1016/j.chemosphere.2023.139264] [PMID: 37348617]

[108] Hosseini S, Sharifi R, Habibi A. Simultaneous removal of aliphatic and aromatic crude oil hydrocarbons by *Pantoea agglomerans* isolated from petroleum-contaminated soil in the west of Iran. Arch Microbiol 2024; 206(3): 98.
[http://dx.doi.org/10.1007/s00203-023-03819-y] [PMID: 38351169]

[109] Chen X, Shan G, Shen J, Zhang F, Liu Y, Cui C. *In situ* bioremediation of petroleum hydrocarbon–contaminated soil: isolation and application of a *Rhodococcus* strain. Int Microbiol 2022; 26(2): 411-21.
[http://dx.doi.org/10.1007/s10123-022-00305-1] [PMID: 36484911]

[110] Elumalai P, Parthipan P, Huang M, *et al.* Enhanced biodegradation of hydrophobic organic pollutants by the bacterial consortium: Impact of enzymes and biosurfactants. Environ Pollut 2021; 289: 117956.
[http://dx.doi.org/10.1016/j.envpol.2021.117956] [PMID: 34426181]

[111] Nhi-Cong LT, Lien DT, Mai CTN, *et al.* Advanced materials for immobilization of purple phototrophic bacteria in bioremediation of oil-polluted wastewater. Chemosphere 2021; 278: 130464.
[http://dx.doi.org/10.1016/j.chemosphere.2021.130464] [PMID: 33845437]

[112] Ortega-González DK, Martínez-González G, Flores CM, *et al. Amycolatopsis sp.* Poz14 isolated from oil-contaminated soil degrades polycyclic aromatic hydrocarbons. Int Biodeterior Biodegradation

2015; 99: 165-73.
[http://dx.doi.org/10.1016/j.ibiod.2015.01.008]

[113] Medić A, Lješević M, Inui H, *et al.* Efficient biodegradation of petroleum *n* -alkanes and polycyclic aromatic hydrocarbons by polyextremophilic *Pseudomonas aeruginosa* san ai with multidegradative capacity. RSC Advances 2020; 10(24): 14060-70.
[http://dx.doi.org/10.1039/C9RA10371F] [PMID: 35498501]

[114] Delegan Y, Sushkova S, Minkina T, *et al.* Diversity and metabolic potential of a PAH-degrading bacterial consortium in technogenically contaminated haplic chernozem, Southern Russia. Processes (Basel) 2022; 10(12): 2555.
[http://dx.doi.org/10.3390/pr10122555]

[115] Zhou M, Liu Z, Wang J, Zhao Y, Hu B. Sphingomonas relies on chemotaxis to degrade polycyclic aromatic hydrocarbons and maintain dominance in coking sites. Microorganisms 2022; 10(6): 1109.
[http://dx.doi.org/10.3390/microorganisms10061109] [PMID: 35744627]

[116] Das A, Das N, Rajkumari J, Pandey P, Pandey P. Exploring the bioremediation potential of *Bacillus spp.* for sustainable mitigation of hydrocarbon contaminants. Environmental Sustainability. 2024; pp. 1-22.

[117] Ghosal D, Chakraborty J, Khara P, Dutta TK. Degradation of phenanthrene *via* meta-cleavage of 2-hydroxy-1-naphthoic acid by *Ochrobactrum sp.* strain PWTJD. FEMS Microbiol Lett 2010; 313(2): 103-10.
[http://dx.doi.org/10.1111/j.1574-6968.2010.02129.x] [PMID: 20964703]

[118] Alegbeleye OO, Opeolu BO, Jackson V. Bioremediation of Polycyclic Aromatic Hydrocarbon (PAH) compounds: (acenaphthene and fluorene) in water using indigenous bacterial species isolated from the Diep and Plankenburg rivers, Western Cape, South Africa. Braz J Microbiol 2017; 48(2): 314-25.
[http://dx.doi.org/10.1016/j.bjm.2016.07.027] [PMID: 27956015]

[119] Nzila A, Ramirez CO, Musa MM, Sankara S, Basheer C, Li QX. Pyrene biodegradation and proteomic analysis in *Achromobacter xylosoxidans*, PY4 strain. Int Biodeterior Biodegradation 2018; 130: 40-7.
[http://dx.doi.org/10.1016/j.ibiod.2018.03.014]

[120] Qin W, *et al.* Anaerobic biodegradation of benzo (a) pyrene by a novel *Cellulosimicrobium cellulans* CWS2 isolated from polycyclic aromatic hydrocarbon-contaminated soil. Braz J Microbiol. 2018; 49: pp. (2)258-68.
[http://dx.doi.org/10.1016/j.bjm.2017.04.014]

[121] Zhu X, Ni X, Waigi M, Liu J, Sun K, Gao Y. Biodegradation of mixed PAHs by PAH-degrading endophytic bacteria. Int J Environ Res Public Health 2016; 13(8): 805.
[http://dx.doi.org/10.3390/ijerph13080805] [PMID: 27517944]

[122] Jiménez-Volkerink SN, Vila J, Jordán M, Minguillón C, Smidt H, Grifoll M. Multi-omic profiling of a newly isolated oxy-PAH degrading specialist from PAH-contaminated soil reveals bacterial mechanisms to mitigate the risk posed by polar transformation products. Environ Sci Technol 2023; 57(1): 139-49.
[http://dx.doi.org/10.1021/acs.est.2c05485] [PMID: 36516361]

[123] Lu Q, Jiang Z, Feng W, *et al.* Exploration of bacterial community-induced polycyclic aromatic hydrocarbons degradation and humus formation during co-composting of cow manure waste combined with contaminated soil. J Environ Manage 2023; 326(Pt B): 116852.
[http://dx.doi.org/10.1016/j.jenvman.2022.116852] [PMID: 36435124]

[124] Al-Hawash AB, Zhang X, Ma F. Removal and biodegradation of different petroleum hydrocarbons using the filamentous fungus *Aspergillus* sp. RFC-1. MicrobiologyOpen 2019; 8(1): e00619.
[http://dx.doi.org/10.1002/mbo3.619] [PMID: 29577679]

[125] Mohammed SA, Omar TJ, Hasan AH. Degradation of crude oil and pure hydrocarbon fractions by some wild bacterial and fungal species. arXiv preprint arXiv:230108715 2023.

[126] Haritash A, Kaushik C. Degradation of low molecular weight polycyclic aromatic hydrocarbons by microorganisms isolated from contaminated soil. Int J Environ Sci 2016; 6(5): 808-19.

[127] Birolli WG, de A Santos D, Alvarenga N, Garcia ACFS, Romão LPC, Porto ALM. Biodegradation of anthracene and several PAHs by the marine-derived fungus *Cladosporium sp.* CBMAI 1237. Mar Pollut Bull 2018; 129(2): 525-33.
[http://dx.doi.org/10.1016/j.marpolbul.2017.10.023] [PMID: 29055563]

[128] Agrawal N, Shahi SK. Degradation of polycyclic aromatic hydrocarbon (pyrene) using novel fungal strain *Coriolopsis byrsina* strain APC5. Int Biodeterior Biodegradation 2017; 122: 69-81.
[http://dx.doi.org/10.1016/j.ibiod.2017.04.024]

[129] Kumar V, Kumar H, Vishal V, Lal S. Studies on the morphology, phylogeny, and bioremediation potential of *Penicillium citrinum* and *Paecilomyces variotii* (Eurotiales) from oil-contaminated areas. Arch Microbiol 2023; 205(1): 50.
[http://dx.doi.org/10.1007/s00203-022-03383-x] [PMID: 36598589]

[130] Hadibarata T, Kristanti RA, Bilal M, Al-Mohaimeed AM, Chen TW, Lam MK. Microbial degradation and transformation of benzo[a]pyrene by using a white-rot fungus *Pleurotus eryngii* F032. Chemosphere 2022; 307(Pt 3): 136014.
[http://dx.doi.org/10.1016/j.chemosphere.2022.136014] [PMID: 35970216]

[131] Varjani SJ. Microbial degradation of petroleum hydrocarbons. Bioresour Technol 2017; 223: 277-86.
[http://dx.doi.org/10.1016/j.biortech.2016.10.037] [PMID: 27789112]

[132] Marghade DT, Chahande AD, Tiwari MS, Patil PD. Microbial degradation of xenobiotic compounds. Recent Adv Microb Degrad. 173-217. 2021; pp.
[http://dx.doi.org/10.1007/978-981-16-0518-5_7]

[133] Singh P, Iyengar L, Pandey A. Bacterial decolorization and degradation of azo dyes. Microbial degradation of xenobiotics. 101-33. 2012; pp.
[http://dx.doi.org/10.1007/978-3-642-23789-8_4]

[134] Costa MF, de Oliveira AM, Oliveira Junior EN. Biodegradation of linear alkylbenzene sulfonate (LAS) by *Penicillium chrysogenum*. Bioresour Technol Rep 2020; 9: 100363.
[http://dx.doi.org/10.1016/j.biteb.2019.100363]

[135] Singh A, Prasad SM, Singh RP. Plant responses to xenobiotics. Springer 2016.
[http://dx.doi.org/10.1007/978-981-10-2860-1]

[136] Phale PS, Sharma A, Gautam K. Microbial degradation of xenobiotics like aromatic pollutants from the terrestrial environments. Pharmaceuticals and personal care products: waste management and treatment technology. Elsevier 2019; pp. 259-78.
[http://dx.doi.org/10.1016/B978-0-12-816189-0.00011-1]

[137] Xu W, Zhao Q, Li Z, Lu X, Han S, Ye Z. Biodegradation of dinitrotoluene sulfonates and other nitro-aromatic compounds by *Pseudomonas sp.* X5 isolated from TNT red water contaminated soil. J Clean Prod 2019; 214: 782-90.
[http://dx.doi.org/10.1016/j.jclepro.2019.01.025]

[138] Singh D, Mishra K, Ramanthan G. Bioremediation of nitroaromatic compounds. Wastewater Treatment Engineering. 51-83. 2015; pp.
[http://dx.doi.org/10.5772/61253]

[139] Arora P, Bae H. Bacterial degradation of chlorophenols and their derivatives. Microb Cell Fact 2014; 13(1): 31.
[http://dx.doi.org/10.1186/1475-2859-13-31] [PMID: 24589366]

[140] Singh PK, Singh RL. Bio-removal of azo dyes: a review. Int J Appl Sci Biotechnol 2017; 5(2): 108-26.
[http://dx.doi.org/10.3126/ijasbt.v5i2.16881]

[141] Sarkar S, Banerjee A, Halder U, Biswas R, Bandopadhyay R. Degradation of synthetic azo dyes of

textile industry: a sustainable approach using microbial enzymes. Water Conservation Science and Engineering 2017; 2(4): 121-31.
[http://dx.doi.org/10.1007/s41101-017-0031-5]

[142] Chekroun KB, Sánchez E, Baghour M. The role of algae in bioremediation of organic pollutants. J Iss ISSN 2014; 2360(8803)

[143] Anasonye F, Winquist E, Räsänen M, *et al.* Bioremediation of TNT contaminated soil with fungi under laboratory and pilot scale conditions. Int Biodeterior Biodegradation 2015; 105: 7-12.
[http://dx.doi.org/10.1016/j.ibiod.2015.08.003]

[144] Misal SA, Humne VT, Lokhande PD, Gawai KR. Biotransformation of nitro aromatic compounds by flavin-free NADH-azoreductase. J Bioremediat Biodegrad 2015; 6(272): 2.

[145] Liu DF, Min D, Cheng L, *et al.* Anaerobic reduction of 2,6-dinitrotoluene by *Shewanella oneidensis* MR-1: Roles of Mtr respiratory pathway and NfnB. Biotechnol Bioeng 2017; 114(4): 761-8.
[http://dx.doi.org/10.1002/bit.26212] [PMID: 27869299]

[146] Hsueh CC, You LP, Li JY, Chen CT, Wu CC, Chen BY. Feasibility study of reduction of nitroaromatic compounds using indigenous azo dye-decolorizers. J Taiwan Inst Chem Eng 2016; 64: 180-8.
[http://dx.doi.org/10.1016/j.jtice.2016.04.015]

[147] Erdem Ö, Türkmen KE, Aracagök D, Erdönmez D, Cihangir N. Decolorization of reactive blue 19 dye by *bacillus megaterium* isolated from soil. Hacet J Biol Chem 2019; 47(2): 193-201.
[http://dx.doi.org/10.15671/hjbc.623786]

[148] El Bouraie M, El Din WS. Biodegradation of Reactive Black 5 by *Aeromonas hydrophila* strain isolated from dye-contaminated textile wastewater. Sustain Environ Res 2016; 26(5): 209-16.
[http://dx.doi.org/10.1016/j.serj.2016.04.014]

[149] Neifar M, Chouchane H, Mahjoubi M, Jaouani A, Cherif A. *Pseudomonas extremorientalis* BU118: a new salt-tolerant laccase-secreting bacterium with biotechnological potential in textile azo dye decolourization. 3 Biotech. 2016; 6: pp. 1-9.

[150] Blánquez A, Rodríguez J, Brissos V, *et al.* Decolorization and detoxification of textile dyes using a versatile *Streptomyces* laccase-natural mediator system. Saudi J Biol Sci 2019; 26(5): 913-20.
[http://dx.doi.org/10.1016/j.sjbs.2018.05.020] [PMID: 31303819]

[151] Sofu A. Investigation of dye removal with isolated biomasses from whey wastewater. Int J Environ Sci Technol 2019; 16(1): 71-8.
[http://dx.doi.org/10.1007/s13762-018-1977-3]

[152] Patel DK, Tipre DR, Dave SR. Enzyme mediated bacterial biotransformation and reduction in toxicity of 1:2 chromium complex AB193 and AB194 dyes. J Taiwan Inst Chem Eng 2017; 77: 1-9.
[http://dx.doi.org/10.1016/j.jtice.2017.02.027]

[153] Guo Z, Yin H, Wei X, Zhu M, Lu G, Dang Z. Effects of methanol on the performance of a novel BDE-47 degrading bacterial consortium QY2 in the co-metabolism process. J Hazard Mater 2021; 415: 125698.
[http://dx.doi.org/10.1016/j.jhazmat.2021.125698] [PMID: 33773249]

[154] Kaur J, Gosal S. Biotransformation of pollutants: a microbiological perspective. Rhizobiont in Bioremediation of Hazardous Waste 2021; pp. 151-62.

[155] Zhang Q, Chen L, Xie X, *et al.* Molecular mechanism triggered by co-metabolic biodegradation of azo dyestuff by *Klebsiella sp.* KL-1: Based on enzymatic and proteomic responsiveness. J Water Process Eng 2024; 62: 105339.
[http://dx.doi.org/10.1016/j.jwpe.2024.105339]

[156] Wang Q, Wang H, Lv M, Wang X, Chen L. Sulfamethoxazole degradation by *Aeromonas caviae* and co-metabolism by the mixed bacteria. Chemosphere 2023; 317: 137882.
[http://dx.doi.org/10.1016/j.chemosphere.2023.137882] [PMID: 36657578]

[157] Huang W, Yin H, Yu Y, Lu G, Dang Z, Chen Z. Co-metabolic degradation of tetrabromobisphenol A by *Pseudomonas aeruginosa* and its auto-poisoning effect caused during degradation process. Ecotoxicol Environ Saf 2020; 202: 110919.
[http://dx.doi.org/10.1016/j.ecoenv.2020.110919] [PMID: 32800254]

[158] Zhu M, Su Y, Wang Y, *et al.* Biodegradation characteristics of p-Chloroaniline and the mechanism of co-metabolism with aniline by *Pseudomonas sp.* CA-1. Bioresour Technol 2024; 406: 131086.
[http://dx.doi.org/10.1016/j.biortech.2024.131086] [PMID: 38977036]

[159] Wang P, Li D, Sun M, Yin J, Zheng T. Microalgae enhanced co-metabolism of sulfamethoxazole using aquacultural feedstuff components: Co-metabolic pathways and enzymatic mechanisms. J Hazard Mater 2024; 470: 134279.
[http://dx.doi.org/10.1016/j.jhazmat.2024.134279] [PMID: 38613960]

[160] Xiong Q, Liu YS, Hu LX, *et al.* Co-metabolism of sulfamethoxazole by a freshwater microalga *Chlorella pyrenoidosa*. Water Res 2020; 175: 115656.
[http://dx.doi.org/10.1016/j.watres.2020.115656] [PMID: 32145399]

[161] Fan Y, Yan D, Chen X, *et al.* Novel insights into the co-metabolism of pyridine with different carbon substrates: Performance, metabolism pathway and microbial community. J Hazard Mater 2024; 465: 133396.
[http://dx.doi.org/10.1016/j.jhazmat.2023.133396] [PMID: 38176261]

CHAPTER 4

Arsenic Bioremediation: A New Paradigm in Microbial Arsenic Clean-up Strategies

Vinod Nandre[1], Aditi Bagade[1], Savita Tapase[1], Sougata Ghosh[2,3] and Kisan Kodam[1,*]

[1] *Department of Chemistry, Division of Biochemistry, Savitribai Phule Pune University (Formerly University of Pune), Pune-411007, Maharashtra, India*

[2] *Department of Physics, Faculty of Science, Kasetsart University, Bangkok-10900, Thailand*

[3] *Department of Microbiology, School of Science, RK University, Rajkot-360020, Gujarat, India*

Abstract: The biosphere is under siege from heavy metal pollution, a dire consequence of human actions. Heavy metals are non-biodegradable, which persist for a long time in the environment, and cause severe water, soil, and air pollution. Green technology, like bioremediation is one of the promising approaches towards hazardous waste. This can be done by reducing bioavailability, mobility, and toxicity by transformation strategies. In the history of heavy metal pollution, arsenic (As) was one of the mass poisoning priorities pollutants extensively studied. In Bangladesh, more than 10 million people suffer from a huge amount of arsenic poisoning, and to date, people there face arsenic pollution in their day-to-day lives. Arsenic is the top carcinogen reported in different studies. This is due to the strong chemical relevance of phosphate as an essential biological moiety in nature and irreversible biochemical interactions with vital proteins. Various strategies have been developed in the last few decades, like physical methods, chemical methods, and phytoremediation, to overcome arsenic poisoning through contaminated water or bioaccumulation of arsenic metalloids in the food chain. Moreover, microbes subjected to continuous arsenic exposure develop several mechanisms to tolerate high arsenic concentrations, such as adsorption, complexation, and biotransformation of arsenic into a less toxic form by enzymatic reduction or by using them as terminal electron acceptors or donors in microbial respiration. Arsenic bioremediation is getting more attention because of its efficiency and cost-effective parameters.

Keywords: Arsenic poisoning, Bioremediation, Heavy metal, Remediation parameters, Toxicity.

*** Corresponding author Kisan Kodam:** Department of Chemistry, Division of Biochemistry, Savitribai Phule Pune University (Formerly University of Pune), Pune - 411007, Maharashtra, India; E-mails: kisankodam@gmail.com, kisan.kodam@unipune.ac.in

INTRODUCTION

"Heavy metals" are defined as elements with an atomic weight between 63.5 and 200.6 and a specific gravity higher than 5.0 [1]. These are elements that are naturally present in the crust of Earth, some of which are biologically necessary that humans have brought into diverse ecosystems through various activities [1 - 4]. The prevalence of heavy metal discharge into wastewater, either directly or indirectly, is rising in developing nations due to the explosive growth of businesses, including metal plating, mining, fertilizer, tanneries, batteries, paper, pesticides, *etc*. Heavy metals are not biodegradable organic contaminants and tend to accumulate in living organisms [5]. Therefore heavy metal contamination of soil in industrialized countries has become a serious problem, especially in dense population regions where land is immensely used [2, 6]. By a variety of exposure pathways, such as consuming crops cultivated on contaminated soils or breathing in dust that clings to plants, these activities pose a health risk to the nearby population [1, 7, 8].

In complex ecosystems, heavy metals transform and transport between environmental compartments (soil/sediments, water, and air) through physical, chemical, and biological processes. Metal speciation deterministically controls these processes [9]. For the last few decades, soils have been considered the definitive sink for heavy metal discharge (Fig. **1**). The difficulty in managing these metals is partly due to their strong binding to soil and the complex nature of soil environments [6, 9, 10].

ARSENIC

Although it is typically described as a heavy metal that exists naturally in the earth's crust, arsenic exists as a semi-metallic element that is frequently found in the environment and is the 20th most prevalent element in the crust [11, 12]. Almost all arsenic species are odorless, water-soluble, and tasteless, which generates an eminent health risk, whereas arsenic is relatively solid and has a grey color in its pure elemental state. In the environment arsenic combines with other elements, and it changes to either white or colorless powder form that is very difficult to differentiate. Due to the absence of smell and taste, it makes arsenic compounds enormously difficult to identify in water, air, or food [12]. While naturally occurring arsenic is often found in very small amounts in soil, several parts of the world have substantial deposits of the element, which are thought to represent elevated levels of arsenic in groundwater. The groundwater sources in these arsenic-rich areas are frequently contaminated, and the local inhabitants regularly get their drinking water from these supplies [12, 13].

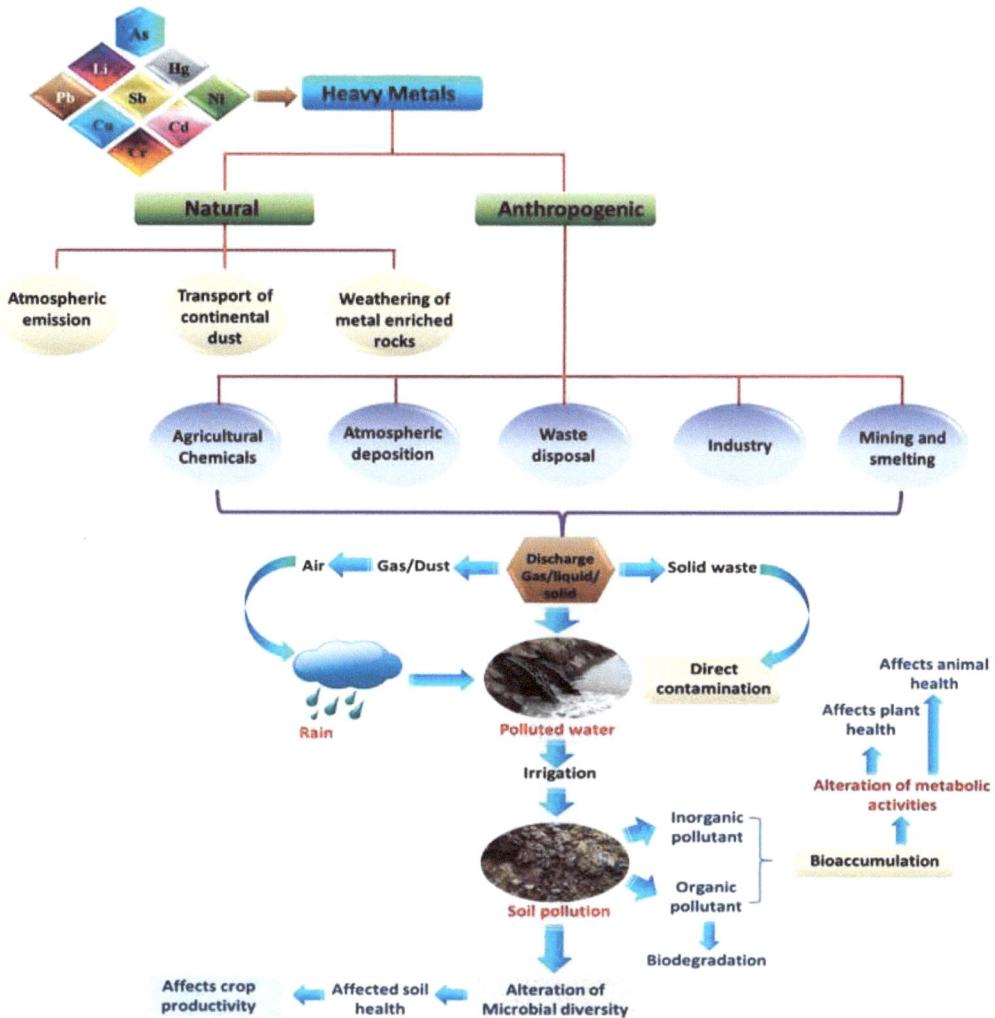

Fig. (1). Origin, sink, and toxic effects of heavy metal soil contamination.

Human activity exacerbates the amount of arsenic that flows into groundwater systems from the earth's crust. In many regions of the world, local communities rely only on groundwater systems, which use pump wells to retrieve water from far beneath the surface of the earth. Water from these wells is used for several purposes, including irrigation [14 - 16]. As a result, excessive water pumping became responsible for increased natural arsenic concentrations in groundwater systems, which contaminate all water sources. Globally, South Asia—mainly Bangladesh, Nepal, and India—is the region where severe groundwater contamination and chronic human consumption have been documented [17].

According to the WHO report, an estimated 35 to 77 million people are at risk of arsenic poisoning in groundwater, which is causing "the largest mass poisoning of a population in history" [17, 18]. The major route through which arsenic enters into the systems of humans and animals is through the ingestion of contaminated food and water. Different arsenic species have been identified as key contributors to human exposure, leading to various health problems. These species include volatile arsenic hydride, methylated arsenic, arsenite, arsenate, and organoarsenic in food [19]. Chronic arsenic exposure can have a number of adverse effects on health, including keratosis, hyperpigmentation, depigmentation, peripheral vascular disease, and others [20 - 23]. Human exposure to arsenic has also been linked to several internal and skin malignancies. In contrast, reports of non-cancer effects include those about the gastrointestinal tract, skin, neurological system, hematological system, hepatic system, respiratory system, and genotoxic effects [8, 19, 24].

Arsenic Speciation in Nature: Aqueous and Gaseous States

Arsenite ($H_3As^{3+}O_3$) and arsenate ($H_3As^{5+}O_4$) are the two important anionic species that form in aqueous conditions. While in most of the natural waters (pH ~ 4-10), arsenite exists as a neutral species (*e.g.*, H_3AsO_3), arsenate can form two different species like H_2AsO^{4-} and $HAsO_4^{2-}$ [25]. However, the range of oxidation potentials (Eh) and pH values that are frequently found in natural environments affect the equilibrium state of arsenic. Thus, arsenite is a stable aqueous form observed under the reducing environment where the Eh value is about +300 mV at pH 4 to –200 mV at pH 9, while in an oxidized environment, arsenate was found to be a very stable form [26, 27]. The common forms of arsenic identified in nature, such as arsenite and arsenate, are briefly described below.

ARSENIC BIOREMEDIATION

A wide range of metalloids and elements with characteristics intermediate between those of metals and non-metals can be changed by a variety of microorganisms. Along with Te, Si, Ge, Sb, and B, other metals like Se and As are also classified under this heading. Microbial oxidation, reduction, and methylation affect the bioavailability and toxicity of numerous metalloids. For instance, inorganic forms of Se are more toxic than volatile methylated derivatives of Se [28]. Prokaryotes may readily metabolize As and Se by contributing to a series of metabolic functions. Because of resistance and respiratory processes *via* microbial oxidation/reduction reactions, the speciation and mobility of arsenic are affected [28, 29].

Necessity of Arsenic Bioremediation

Arsenic has been a controversial element since its initial isolation by Albertus Magnus in 1250 A.D. Since then, it has been used in diverse applicative fields, namely medication, farming, electronics, industry, and metallurgy [30]. Many studies over the past decades have accepted that arsenic consumption leads to carcinogenesis, even at low levels [8, 12, 30]. The most severe documented contamination of arsenic lies in South Asia, mainly in Nepal, India, and Bangladesh. WHO reports Bangladesh to be facing a massive mass population poisoning where approximately 35 to 77 million people have been at risk due to arsenic contamination [30]. Arsenic cannot be destroyed, and it can interchangeably spread from soil, air, and water. Once arsenic is released from the earth's crust, it forms a complex with hydrogen and oxygen, which helps it to establish itself in soil, and water sediments, contaminating drinking water and agricultural products, which results in human health risks [31, 32].

Groundwater contamination is a serious problem caused by man-made industrial activities leaching arsenic from the earth's crust into groundwater systems [13]. Local communities throughout the world rely on groundwater for drinking and irrigation in agricultural fields [33]. Digging deep and overpumping wells are responsible for increased natural arsenic in groundwater systems, which further contaminates the whole soil [13].

However, the major contamination of arsenic comes from the minerals where most of the arsenic minerals are in their sulfide form, namely *Arsenopyrite* (FeAsS). It is the most abundant mineral ore for arsenic, which, on oxidation, releases arsenite As (III) and arsenate As (V). Similar oxidation products are produced by yellow-colored *Orpiment* (As_2S_3), while conversely, *Energite* ($Cu_3 AsS_4$) produces As (V) on oxidation. The other mineral ores for arsenic are *Realgar* (As_2S_2), *Loellingite* ($FeAs_2$), *and Seligmannite* ($PbCuAsS_3$). The minerals having an elemental form of arsenic as the mineral group are *Pararsenolamprite* As (0), *Arsenolamprite,* and *Rhombohedral* arsenic, and all these originate in the hydrothermal deposits. *Arsenolite* and *Claudite* with the molecular formula As_2O_3 are As (III) containing minerals originating from the weathering products, while *Scorodite* ($FeAsO_4.2H_2O$) is a mineral containing As (V) [34].

Human populations are exposed to arsenic contamination through the most common routes, such as ingestion, drinking, and irrigation. Arsenic toxicity has gained much international attention due to the widespread and significant human health issues [13, 35]. The areas in the flood plains of rivers Ganga, Brahmaputra, and Imphal have groundwater arsenic contamination above the permissible limit of 10 μg L^{-1}. These areas include West Bengal, Jharkhand, Bihar, Uttar Pradesh,

Assam, Manipur along with Rajnandgaon village in Chhattisgarh [24, 36, 37]. Numerous unknown factors are contributing to the constant rise in the number of people suffering from As toxicity [13, 31]. The most affected states in India are West Bengal, Bihar, Assam, and Chhattisgarh, with increasing health problems. The Bengal Delta Plain, spanning West Bengal, India, and Bangladesh, is facing one of the most significant environmental health crises of the 21st century. The widespread contamination of groundwater with arsenic has put at least 50 million people at risk of developing cancer and other debilitating diseases.

Arsenic Contamination Scenario

There are many places in and around India contaminated with groundwater arsenic, which has become a major public health problem. The WHO reports that across 70 countries headcount of people exposed to drinking water with arsenic is around 140 million [38]. In 2010, the Food and Agricultural Organization of the United Nations and the World Health Organization Joint Expert Committee on Food Additives (JECFA) re-evaluated the consequences of arsenic on human health together. The joint venture concluded that certain regions of the world with inorganic arsenic concentrations exceeding 50–100 µg L^{-1} had some evidence of having adverse effects. Whereas, in the areas where arsenic in water was elevated to around 10–50 µg L^{-1}, there was a lower chance of adverse effects that would have been difficult to detect in epidemiological studies. Considering the global scenario, Bangladesh is heavily arsenic-contaminated, followed by India > the United States > China, and Nigeria.

Available Remediation Methods for Arsenic

Exposure to arsenic from contaminated agricultural fields causing severe health hazards needs to be curbed strategically [31, 32]. Removing arsenic by oxidizing As (III) to As (V) appears to be a decontamination strategy. Numerous remediation processes for arsenic removal have been described, often involving agents like iron and aluminum salts that cause its coagulation and precipitation. Other strategies for removal involve activated alumina-directed adsorption, ion exchange, and reverse osmosis. Fig. (**2**) illustrates the schematic representation of different techniques used for the removal of arsenic from soil and water. Recent technologies are based on oxidation and adsorption using green sand filtration, ferruginous manganese ore, clay minerals, and zero-valent iron were used for arsenic removal and are inefficient for As (III) removal [39, 40]. The generation of additional pollution and the treatment costs are the foremost disadvantages of these developments, which have led to biological oxidation-based alternative arsenic remediation methods.

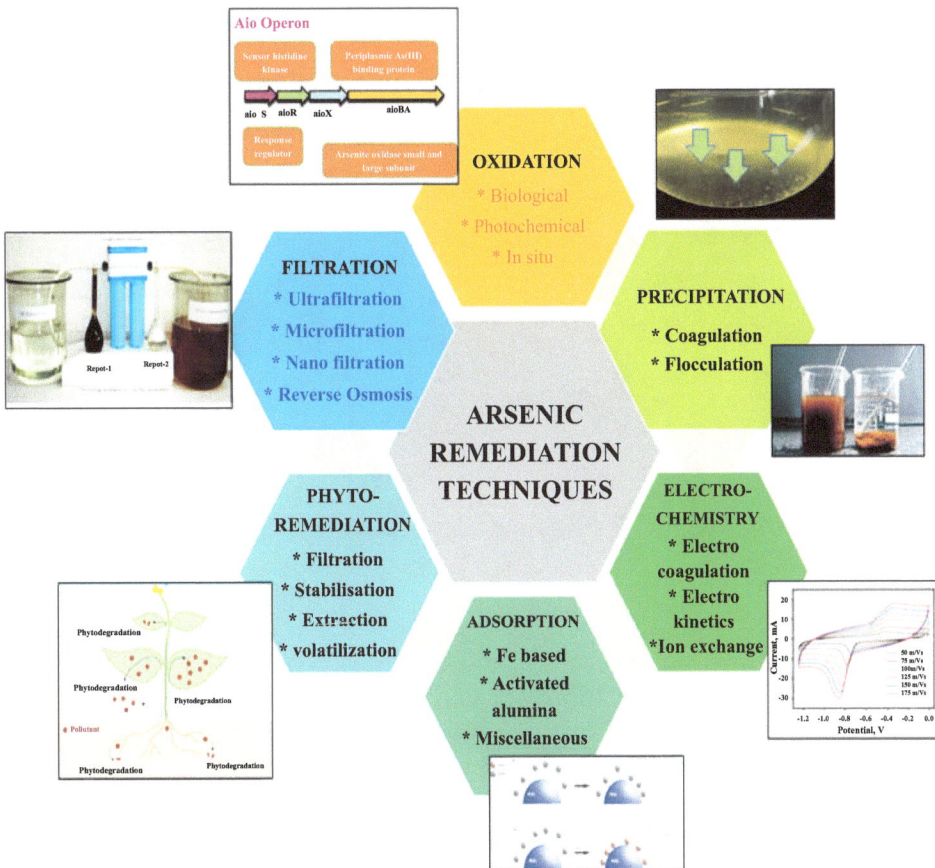

Fig. (2). Different techniques used for the removal of arsenic from soil and water.

BACTERIAL ARSENIC REMEDIATION

Microbes have evolved resistance mechanisms to cope with arsenic toxicity, allowing them to thrive in arsenic environments. As (V) enters through phosphate transport proteins in microbial cells, while As (III) enters *via* aquaglyceroporins at neutral pH. The bacterial arsenic resistance includes detoxification and redox reactions concerning As (V) reduction or As (III) oxidation that conserve the energy gained by the reactions to provide for cell growth. The *ars* operon is considered to be the most widespread resistance mechanism found in both Gram-negative and Gram-positive bacteria located either on chromosomes or plasmid. Ars operon consists of either 3-5 genes which confer arsenic resistance or detoxification. This multifaceted operon provides resistance against 3 ions, namely As (III), As (V), and antimony [34]. *Escherichia coli* Ars operon presents 3 open reading frames with the repressor *arsR*, As (III) efflux pump *arsB,* and arsenic reductase *arsC* Fig. (**3**).

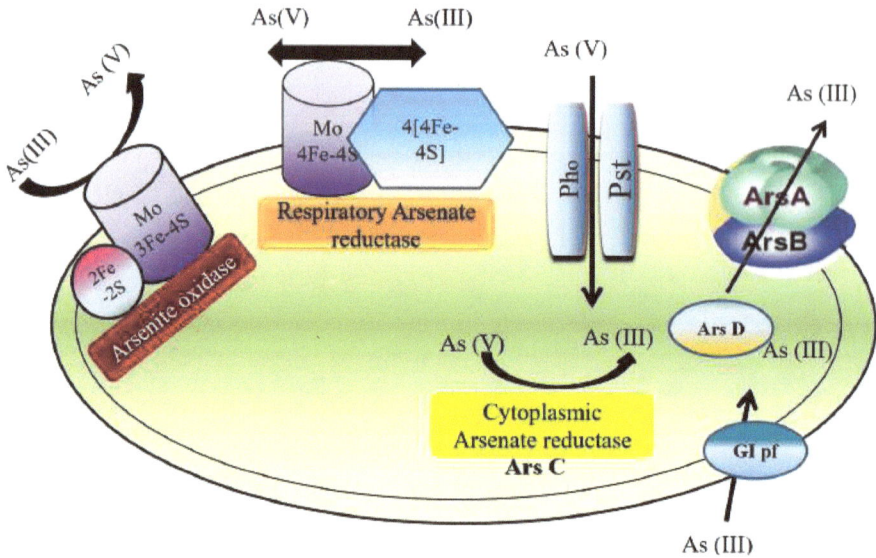

Fig. (3). Different arsenic transformation mechanisms in bacteria governed by the *ars* operon and *aio* operon.

Arsenic toxicity in all living things is defined by its state of oxidation and its chemical form. All organic arsenic derivatives are less toxic forms, while, amongst the inorganic forms, As (III) is a hundred times more toxic than As (V), the less mobile form of arsenic [41, 42]. Several As (III) oxidizing bacteria have been isolated from diverse environments [43]. *Alcaligenes faecalis* is the most common arsenite-oxidizing heterotrophic bacteria, which requires organic matter for growth. The heterotrophic nature of arsenite oxidizing bacteria is contemplated as a strategy for the removal of toxic arsenic rather than generating energy or supporting growth; however, *Pseudomonas arsenitoxidans* grow chemolithoautotrophically in the presence of arsenite, oxygen, and carbon dioxide to gain energy [29]. This gold mine-derived novel bacterium utilizes a carbon source in the form of carbon dioxide or bicarbonate to donate electrons to oxygen being an electron acceptor demonstrating distinctive metabolic capabilities. These important findings help in understanding the factors that affect the bioremediation of arsenic [44].

Role of Arsenite Oxidase in Arsenic Remediation

The first scientist to identify arsenite oxidase was Green in 1918 from a cattle dipping site in South Africa, where As (III) was added to protect them against insect bites. Oxidation was carried out by *Bacillus arsenoxidans*. This finding was ignored until Turner, in 1949, found the same bacteria along with *Pseudomonas, Xanthomonas,* and *Achromobacter* in the cattle dipping site in Queensland,

Australia. Later, Turner and Legge showed for the first time that As (III) oxidation was maximum with 3-4 days old cells and that oxidation was done only with those cells that were previously exposed to As (III). This later laid the foundation for the washed cell assay where previously As (III) exposed cells are taken for higher oxidation efficiency and cells are devoid of growing media. Maximum organisms experience detrimental effects of the inorganic arsenite [As (III)] and arsenate [As (V)]. To support growth and survival, few bacteria with phylogenetically diverse backgrounds make use of the reduction potential of the arsenite/ arsenate couple of +60 mV, where arsenite acts as an electron donor or arsenate as a terminal electron acceptor facilitating the metabolic processes [29, 45].

Studying the phylogenetic relationship reveals that arsenite oxidase (Aio) had a universal common ancestor before bacteria and archaea diverged [46]. Aio plays an important role in the catalysis of As (III) oxidation to form arsenate As (V), with the coupling of the reduction of oxygen to water generating ATP with arsenite oxidation. In some cases it also pairs with NADH for carbon dioxide fixation [47]. Azurin or c-type cytochromes seemed to be the physiological electron acceptors for Aio [44, 48]. Aio is a member of the Dimethyl Sulfoxide Reductase (DMSOR) superfamily of molybdoenzymes, which all contain two equivalents of an organic pyranopterin cofactor coordinated to the molybdenum, usually present as the dinucleotide of guanine and termed as MGD (for Molybdopterin Guanine Dinucleotide).

As a member of the DMSOR superfamily, Aio contains a unique molybdenum (Mo) centre with 2 equivalents of pyranopterin cofactor, coordinated to Mo as a dinucleotide of guanine [46]. Unlike other DMSOR members, the molybdenum in Aio is not coordinated with the protein by an amino acid side chain. The molybdenum center of Aio exhibits highly cooperative two-electron transfer, with the intermediate Mo (V) oxidation state rarely observed until recently in a mutant with altered hydrogen bonding to the MGD. Additionally, Aio contains a high-potential 3Fe-4S cluster in the large catalytic subunit (AioA) and a Rieske center in the small subunit (AioB), homologous to the Rieske protein in bc1 and b6f complexes [42-44]. It has been suggested that the electrons from arsenite oxidation pass to the molybdenum center, to the 3Fe-4S cluster, the Rieske cluster, and finally to an electron acceptor [46, 48, 49]. The final product of oxidation was further used to remediate from the environment by using different metal affinity-based adsorption matrices, *e.g.*, Fe, Al, and zeolite. The rate of As (V) adsorption is high as compared to As (III) because of its less reactivity and mobility.

SCIENCE BEYOND RESEARCH ARTICLES

It has long been known that science is restricted only to peer-reviewed manuscripts. The need to bridge the gap between lab research and the common man is still a matter of concern. Huge amounts of government subsidies and money are channeled for human welfare, limiting the output. Despite country-wide mass awareness programs by government and NGOs, the output remains below a satisfactory level. The real help to the affected people should include providing easy and handy arsenic testing kits to check whether the water supplied to them is contaminated. The next approach is to connect the arsenic removal system suppliers with the villagers to nullify the post-installation problems like changing filters or their repairing and to address the downstream problems experienced after a few years of installation. Also, proper pricing of these arsenic units is to be done to eliminate the abandonment due to financial reasons. Arranging science clubs in schools and laboratory facilities at colleges to increase awareness can be the next approach. Rainwater harvesting can be a remedy for the current groundwater arsenic contamination problem, where proper storage of rainwater can eliminate the possibility of excess surface water being stored in groundwater basins during wet seasons. Providing proper treatment and cheap, easily available medications for the already-affected victims of arsenicosis is also an important factor to be considered.

CONCLUSION AND FUTURE PERSPECTIVES

Underground arsenic contamination and its related toxicity is a worldwide issue. Different technologies have also been developed based on the chemo-physical properties of arsenic species present in the ecosystem. However, microbial-mediated arsenic remediation is an all-time best and very efficient technology that saves energy and is an environment-friendly approach. The complexity and secondary pollution caused by the present bioremediation techniques warrant the need for a greener and faster technology. However, biological routes for remediation have several disadvantages: they are time-consuming, exhibit slow enzyme activities (*e.g.*, arsenic oxidase or methylase), face issues with the shelf life of the organisms used, pose reusability challenges, and in some cases, lead to the production of harmful by-products (*e.g.*, arsenic species generated by arsenate reductase activity). Hence, it is not necessary that in all bioremediation cases, the microbes break toxic metals in a harmless form. Hence, to overcome such difficulties different ways should be indulged. Absorbing the bacterial cells into a permeable matrix can solve the reusability problems to some extent. Pre-induction of enzymes in detoxification and overexpression of proteins can fasten the rate of bioremediation. Modifying the permeability of the bacterial membrane allows for a quicker accumulation of toxic arsenic inside the bacteria. Future studies should focus on genetically modified organisms having improved bioremediation capacity. This can be achieved by incorporating more than one arsenic operon, overexpression of arsenic influx proteins, and upregulation of proteins in *aio* operon, multiple heavy metal resistant organisms to provide a green and clean environment in the near future.

ACKNOWLEDGEMENTS

The authors thank the University Grants Commission (UGC), New Delhi, India, and UPE-II, SPPU, Pune, for the research fellowship and financial support.

REFERENCES

[1]   Ali H, Khan E, Sajad MA. Phytoremediation of heavy metals—Concepts and applications. Chemosphere 2013; 91(7): 869-81.
[http://dx.doi.org/10.1016/j.chemosphere.2013.01.075] [PMID: 23466085]

[2]   Krishna AK, Govil PK. Assessment of heavy metal contamination in soils around Manali industrial area, Chennai, Southern India. Environ Geol (Berl) 2008; 54(7): 1465-72.
[http://dx.doi.org/10.1007/s00254-007-0927-z]

[3]   Rasool A, Xiao T, Farooqi A, *et al.* Arsenic and heavy metal contaminations in the tube well water of Punjab, Pakistan and risk assessment: A case study. Ecol Eng 2016; 95: 90-100.
[http://dx.doi.org/10.1016/j.ecoleng.2016.06.034]

[4]   Zhang J, Zhou W, Liu B, He J, Shen Q, Zhao FJ. Anaerobic arsenite oxidation by an autotrophic arsenite-oxidizing bacterium from an arsenic-contaminated paddy soil. Environ Sci Technol 2015; 49(10): 5956-64.
[http://dx.doi.org/10.1021/es506097c] [PMID: 25905768]

[5]   Fu F, Wang Q. Removal of heavy metal ions from wastewaters: A review. J Environ Manage 2011; 92(3): 407-18.
[http://dx.doi.org/10.1016/j.jenvman.2010.11.011] [PMID: 21138785]

[6]   Al-Khashman OA. Heavy metal distribution in dust, street dust and soils from the work place in Karak Industrial Estate, Jordan. Atmos Environ 2004; 38(39): 6803-12.
[http://dx.doi.org/10.1016/j.atmosenv.2004.09.011]

[7]   Wu C, Huang L, Xue SG, *et al.* Oxic and anoxic conditions affect arsenic (As) accumulation and arsenite transporter expression in rice. Chemosphere 2017; 168: 969-75.
[http://dx.doi.org/10.1016/j.chemosphere.2016.10.114] [PMID: 27817896]

[8]   Xue S, Shi L, Wu C, *et al.* Cadmium, lead, and arsenic contamination in paddy soils of a mining area and their exposure effects on human HEPG2 and keratinocyte cell-lines. Environ Res 2017; 156: 23-30.
[http://dx.doi.org/10.1016/j.envres.2017.03.014] [PMID: 28314151]

[9]   Dang Z, Liu C, Haigh MJ. Mobility of heavy metals associated with the natural weathering of coal mine spoils. Environ Pollut 2002; 118(3): 419-26.
[http://dx.doi.org/10.1016/S0269-7491(01)00285-8] [PMID: 12009140]

[10] Obiajunwa EI, Pelemo DA, Owolabi SA, Fasasi MK, Johnson-Fatokun FO. Characterisation of heavy metal pollutants of soils and sediments around a crude-oil production terminal using EDXRF. Nucl Instrum Methods Phys Res B 2002; 194(1): 61-4.
[http://dx.doi.org/10.1016/S0168-583X(02)00499-8]

[11] Bahar MM, Megharaj M, Naidu R. Kinetics of arsenite oxidation by Variovorax sp. MM-1 isolated from a soil and identification of arsenite oxidase gene. J Hazard Mater 2013; 262: 997-1003.

[http://dx.doi.org/10.1016/j.jhazmat.2012.11.064] [PMID: 23290483]

[12]   Wang S, Mulligan CN. Occurrence of arsenic contamination in Canada: Sources, behavior and distribution. Sci Total Environ 2006; 366(2-3): 701-21.
[http://dx.doi.org/10.1016/j.scitotenv.2005.09.005] [PMID: 16203025]

[13]   Bernhardt A, Gysi N. The Toxics beneath Our Feet. Blacksmith Institute 2016.

[14]   Jain H. From pollution to progress: groundbreaking advances in clean technology unveiled. Innov Green Dev 2024; 3(2): 100143.
[http://dx.doi.org/10.1016/j.igd.2024.100143]

[15]   Jain H, Dhupper R, Shrivastava A, Kumari M. Enhancing groundwater remediation efficiency through advanced membrane and nano-enabled processes: a comparative study. Groundw Sustain Dev 2023; 23: 100975.
[http://dx.doi.org/10.1016/j.gsd.2023.100975]

[16]   Shah MP, Shah N. Environmental approach to remediate refractory pollutants from industrial wastewater treatment plant. Elsevier 2024.

[17]   Edmunds WM, Ahmed KM, Whitehead PG. A review of arsenic and its impacts in groundwater of the Ganges–Brahmaputra–Meghna delta, Bangladesh. Environ Sci Process Impacts 2015; 17(6): 1032-46.
[http://dx.doi.org/10.1039/C4EM00673A] [PMID: 25683650]

[18]   Harris J, McCartor A. The world's worst toxic pollution problems report 2011. Blacksmith Institute 2011; pp. 1-76.

[19]   Hossain MF. Arsenic contamination in Bangladesh—An overview. Agric Ecosyst Environ 2006; 113(1-4): 1-16.
[http://dx.doi.org/10.1016/j.agee.2005.08.034]

[20]   Agarwal S, Jain H. Investigating the potential of floral waste as a vermicompost and dual-functional biosorbent for sustainable environmental management. Water Air Soil Pollut 2024; 235(5): 322.
[http://dx.doi.org/10.1007/s11270-024-07144-y]

[21]   Jain H. Data analytics enabled by the Internet of Things and artificial intelligence for the management of Earth's resources. In: Kumar D, Tewary T, Shekhar S, Eds. Data Analytics and Artificial Intelligence for Earth Resource Management. Elsevier 2025; pp. 19-36.
[http://dx.doi.org/10.1016/B978-0-443-23595-5.00002-4]

[22]   Jain H, Dhupper R, Shrivastava A, Kumar D, Kumari M. Leveraging machine learning algorithms for improved disaster preparedness and response through accurate weather pattern and natural disaster prediction. Review. 2023; 11.
[http://dx.doi.org/10.3389/fenvs.2023.1194918]

[23]   Shah MP. Emerging innovative trends in the application of biological processes for industrial wastewater treatment. Elsevier 2024.

[24]   Patel KS, Shrivas K, Brandt R, Jakubowski N, Corns W, Hoffmann P. Arsenic contamination in water, soil, sediment and rice of central India. Environ Geochem Health 2005; 27(2): 131-45.
[http://dx.doi.org/10.1007/s10653-005-0120-9] [PMID: 16003581]

[25]   Cullen WR, Reimer KJ. Arsenic speciation in the environment. Chem Rev 1989; 89(4): 713-64.
[http://dx.doi.org/10.1021/cr00094a002]

[26]   Inskeep W P, McDermott T R, Fendorf S. Arsenic (v)/(iii) cycling in soils and natural waters: chemical and microbiological processes. Environ Chem Arsenic 2002; 391

[27]   Nordstrom DK, Archer DG. Arsenic thermodynamic data and environmental geochemistry: an evaluation of thermodynamic data for modeling the aqueous environmental geochemistry of arsenic. Arsenic in Ground Water: Geochemistry and Occurrence. Springer 2003; pp. 1-25.
[http://dx.doi.org/10.1007/0-306-47956-7_1]

[28]   Gadd GM. Metals, minerals and microbes: geomicrobiology and bioremediation. Microbiology

(Reading) 2010; 156(3): 609-43.
[http://dx.doi.org/10.1099/mic.0.037143-0] [PMID: 20019082]

[29]   J S, P B, R O. Microbial transformation of elements: the case of arsenic and selenium. Int Microbiol 2002; 5(4): 201-7.
[http://dx.doi.org/10.1007/s10123-002-0091-y] [PMID: 12497186]

[30]   Mandal BK, Suzuki KT. Arsenic round the world: a review. Talanta 2002; 58(1): 201-35.
[http://dx.doi.org/10.1016/S0039-9140(02)00268-0] [PMID: 18968746]

[31]   Bhattacharya P, Samal AC, Majumdar J, Santra SC. Arsenic contamination in rice, wheat, pulses, and vegetables: a study in an arsenic affected area of West Bengal, India. Water Air Soil Pollut 2010; 213(1-4): 3-13.
[http://dx.doi.org/10.1007/s11270-010-0361-9]

[32]   Eisler R. Mercury hazards from gold mining to humans, plants, and animals. Rev Environ Contam Toxicol 2004; 181: 139-98.
[PMID: 14738199]

[33]   Brammer H. Mitigation of arsenic contamination in irrigated paddy soils in South and South-east Asia. Environ Int 2009; 35(6): 856-63.
[http://dx.doi.org/10.1016/j.envint.2009.02.008] [PMID: 19394085]

[34]   Henke K. Arsenic: environmental chemistry, health threats and waste treatment. John Wiley & Sons 2009.
[http://dx.doi.org/10.1002/9780470741122]

[35]   Adeniji A. Bioremediation of arsenic, chromium, lead, and mercury. National network of enviromental management studies fellow for US Enviromental Protection Agency Office of Solid Waste and Emergency Response Technology Innovation Office. Washington, DC 2004.

[36]   Chakraborti D, Mukherjee SC, Pati S, *et al.* Arsenic groundwater contamination in Middle Ganga Plain, Bihar, India: a future danger?. Environ Health Perspect 2003; 111(9): 1194-201.
[http://dx.doi.org/10.1289/ehp.5966] [PMID: 12842773]

[37]   Chowdhury UK, Biswas BK, Chowdhury TR, *et al.* Groundwater arsenic contamination in Bangladesh and West Bengal, India. Environ Health Perspect 2000; 108(5): 393-7.
[http://dx.doi.org/10.1289/ehp.00108393] [PMID: 10811564]

[38]   Ravenscroft P, Brammer H, Richards K. Arsenic pollution: a global synthesis. John Wiley & Sons 2011.

[39]   Ratna Kumar P, Chaudhari S, Khilar KC, Mahajan SP. Removal of arsenic from water by electrocoagulation. Chemosphere 2004; 55(9): 1245-52.
[http://dx.doi.org/10.1016/j.chemosphere.2003.12.025] [PMID: 15081765]

[40]   Mohan D, Pittman CU Jr. Arsenic removal from water/wastewater using adsorbents—A critical review. J Hazard Mater 2007; 142(1-2): 1-53.
[http://dx.doi.org/10.1016/j.jhazmat.2007.01.006] [PMID: 17324507]

[41]   Muller D, Lièvremont D, Simeonova DD, Hubert JC, Lett MC. Arsenite oxidase aox genes from a metal-resistant β-proteobacterium. J Bacteriol 2003; 185(1): 135-41.
[http://dx.doi.org/10.1128/JB.185.1.135-141.2003] [PMID: 12486049]

[42]   Santini JM, vanden Hoven RN. Molybdenum-containing arsenite oxidase of the chemolithoautotrophic arsenite oxidizer NT-26. J Bacteriol 2004; 186(6): 1614-9.
[http://dx.doi.org/10.1128/JB.186.6.1614-1619.2004] [PMID: 14996791]

[43]   Duval S, Santini JM, Lemaire D, *et al.* The H-bond network surrounding the pyranopterins modulates redox cooperativity in the molybdenum- bis PGD cofactor in arsenite oxidase. Biochim Biophys Acta Bioenerg 2016; 1857(9): 1353-62.
[http://dx.doi.org/10.1016/j.bbabio.2016.05.003] [PMID: 27207587]

[44]   Santini JM, Kappler U, Ward SA, Honeychurch MJ, vanden Hoven RN, Bernhardt PV. The NT-26 cytochrome c552 and its role in arsenite oxidation. Biochim Biophys Acta Bioenerg 2007; 1767(2): 189-96.
[http://dx.doi.org/10.1016/j.bbabio.2007.01.009]

[45]   Santini JM, Sly LI, Schnagl RD, Macy JM. A new chemolithoautotrophic arsenite-oxidizing bacterium isolated from a gold mine: phylogenetic, physiological, and preliminary biochemical studies. Appl Environ Microbiol 2000; 66(1): 92-7.
[http://dx.doi.org/10.1128/AEM.66.1.92-97.2000] [PMID: 10618208]

[46]   Warelow TP, Oke M, Schoepp-Cothenet B, *et al.* The respiratory arsenite oxidase: structure and the role of residues surrounding the rieske cluster. PLoS One 2013; 8(8): e72535.
[http://dx.doi.org/10.1371/journal.pone.0072535] [PMID: 24023621]

[47]   Bryan CG, Marchal M, Battaglia-Brunet F, *et al.* Carbon and arsenic metabolism in Thiomonas strains: differences revealed diverse adaptation processes. BMC Microbiol 2009; 9(1): 127.
[http://dx.doi.org/10.1186/1471-2180-9-127] [PMID: 19549320]

[48]   Watson C, Niks D, Hille R, *et al.* Electron transfer through arsenite oxidase: Insights into Rieske interaction with cytochrome c. Biochim Biophys Acta Bioenerg 2017; 1858(10): 865-72.
[http://dx.doi.org/10.1016/j.bbabio.2017.08.003] [PMID: 28801050]

[49]   Ellis PJ, Conrads T, Hille R, Kuhn P. Crystal structure of the 100 kDa arsenite oxidase from *Alcaligenes faecalis* in two crystal forms at 1.64 A and 2.03 A. Structure 2001; 9(2): 125-32.
[http://dx.doi.org/10.1016/S0969-2126(01)00566-4] [PMID: 11250197]

CHAPTER 5

Microbes in Green Nanotechnology and Energetics

Mohammad H. El-Zmrany[1], Mohamed Ebrahim[2], Samah H. Abu-Hussien[3], Ziad Samy[1], Kinzey M. Abohussein[1], Maria A. Farag[1] and Muhammad Aslam Khan[4,*]

[1] *Biotechnology Program, New Programs Administration, Faculty of Agriculture, Ain Shams University, Cairo, Egypt*

[2] *Department of Plant Pathology, Faculty of Agriculture, Ain Shams University, Cairo, Egypt*

[3] *Department of Agricultural Microbiology, Faculty of Agriculture, Ain Shams University, Cairo, Egypt*

[4] *Department of Biological Sciences, Faculty of Sciences, International Islamic University (IIU), Islamabad, Pakistan*

Abstract: Green nanotechnology is an emerging field that uses eco-friendly methods for synthesizing nanomaterials, offering sustainable alternatives for pollution control, resource recovery, and renewable energy production. This chapter delves into the innovative use of microorganisms, including bacteria, yeasts, fungi, plants, and algae, to biosynthesize nanomaterials as a green alternative to traditional chemical and physical synthesis techniques. Microbial synthesis, often termed "green" nanomanufacturing, eliminates toxic byproducts, paving the way for applications in environmental remediation, biomedicine, and sensor development. For example, bacteria and microalgae produce unique nanostructures, such as bacterial nanocellulose, exopolysaccharides, and biomineralized materials, which have significant applications in biomedical devices, sensors, plant enhancement, and environmental monitoring. Yeast and molds facilitate extracellular synthesis, enabling culture reuse and reducing purification demands, making microbial systems both adaptable and scalable for industrial production [1]. This chapter also reviews developments from the past decade, highlighting microbial biosynthesis capabilities and challenges, including standardization issues and the role of genetic engineering in enhancing nanoparticle consistency. The chapter emphasized that in fields such as agriculture, nanofertilizers and nanopesticides derived from microbial sources improve nutrient delivery and pest resistance, minimizing chemical inputs. In energy applications, microbial nanomaterials are integrated into solar cells and hydrogen production processes, providing cleaner and more sustainable energy solutions. By integrating green chemistry principles, microbial biosynthesis offers an environmentally friendly pathway for producing nanomaterials with broad applicability across healthcare, agriculture, and clean energy sectors. As research advances, the

---

* **Corresponding author Muhammad Aslam Khan:** Department of Biological Sciences, Faculty of Sciences, International Islamic University (IIU), Islamabad, Pakistan;
E-mail: muhammadaslamkhanmarwat@gmail.com

Harshita Jain & Maulin P. Shah (Eds.)

standardization of microbial nanomaterial production is expected to enable the scale-up of these promising "nanofactories" for widespread industrial use, significantly contributing to sustainability and aligning with global environmental goals.

Keywords: Biogenic nanomaterials, Environmental remediation, Green nanotechnology, Microbial biosynthesis, Nano-fertilizers, Nano-pesticides, Sustainable energy solutions.

INTRODUCTION

Green science and technology have increasingly shaped the way we approach industrial processes and resource management, creating avenues for sustainable practices across numerous sectors. This field of science involves not only a single discipline but an integration of various approaches to reduce the ecological footprint of human activity. By prioritizing renewable resources, minimizing waste, and reducing reliance on hazardous chemicals, green science has traditionally transformed resource-intensive and polluting industries. For example, innovations in green methodologies have contributed to reductions in the release of hazardous materials used in manufacturing and cleaning processes. These reductions illustrate how shifting to eco-friendly practices is vital not only for industrial progress but also for preserving natural ecosystems and public health. Today, green technologies not only enhance energy efficiency but also pave the way for advanced solutions such as improved solar cells, fuel cells, and high-performance batteries that sustainably store energy, fuelling cleaner and more reliable energy options for future generations.

A notable development within green science is the emergence of green nanotechnology, an area that combines the principles of environmental sustainability with the power of nanoscale materials. Nanotechnology itself, a field that studies materials and processes at the atomic and molecular scales, has opened new possibilities for improving efficiency and functionality in countless applications. Central to nanotechnology are ultrasmall nanoparticle materials, often between 1 and 100 nanometres, with properties that differ significantly from those of their bulk counterparts. These particles, whether composed of metals such as silver, copper, or zinc, have unique optical, electrical, and catalytic characteristics that make them valuable in fields ranging from electronics to medicine. Importantly, nanotechnology's potential extends beyond product innovation; it has become a key element in environmental science. By enabling the creation of materials with minimal environmental impact, green nanotechnology supports the development of nanoscale products, such as nanosensors and nanocatalysts, that aid in waste reduction, pollution monitoring, and resource-efficient manufacturing. Researchers have discovered that nanoparticles can be synthesized through biological methods that are both cost-effective and environmentally safe and rely on microorganisms such as bacteria, fungi, yeast, and even plants. This bioinspired approach reduces the need for toxic reagents traditionally used in nanoparticle synthesis, thus aligning nanotechnology more closely with green principles.

In the future, green nanotechnology holds significant promise for addressing some of society's most pressing environmental and energy challenges. The future of this field lies in its potential to deliver solutions that are not only technologically advanced but also sustainable. In the coming years, advancements in microbial synthesis methods could lead to the creation of specialized nanomaterials for targeted applications, such as water purification, soil remediation, and efficient energy storage. By harnessing the capabilities of green nanotechnology, researchers aim to reduce our dependence on limited resources and foster sustainable practices that support both economic growth and environmental health. The continued development of these technologies promises to enhance clean energy initiatives, provide new ways of combating pollution, and inspire innovative approaches to preserving ecosystems. With an emphasis on renewable inputs, minimal energy consumption, and the elimination of toxic byproducts, green nanotechnology stands as a critical tool in our journey toward a more sustainable future. Through collaborative research, technological innovation, and a strong commitment to environmental stewardship, green nanotechnology is poised to have a lasting impact on industries and communities around the world, contributing to a healthier, more resilient planet.

This synthesis of green science, biotechnology, and nanotechnology underscores a pivotal shift in how we approach industrial development. As green nanotechnology continues to evolve, its applications will undoubtedly play a fundamental role in shaping a future where sustainable choices are woven seamlessly into the fabric of modern life and industry.

Production of Nanoparticles using Microbes and their Applications

Green nanotechnology utilizes the ability of microorganisms and plants to produce nanomaterials through environmentally friendly and sustainable methods. This approach minimizes harmful chemical usage and offers a more eco-friendly alternative to traditional nanoparticle synthesis. Among the various organisms utilized in green nanotechnology, bacteria, fungi, yeast, algae, actinomycetes, and plants play significant roles in nanoparticle synthesis, each with unique mechanisms and applications (Table **1**).

Bacteria

Bacterial nanoparticles have shown exceptional potential in various applications because of their ability to interact with bacterial structures in multiple ways. One of the primary mechanisms through which these nanoparticles exert antimicrobial effects is by directly interacting with the bacterial cell membrane. NPs such as silver and titanium dioxide bind to the phospholipid bilayer of bacterial cells, destabilizing the membrane and causing the leakage of vital intracellular components. This disruption compromises the integrity of bacteria, leading to a halt in cellular processes and ultimately preventing bacteria from surviving and replicating (Fig. 1) [2].

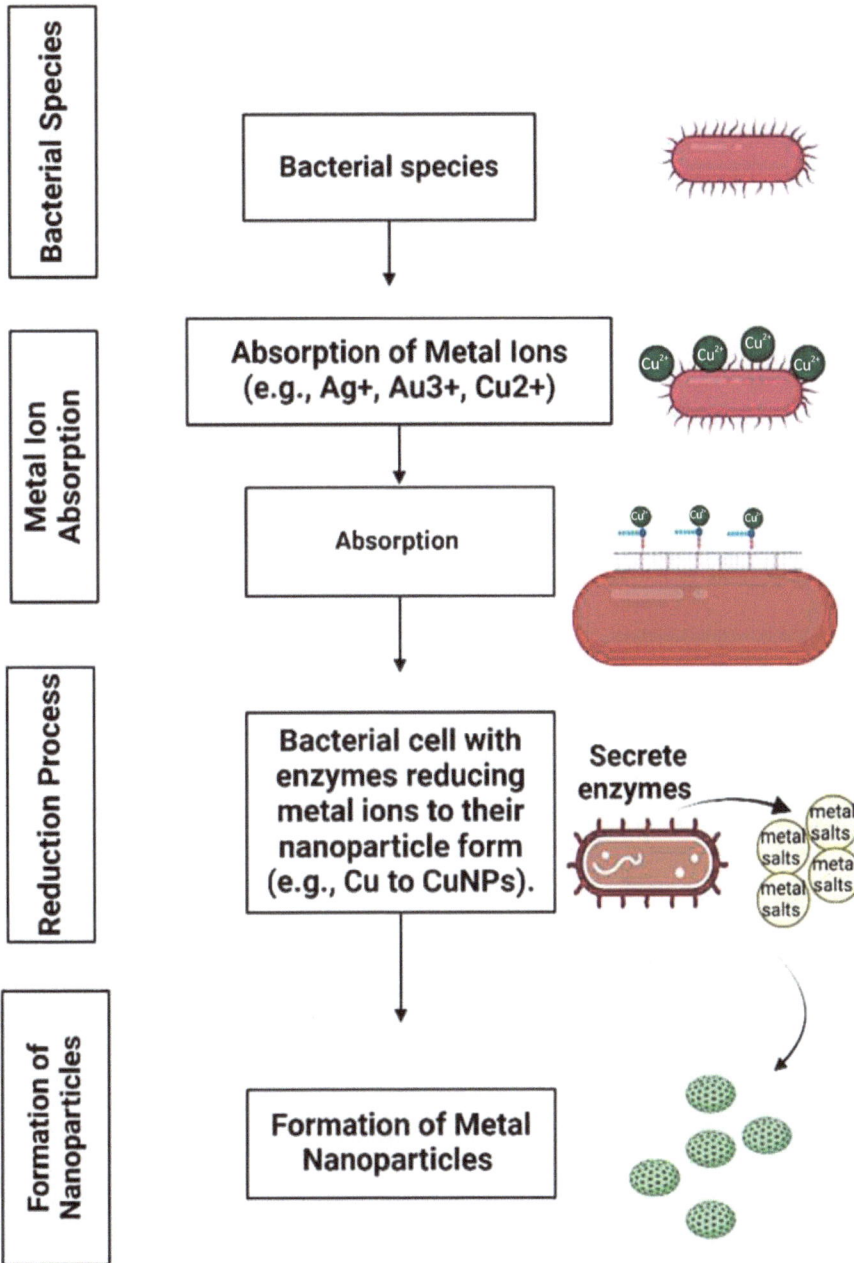

Fig. (1). The role of bacteria in the production of nanoparticles (NPs).

Additionally, these nanoparticles can generate Reactive Oxygen Species (ROS), such as superoxide radicals, hydrogen peroxide, and hydroxyl radicals. These

ROS target crucial cellular components, including proteins, lipids, and DNA, leading to oxidative stress that disrupts bacterial function and promotes cell death [3]. In addition to membrane disruption and ROS generation, nanoparticles can directly interact with bacterial DNA and proteins. Once inside the bacterial cell, the nanoparticles may bind to the DNA, inhibiting its replication and transcription, thereby impairing the ability of the bacteria to propagate. NPs can also bind to essential enzymes and proteins, altering their structure and inhibiting their functions, which has a detrimental effect on bacterial metabolism and growth [4].

Another important aspect of bacterial nanoparticle action is their ability to disrupt biofilm formation. Biofilms are clusters of bacteria encased in a self-produced extracellular matrix, and the formation of these biofilms is regulated by quorum sensing—a bacterial communication system. NPs can interfere with this system, preventing biofilm formation and disrupting the biofilm structure, which increases the susceptibility of bacteria to antibiotics and immune system attacks [5]. The ability of nanoparticles to enhance antibiotic effectiveness also holds significant promise. When combined with conventional antibiotics, nanoparticles can increase their antimicrobial activity, helping to overcome bacterial resistance mechanisms, such as efflux pumps and enzymatic inactivation. This synergistic effect is particularly valuable in fighting multidrug-resistant bacterial strains [6].

Despite these advantages, there are challenges, such as the potential for bacterial adaptation to nanoparticle exposure. The prolonged use of nanoparticles may lead to bacterial resistance, with bacteria developing mechanisms such as the production of efflux pumps or enzymes that neutralize nanoparticles. This highlights the need for careful regulation of nanoparticle use to minimize resistance development [7]. While bacterial nanoparticles are powerful tools for combating infections, further research is needed to understand their long-term effects and mitigate environmental risks.

In addition to their antimicrobial applications, bacterial nanoparticles, especially Exopolysaccharides (EPSs), have diverse uses across various fields. Exopolysaccharides are high-molecular-weight polymers secreted by bacteria into their surrounding environment. These EPS molecules have been widely studied for their role in biofilm formation and their protective properties. In addition to biofilm management, EPS can also be used in medical applications, such as wound healing, where they can enhance tissue regeneration and reduce inflammation. Furthermore, EPS has been explored for its potential in drug delivery systems, where it can serve as a biodegradable carrier for the controlled release of therapeutics. In agriculture, EPS-based nanoparticles have been applied to increase plant growth and provide protection against environmental stress.

The versatility of bacterial exopolysaccharides also extends to the food industry, where they serve as natural stabilizers, emulsifiers, and thickening agents. The biocompatibility and sustainability of these materials make them valuable alternatives to synthetic additives. In environmental applications, EPS can be used for bioremediation purposes, where they help trap heavy metals and toxic compounds, thereby facilitating their removal from contaminated sites. As research continues, the potential applications of bacterial nanoparticles and EPS are expected to expand further, offering sustainable and eco-friendly solutions to a wide range of challenges.

Fungi

Fungi have emerged as powerful tools for the biosynthesis of Nanoparticles (NPs), offering a range of advantages for the fabrication of metal-based nanomaterials, as shown in Fig. (**2**). The ability of these compounds to secrete potent enzymes and proteins makes them highly efficient at synthesizing a wide variety of NPs. Additionally, fungi are easily manipulated in laboratory settings, and their ability to produce significant amounts of proteins enhances their suitability for large-scale nanoparticle production. Fungi are also cost-effective and scalable, providing a feasible solution for industrial applications. The mycelial structure of these materials offers a large surface area, which is beneficial for nanoparticle formation, and their inherent tolerance to metals and bioaccumulation capabilities are critical for metal nanoparticle synthesis in green nanotechnology.

Several fungal species have been identified for their role in the biosynthesis of various metallic nanoparticles. For example, *Phanerochaete chrysosporium, Pleurotus sajor-caju, Coriolus versicolor*, and *Schizophyllum commune* are widely used for the production of gold and silver nanoparticles. On the other hand, species such as *Aspergillus niger, Fusarium keratoplasticum*, and *Alternaria alternata* are instrumental in the production of zinc oxide and iron oxide nanoparticles. *Fusarium oxysporum* is particularly notable for its ability to produce other nanomaterials, such as zinc sulfide, lead sulfide, and cadmium sulfide nanoparticles. This diversity in fungal species has significantly advanced the field of myconanotechnology, a subdiscipline at the intersection of mycology and nanotechnology, which holds immense potential for applications in fields such as medicine, environmental remediation, and nanomaterial engineering.

The mode of action behind nanoparticle synthesis in fungi involves both extracellular and intracellular processes. Fungal enzymes, such as reductases and oxidoreductases, are central to the reduction of metal ions to their corresponding nanoparticle forms. This reduction can occur in both solid-state fermentation and liquid culture media, providing flexibility in nanoparticle production under controlled conditions. The biomineralization process, in which fungi act as natural scaffolds for metal ion deposition, further aids in the formation of nanoparticles. These environmentally friendly processes are sustainable and provide alternatives to traditional chemical synthesis methods, making fungi valuable resources for the green production of metal nanoparticles.

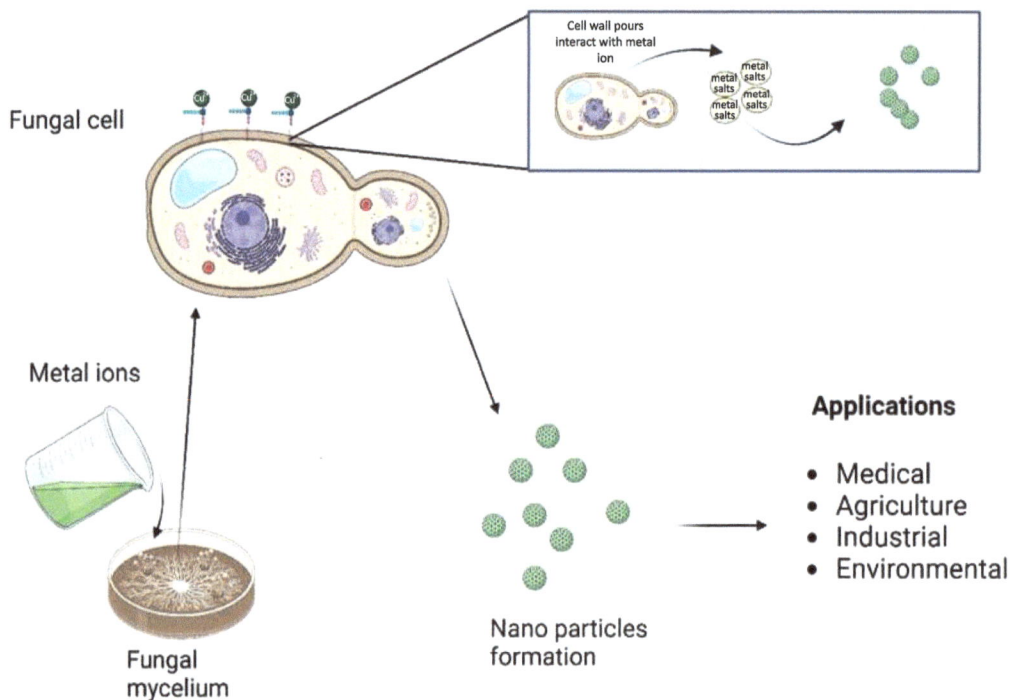

Fig. (2). The role of fungi in the synthesis of nanoparticles (NPs).

The applications of fungal-biosynthesized nanoparticles are wide-ranging. In medicine, they are being explored for drug delivery systems, biosensors, and diagnostic tools. For example, silver and gold nanoparticles have antimicrobial properties and are used in wound healing, antibacterial coating, and infection treatment. Fungal-derived nanoparticles are also used in environmental applications, such as the remediation of heavy metals and toxic pollutants. Their potential for use in agriculture for pest control and crop protection is another area of growing interest. Furthermore, their scalability and environmentally friendly production methods make them ideal for large-scale industrial applications.

Yeast

Yeasts have garnered significant attention in the field of nanotechnology because of their remarkable ability to synthesize nanoparticles (NPs) through both intracellular and extracellular processes (Fig. **3**). They are highly advantageous for nanoparticle synthesis because of their simple cultivation requirements, rapid growth, and ability to produce a variety of nanoparticles *via* eco-friendly, low-cost methods. Yeast species such as *Saccharomyces cerevisiae, Candida albicans*, and *Candida utilis* are recognized as efficient producers of metal nanoparticles, including gold (Au), silver (Ag), and zinc oxide (ZnO). The ability of yeasts to produce large amounts of biomass, coupled with their eukaryotic nature, enables genetic modifications that increase nanoparticle synthesis, making them promising organisms for use in green nanotechnology [8] Table **1**.

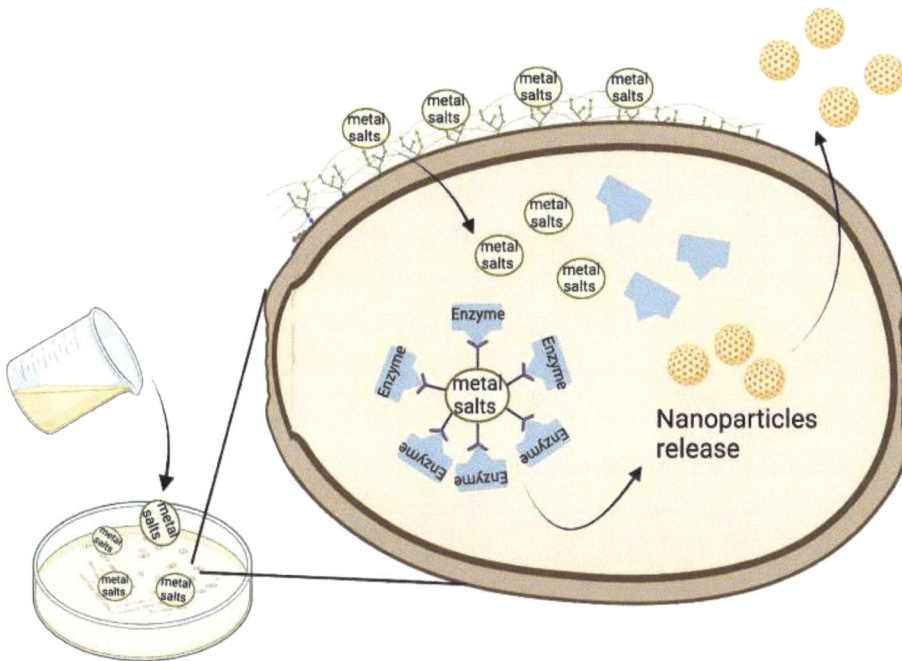

Fig. (3). Yeast cells facilitating the absorption and reduction of metal ions into nanoparticles (NPs).

Table 1. The uses of nanoparticles (NPs) biomineralized from metal ions by different microorganisms.

Metal ion	Classification	NPs	Uses	Refs.
Bacteria	*Pseudomonas sp.*	Ag	Antibacterial activity	[9]
	Bacillus thuringiensis	Ag	Larvicidal activity against *Culex quinquefasciatus* and *Aedes aegypti*	[10]
	Bacillus licheniformis	Ag		[11]
	Ochrobactrum anthropi	Ag	Antibacterial activity	[12]
	Bacillus spp.	Ag	Antimicrobial activity, antiviral activity	[13]
	Pantoea ananatis	Ag	Antibacterial against multidrug resistant	[14]
	Bacillus brevis NCIM2533	Ag	Antibacterial activity	[15]
	Bacillus mojavensis BTCB15	Ag	Antibacterial activity against multidrug resistant	[16]
	Actinobacteria	Ag	Antibacterial activity	[17]
	Sinomonas mesophila	Ag	Antimicrobial activity	[18]
	Bacillus endophyticus	Ag	Antimicrobial activity	[19]
	Bacillusbrevis	Ag	Antibacterial activity	[20]
	Bacillus licheniformis Dahb1	Ag	Antibiofilm activity	[21]
	Bacillus methylotrophicus DC3	Ag	Antimicrobial activity	[22]
	Stenotrophomonas GSG2	Au–Ag		[23]
	Kocuriaflava	Cu		[24]
	Shewanella loihica	Cu	Antibacterial activity	[22]
	Shewanella oneidensis MR-1	Cu	Biocatalysts	[25]
	Shewanella loihica	Pt	Decolorization of dyes	[26]
	Shewanella oneidensis MR-1	Pt	Biocatalysts for reduction of 4-nitrophenol	[27]
	Shewanella loihica	Pd	Degradation of dyes	[28]
	Shewanella oneidensis MR-1	Pd	Electrocatalysts	[29]
	Shewanella loihica	Pd–Pt	Degradation of dyes	[28]
	Ochrobactrum sp	TeO_3–SeO_3	Reduce of toxic substances	[30]

(Table 1) cont.....

-	Pseudomonas aeruginosa	Au		[31]
	Rhodopseudomonas capsulata	Au		[32]
	Escherichia coli DH5α	Au	Direct electrochemistry of hemoglobin	[33]
	Bacillus subtilis	Au	Degradation of dyes	[34]
	Shewanella loihica	Au	Degradation of dyes	[35]
	Micrococcus yunnanensis	Au	Antibacterial and anticancer activity	[36]
	Mycobacterium sp.	Au	Anticancer activity	[37]
	Halomonas salina	Au		[38]
	Shewanella oneidensis MR-1	Au	Biocatalysts for reduction of nitroaromatic compounds	[39]
	E.coli	CdS		[40]
	Pseudomonas aeruginosa	CdS	Removal of heavy metas suchl as cadmium	[41]
	Bacillus mycoides	TiO_2	Used in solar cells	[42]
	Bacillus amyloliquefaciens	TiO_2	Photocatalytic for dye removal	[43]
	Aeromonas hydrophila	TiO_2	Antibacterial activity	[44]
	Lactobacillus sp.	TiO_2		[45]
	Shewanella baltica	Te	Photocatalytic activity	[46]
	Bacillus subtilis	Co_3O_4		[47]
	Lysinibacillus sp. ZYM-1	Se	Photocatalytic activity	[48]
	Bacillus subtilis	Se	H2O2 sensoristic device	[49]
	Bacillus megaterium NCIM2326	ZnO	Antimicrobial activity	[15]
	Halomonas elongata IBRC - M10214	ZnO	Antimicrobial activity	[50]
	Sphingobacterium thalpophilum	ZnO	Antimicrobial activity	[51]
	Staphylococcus aureus	ZnO	Antimicrobial activity	[52]
Yeast				

(Table 1) cont.....

Rhodotorula sp. ATL72	Ag	Antimicrobial activity	[53]
Saccharomyces cerevisiae	Ag		[54]
Cryptococcus laurentii	Ag	Antifungal against plant pathogen	[55]
Rhodotorula glutinis	Ag	Antifungal activity, reduction of nitrophenol compound, and dye degradation	[56]
Rhodotorula mucilaginosa	Ag	Bioremediation	[57]
Commercial yeast	Au–Ag alloy	Electrochemical sensor	[58]
Phaffia rhodozyma	Ag and Au	Antifungal activity	[59]
Saccharomyces cerevisiae	Au	Enhancement of surface plasmon applications	[60]
Magnusiomyces ingens LHF1	Au	Reduction of nitrophenol compounds	[61]
Saccharomyces cerevisiae	Pd	Dye degradations	[62]
Pichia kudriavzevii	ZnO	Antimicrobial and antioxidant activities	[63]
Magnusiomyces ingens LHF1	Se	Antibacterial activity	[64]
Saccharomyces cerevisiae	CdTe	Applications in bioimaging and biolabelling	[65]
Fungi			

(Table 1) cont.....

Rhizopus stolonifera	Ag		[66]
Candida glabrata	Ag	Antibacterial activity	[67]
Trametes trogii	Ag		[68]
Trichoderma longibrachiatum	Ag	Antimicrobial against phytopathogen	[69]
Fusarium oxysporum	Ag	Antibacterial activity	[70]
Aspergillus terreus	Ag	Antibacterial activity	[71]
Ganoderma sessiliforme	Ag	Antibacterial, antioxidant, and anticancer activities	[72]
Rhodotorula glutinis	Ag	Antifungal, cytotoxicity, and dye degradation	[73]
Aspergillus sp.	Ag	Antibacterial and cytotoxicity activities	[74]
Fusarium keratoplasticum A1-3	Ag	Increasing antibacterial activity of cotton fabrics	[75]
Arthroderma fulvum	Ag	Antifungal activity	[76]
Fusarium oxysporum 405	Ag	Colloidal stability	[77]
Fusarium oxysporum	Ag	Antibacterial and antitumour activities	[78]
Metarhizium anisopliae	Ag	Larvicidal activity	[79]
Trichodermah arzianum	Ag	Antifungal activity	[80]
Fusarium oxysporum	Ag	Antibacterial activity	[78]
Candida albicans ATCC10231	Ag		[81]
Macrophomina phaseolina	Ag/AgCl	Antibacterial activity	[82]

(Table 1) cont.....

Cladosporium cladosporioides	Au	Antibacterial and antioxidant activities	[83]
Trichoderma harzianum	Au	Dye degradation; antibacterial activity	[84,85]
Pleurotus ostreatus	Au	Antimicrobial, anticancer activities	[86]
Aspergillus sp.	Au	Reduction of nitrophenol compounds	[87]
Rhizopus oryzae	Au	Hemocompatible activity	[88]
Penicillium chrysogenum	Pt	Cytotoxicity	[89]
Aspergillus niger	ZnO	Dye degradation; antibacterial activity	[90]
Candida albicans	ZnO	Synthesis of steroidal pyrazolines	[91]
Fusarium keratoplasticum A1-3	ZnO	Antibacterial, cytotoxicity activities, and loaded on textile	[75]
Aspergillus niger G3-1	ZnO	Antibacterial, cytotoxicity activities and medical textile	[75]
Aspergillus terreus	ZnO	Antibacterial, cytotoxicity, medical textile and UV protection	[92]
Pichia kudriavzevii	ZnO	Antibacterial and antioxidant activities	[63]
Aspergillus welwitschiae	Te	Antibacterial activity against MRSA	[63]
Aspergillus flavus	ZnS; ZnS-Gd	Detection of heavy metals in water	[93]
Alternaria alternata	Fe_2O_3		[94]
Colletotrichum sp.	Al_2O_3	Antimicrobial activity	[95]
Aspergillus nidulans	CoO		[96]
Actinomycetes			

(Table 1) cont.....

Streptomyces spp.	Ag	Antimicrobial, antioxidant, and larvicidal activities	[92]
Nocardiopsis sp. MBRC-1	Ag	Antimicrobial activity, in vitro cytotoxicity against the HeLa cell line	[97]
Streptacidiphilus durhamensis	Ag	Antimicrobial activity	[98]
Streptomyces rochei MHM13	Ag	Antimicrobial activity and enhancement of antibiotic action	[99]
Streptomyces spp.	Ag		[100]
Saccaropolyspora hirsuta	Ag	Antimicrobial activity	[101]
Streptomyces parvulus	Ag	Antimicrobial activity	[102]
Streptomyces seoulensis	Ag	Antimicrobial activity	[103]
Streptomyces owasiensis	Ag	Antimicrobial activity	
Nocardiopsis flavascens	Ag	Cytotoxicity	[104]
Streptomyces fradiae	Ag	Antioxidant activity	[105]
Streptomyces griseoplanus	Ag	Antifungal against plant pathogen	[96]
Rhodococcus sp.	Ag	Antimicrobial activity	[106]
Streptomyces sp. Al-Dhabi-87	Ag	Antimicrobial activity, antibacterial activity against multidrug-resistant bacteria	[107]
Streptomyces spp	Ag–Au	Antibacterial activity	[108]
Gordonia amicalis HS-11	Ag–Au		[109]
Streptomyces spp.	CuO	Antibacterial activity	[110]
Streptomyces capillispiralis Ca-1	Cu	Antimicrobial, biocontrol of phytopathogen, and larvicidal activities	[111]
Streptomycetes viridogens HM10	Au	Antibacterial activity	[112]
Streptomyces spp.	Au	Antifungal activity	[113]
Nocardiopsis sp. MBRC-48	Au	Antimicrobial and cytotoxicity activities	[114]
Streptomyces griseoruber	Au	Dye degradation	[115]
Rhodococcus sp.	Au		[116]
Streptomyces hygroscopicus	Au		[117]
Streptomyces spp.	Au		[118]
Streptomyces minutiscleroticus M10A62	Se	Antibiofilm, antiviral; antioxidant activity, antiproliferative activity	[119]
Streptomyces bikiniensis Ess_amA-1	Se	Anticancer activity	[120]
Streptomyces spp.	ZnO	Antimicrobial activity	[121]
Algae			

(Table 1) cont.....

Red algae *Portieria*	Ag	Antibacterial activity against fish pathogens	[122]
Marine macroalgae *Padina sp*	Ag	Antibacterial and antioxidant activities	[123]
Microchaete NCCU-342	Ag	Dye decolorization ability	[124]
Macroalgae (*Ulva lactuca L.*)	Ag	Cancer therapy	[125]
Brown alga *Padina pavonia*	Ag	One-pot method for synthesis	[126]
Gelidium amansii	Ag	Antimicrobial property	[127]
Caulerpa serrulata	Ag	Catalytic and antibacterial activities	[128]
Acanthophora specifera	Ag	Antimicrobial activity	[129]
Gracilaria birdiae	Ag	Antibacterial activity	[130]
Sargassum muticum	Ag	Control tool against mosquito vectors and bacterial pathogens	[131]
Anabaena flos-aquae	Ag	Anticancer and cytotoxic activity against T47D cell lines	[132]
Polysiphonia algae	Ag	Anticancer activity against MCF-7 cell line	[133]
Marine algae Gelidiella acerosa	Au	Biological uses	[134]
Macroalgae (*Ulva lactuca L*)	Au	Cancer therapies	[125]
Marine algae extract	Au	One-pot method for synthesis	[135]

(Table 1) cont.....

Brown algae *Cystoseira baccata*	Au	Cancer therapies	[136]
Pithophora oedogonia	Au	Determination of carbendazim molecules in soil	[137]
Sargassum tenerrimum	Au	Evaluation of their catalytic activity	[138]
Spirulina platensis	Au	Antibacterial efficacy	[139]
Chlorella vulgaris	Pd	Catalytic activity	[140]
Sargassum bovinum	Pd		[141]
Brown alga *Cystoseira trinodis*	CuO	Photocatalytic and antibacterial activities	[142]
Green alga *Botryococcus braunii*	CuO	Antimicrobial activity	[143]
Brown algae *Sargassum polycystum*	CuO	Antimicrobial and anticancer activities	[144]
Bifurcaria bifurcata	CuO	Antimicrobial activity	[145]
Brown seaweed extract	Fe_3O_4	Antimicrobial potency	[146]
Brown seaweed	Fe_3O_4	Bioremediation	[147]
Sargassum muticum	Fe_3O_4	High functional bioactivity	[148]
Chlorella vulgaris	$CuFe_2O$/Ag	Antibacterial activity, antibiofilm activity, inhibit efflux pump genes in Staphylococcus	[149]
Spirulina platensis	Fe_3O_4/Ag	Effect on the expression of norA and norB genes in *Staphylococcus aureus* & Anticancer activity	[150]

The synthesis of nanoparticles in yeasts involves the reduction of metal ions into their corresponding metallic forms through enzymatic processes. For example, Saccharomyces cerevisiae utilizes enzymes such as reductases to reduce silver ions (Ag^+) to silver nanoparticles (Ag-NPs). Similarly, other species, such as *Candida albicans* and *Candida utilis*, are used for the biosynthesis of gold nanoparticles (Au-NPs) *via* similar enzymatic mechanisms. Yeasts also produce zinc oxide nanoparticles (ZnO-NPs) through bioreduction and mineralization processes, where metabolic products such as proteins and organic acids interact with metal ions, facilitating nanoparticle formation. Yeasts are versatile in their ability to produce nanoparticles both in the cell wall and extracellularly, offering flexibility in nanoparticle production.

The mode of action for nanoparticle synthesis in yeasts includes both passive and active processes. In the passive mode, metal ions are absorbed by the yeast cell wall and reduced into nanoparticles through interactions with cell surface proteins. In the active mode, intracellular enzymes catalyze the reduction of metal ions inside yeast cells, leading to the formation of nanoparticles. Factors such as metal ion concentration, pH, temperature, and incubation time play crucial roles in influencing nanoparticle synthesis in yeasts. Yeast-based nanoparticle synthesis is a sustainable and low-cost alternative to traditional chemical synthesis methods, and it holds significant promise for various applications in fields such as medicine, agriculture, and environmental remediation.

The applications of yeast-based nanoparticles include their use in medicine for drug delivery systems, where the unique properties of the nanoparticles enhance the bioavailability and targeting of drugs. Silver and gold nanoparticles synthesized by yeasts have demonstrated antimicrobial properties, making them effective in wound healing, antibacterial coating, and infection treatment. Additionally, yeast-derived nanoparticles have been explored in biosensors for the detection of pathogens, toxins, and heavy metals, highlighting their potential in environmental monitoring. In agriculture, these nanoparticles are being investigated for use in crop protection and pest control, offering eco-friendly alternatives to conventional chemical pesticides. Furthermore, yeast-based nanoparticle synthesis is scalable, which makes it suitable for industrial applications, offering a sustainable, efficient method for large-scale nanoparticle production.

Algae

Algae, particularly microalgae and macroalgae, have gained significant attention for their ability to biosynthesize metallic nanoparticles (NPs) through various bioreduction and biosorption mechanisms. One of the key mechanisms involved in nanoparticle synthesis is the reduction of metal ions into their metallic form (Fig. **4**). This process is often mediated by enzymatic reactions, where algae such as *Chlorella vulgaris* can reduce tetrachloroaurate ions ($AuCl_4^-$) to form gold nanoparticles (Au-NPs) through extracellular reduction. Enzymes such as reductases are often responsible for this reduction, and Reactive Oxygen Species (ROS) generated by algae also contribute to the reduction process, increasing the efficiency of NP production. In some algal species, such as *Phaeodactylum tricornutum*, the metal ions are reduced into nanocrystals within the algal cells, often triggered by exposure to metals such as cadmium (Cd). The algae produce biomolecules such as phytochelatins, which bind to the metal ions and facilitate the formation of nanocrystals such as cadmium sulfide (CdS) inside the cells.

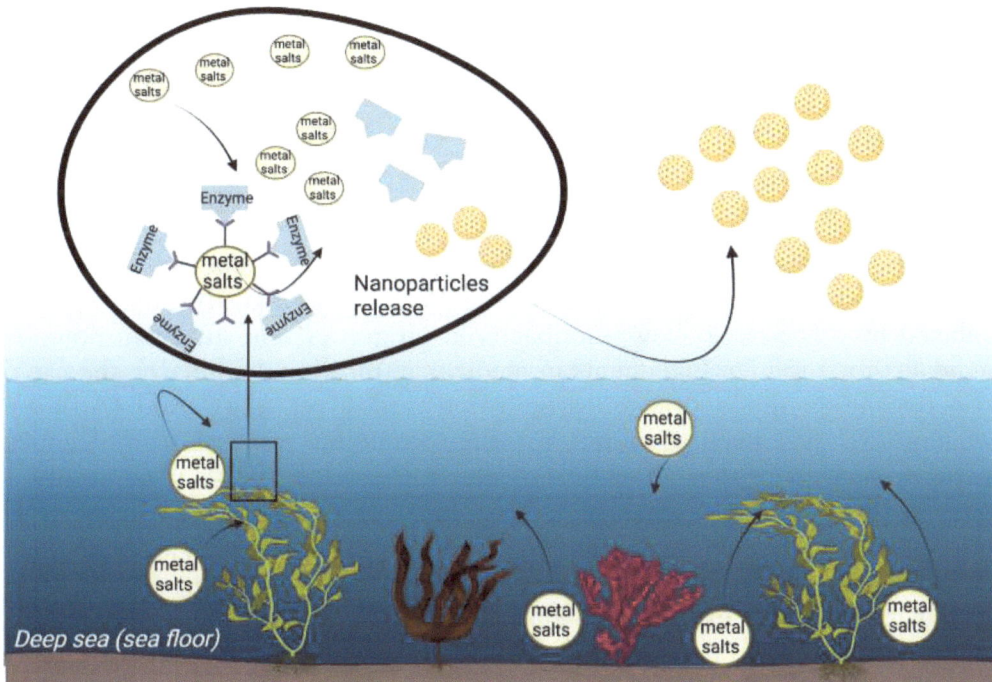

Fig. (4). The absorption of metal ions from water and their reduction into nanoparticles (NPs) by algae.

In addition to being involved in reduction, algae also play a role in biosorption, where metal ions are absorbed by the cell wall. *Fucus vesiculosus*, a brown alga, is known for its ability to biosorb metal ions that form metallic nanoparticles on the surface of the algae. The polysaccharide and protein components of the cell wall play critical roles in binding metal ions through electrostatic interactions. This biosorption mechanism allows the formation of NPs at the surface of the algae, which can be further stabilized by the cell wall components. The extracellular synthesis of nanoparticles is another widely recognized mechanism, particularly in marine algae such as *Sargassum wightii*, where NPs are synthesized outside the algal cells in the culture medium. This extracellular process is facilitated by the release of enzymes, reductases, and other biomolecules that help reduce metal ions and stabilize the NPs, which allows for easier recovery and scalability in industrial applications.

Algae also show a remarkable ability to produce biopolymers or peptides that stabilize nanoparticles after synthesis. These biopolymers prevent the agglomeration of nanoparticles and control their size, which is essential for their effectiveness in various applications. One prominent example is the production of biocompatible and water-soluble cadmium telluride quantum dots by the yeast

Saccharomyces cerevisiae, which have been shown to have size-tunable properties and are used in bioimaging and biolabelling applications. Similarly, the marine alga Shewanella has demonstrated the ability to reduce aqueous platinum chloride ($PtCl_6$) to elemental platinum (Pt), forming biogenic platinum nanoparticles approximately 5 nm in size. This process occurs within the periplasm and is preferable for easy recovery, making it suitable for various nanotechnology applications, such as catalysis and electronics.

The applications of algae-synthesized nanoparticles are vast and include medical, environmental, and industrial uses. For example, the gold and silver nanoparticles produced by algae have shown antibacterial properties, making them potential candidates for antimicrobial agents in medical applications. Similarly, the use of algae for the synthesis of nanoparticles such as silver (Ag), gold (Au), and platinum (Pt) has been explored for their catalytic properties in chemical reactions, such as in the fabrication of electrochemical sensors. The ability of algae to biosynthesize these nanoparticles has opened new avenues in environmental remediation, particularly in the removal of heavy metals from water. The algae-based biosynthesis of nanoparticles, such as zinc sulfide (ZnS) and silver nanoparticles, has also been used in optical detection systems for monitoring toxic metals such as arsenic in water. Moreover, algae-based biosynthesis holds promise in nanomedicine, with algae-derived nanoparticles being utilized for drug delivery systems, wound healing, and cancer therapy, highlighting their potential in therapeutic applications.

These processes highlight the diverse and innovative approaches algae employ in the synthesis of nanoparticles, offering sustainable, eco-friendly alternatives to traditional chemical synthesis methods while providing a wide range of applications in medicine, environmental protection, and industry.

Actinomycetes

Actinomycetes, a group of bacteria, are gaining significant recognition for their ability to biosynthesize nanoparticles (NPs) with desirable sizes and surface characteristics. These microorganisms are capable of producing metallic nanoparticles through both intracellular and extracellular pathways. However, extracellular synthesis has garnered increasing commercial attention because of its advantages, such as the ability to produce nanoparticles with polydispersity, a feature that is important for certain industrial applications. The literature contains many reports detailing the intra- and extracellular synthesis of various metallic nanomaterials by actinomycetes, making them promising options in nanotechnology.

Several species of actinomycetes, including *Rhodococcus sp.,* *Thermoactinomycete sp., Streptomyces viridogens, S. hygroscopicus, Nocardia farcinica,* and *Thermomonsora,* have been identified as effective producers of gold nanoparticles. In addition to gold, actinomycetes are also capable of synthesizing other metallic nanoparticles, such as silver, copper, zinc, and manganese. These nanoparticles, which are synthesized by species such as *Streptomyces spp.*, have a wide range of applications owing to their unique properties, such as antimicrobial effects, catalytic activity, and potential for use in drug delivery systems.

The mechanism of nanoparticle biosynthesis in actinomycetes typically involves biomineralization, where metal ions are reduced into their nanoparticle forms. This process occurs extracellularly in many cases, with actinomycetes secreting enzymes such as reductases that facilitate the reduction of metal ions. The resulting nanoparticles are often stable and well dispersed, making them suitable for various applications. The synthesis of nanoparticles by actinomycetes is not only eco-friendly but also efficient and scalable, which makes these microorganisms particularly valuable for industrial-scale nanoparticle production (Fig. **5**).

The applications of actinomycete-derived nanoparticles are diverse and include antimicrobial treatments, environmental remediation, and drug delivery. Owing to their ability to produce nanoparticles with high stability and biocidal properties, these materials are ideal candidates for controlling pathogenic microorganisms. In environmental applications, these nanoparticles can be used to remove heavy metals from polluted environments, contributing to sustainable environmental cleanup. In medicine, actinomycete-derived nanoparticles hold promise for drug delivery systems, where they can deliver therapeutic agents in a controlled and targeted manner. The wide range of applications of actinomycete-derived nanoparticles is highlighted by their potential in fields as diverse as agriculture, biotechnology, and nanomedicine.

Methods of Green Nanoparticle Synthesis

Bioreduction in Nanoparticle Synthesis

Bioreduction is a process in which microorganisms reduce metal ions into their corresponding nanoparticles, facilitated by biomolecules such as enzymes, proteins, and carbohydrates. During bioreduction, microorganisms such as bacteria, fungi, and algae interact with metal ions and reduce them into metallic nanoparticles, either intracellularly or extracellularly. For example, *Pseudomonas aeruginosa* can reduce silver ions (Ag^+) to silver nanoparticles (Ag-NPs) through the action of microbial enzymes such as nitrate reductase and hydrogenase. The process involves electron transfer, where these enzymes donate electrons to the metal ions, thereby reducing them into stable nanoparticles. The applications of bioreduction are vast and include antimicrobial treatments, drug delivery systems, and environmental applications such as biosensors and bioremediation, where nanoparticles can help with heavy metal removal from contaminated environments (Fig. **6**) [151].

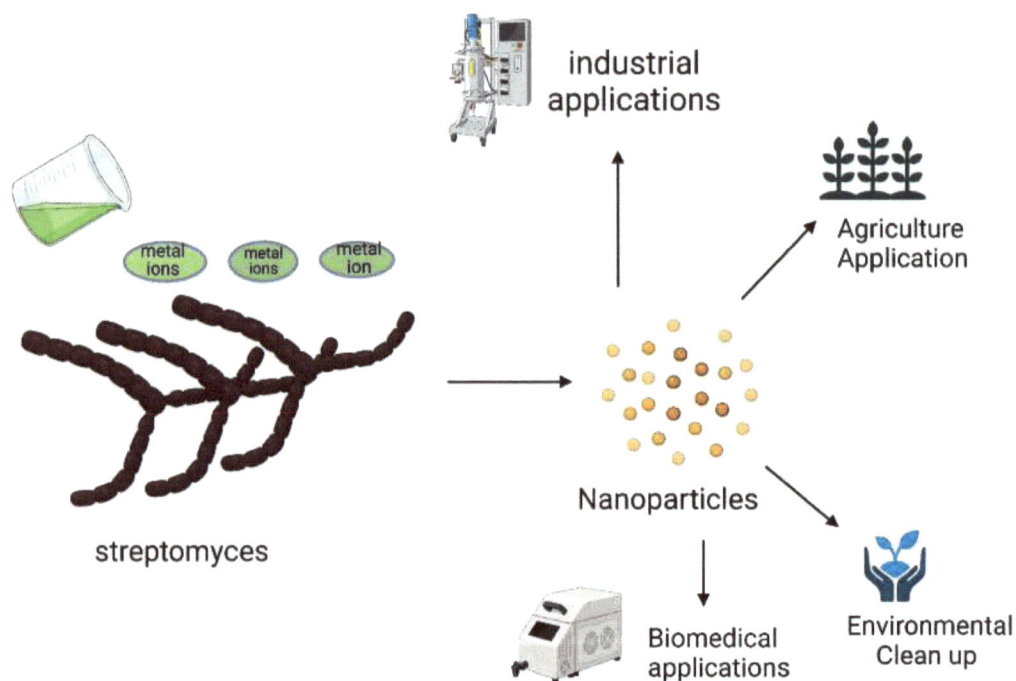

Fig. (5). The biomineralization process of metal ions into nanoparticles (NPs) by Streptomyces.

Microbial Nanoparticle Synthesis: Bioreduction and Biomineralization

Microbial nanoparticle synthesis involves processes such as bioreduction and biomineralization, where microorganisms reduce metal ions into nanoparticles. In bioreduction, metal ions such as Ag^+, Au^{3+}, and Zn^{2+} are reduced to their respective nanoparticles through microbial enzymes or cellular mechanisms. For example, *Pseudomonas aeruginosa* utilizes enzymes to reduce silver ions into silver nanoparticles, which have applications as antimicrobial agents and in wound healing [152]. In biomineralization, microorganisms, particularly fungi such as *Fusarium oxysporum*, reduce metal ions and form nanoparticles through the action of extracellular proteins. These nanoparticles are formed in the extracellular space, and their properties make them suitable for applications in wastewater treatment and as catalysts in chemical reactions. The nanoparticle sizes and uniformity are typically smaller and more controlled than those synthesized chemically, making them more efficient for industrial use [153].

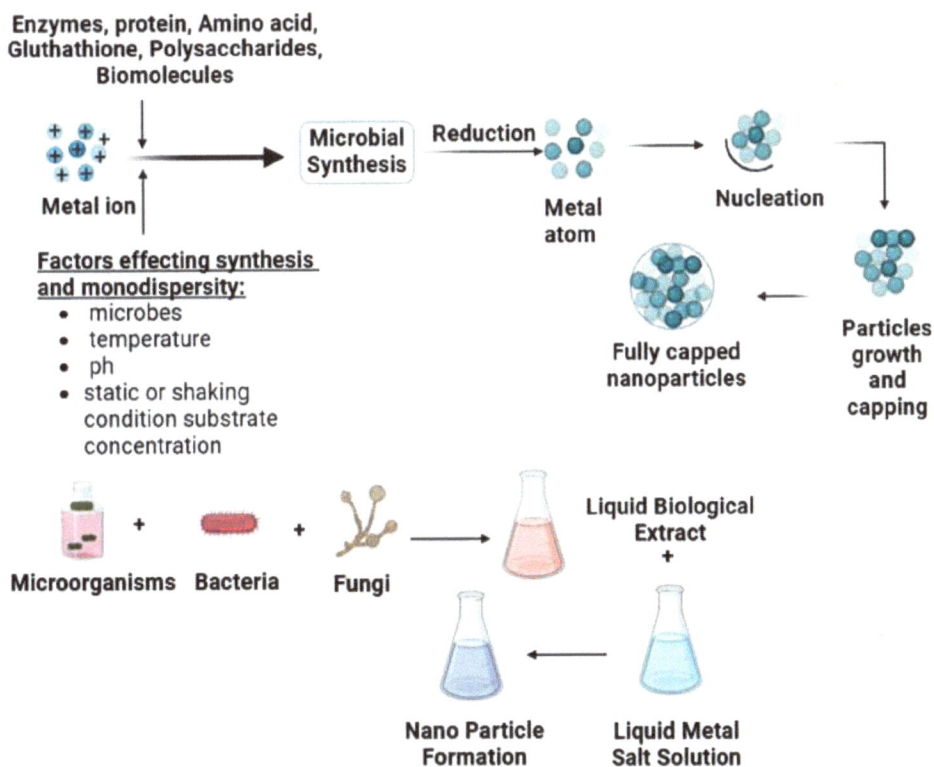

Fig. (6). Production of green nanoparticles through the reduction method using biological agents.

Algae-mediated Synthesis of Nanoparticles

Owing to their rich content of biocompounds such as proteins, polysaccharides, and lipids, algae, particularly microalgae, are efficient at synthesizing nanoparticles. These compounds act as reducing agents, helping to convert metal ions into nanoparticles. Microalgae such as *Spirulina platensis* and *Chlorella vulgaris* have been used to synthesize silver (Ag-NPs) and gold (Au-NPs) nanoparticles. In this process, algal enzymes facilitate the reduction of metal ions, whereas polysaccharides and proteins stabilize the nanoparticles, preventing aggregation. The nanoparticles produced by algae have applications in various fields, such as biomedical treatments, water purification, and environmental cleanup. They can also be used in the creation of biosensors for detecting harmful substances or pathogens in the environment [154].

Biosorption for Nanoparticle Formation

Biosorption refers to the ability of biological materials such as microorganisms, algae, and plants to adsorb metal ions from their surroundings, which are then reduced to form nanoparticles. For example, fungi such as *Aspergillus niger* can adsorb copper (Cu^{2+}) and silver (Ag^+) ions from their environment and reduce them into nanoparticles. During biosorption, the metal ions bind to functional groups on the surface of the microbial biomass, and these ions are subsequently reduced by the microbial metabolites to form nanoparticles. This process can occur either extracellularly or intracellularly, depending on the organism. The applications of biosorption are widespread in environmental cleanup, including the removal of heavy metals from polluted water, as well as in the production of nanomaterials for catalytic, biomedical, and industrial uses [155].

Enzyme-Catalyzed Synthesis of Nanoparticles

Enzyme-catalyzed synthesis involves the use of specific enzymes to catalyze the reduction of metal ions into nanoparticles. Enzymes such as nitrate reductase, hydrogenase, and other reductases can be isolated from microorganisms and used to facilitate nanoparticle formation. For example, *Trichoderma harzianum* produces nitrate reductase, which can reduce silver ions (Ag^+) to silver nanoparticles (Ag-NPs). This enzymatic process offers controlled nanoparticle synthesis with high precision, enabling the production of nanoparticles with specific sizes and properties. Enzyme-catalyzed nanoparticle synthesis is particularly useful in the development of nanomaterials for applications in drug delivery, biosensing, and environmental monitoring [156].

Factors Affecting Nanoparticle Biosynthesis

The synthesis of nanoparticles (NPs) *via* biological systems is influenced by a variety of factors that significantly affect the final properties, such as size, shape, and stability. These factors include the type of biomass, temperature, precursor concentration, pH, exposure time, and presence of enzymes. Each of these elements can be adjusted to optimize the synthesis process, making it suitable for various applications in fields such as medicine, electronics, and environmental science [2].

The type of biomass used in nanoparticle biosynthesis, whether bacteria, fungi, or plant extracts, determines how effectively metal ions are reduced and stabilized. Different organisms possess unique enzymatic activities and metabolic pathways that influence nanoparticle formation. For example, *Aspergillus* niger is particularly efficient at synthesizing gold nanoparticles (Au-NPs), whereas plant extracts from *Azadirachta indica* (neem) and *Coriandrum sativum* (coriander) contain bioactive compounds that act as reducing agents to produce silver and gold nanoparticles. The selection of biomass directly impacts not only the synthesis rate but also the potential application of the nanoparticles, owing to the specific properties imparted by each organism.

Temperature plays a pivotal role in regulating the synthesis rate of nanoparticles. Elevated temperatures accelerate the reduction of metal ions, accelerating nanoparticle formation. However, excessive heat can lead to aggregation, causing the nanoparticles to form irregular shapes or become unstable. For example, when silver nanoparticles are synthesized using *Trichoderma harzianum*, higher temperatures result in larger particles, whereas lower temperatures help produce more uniform, smaller nanoparticles. Conversely, colder conditions may enhance nanoparticle stability, which is particularly beneficial when synthesizing Zinc Oxide (ZnO) nanoparticles for various applications, including UV protection [157].

The concentration of the metal precursor is another crucial factor that influences the size and morphology of the nanoparticles. Higher concentrations typically result in larger nanoparticles, whereas lower concentrations favour the formation of smaller ones. For example, gold nanoparticle synthesis using *Pseudomonas aeruginosa* yields larger particles at higher precursor concentrations, whereas diluted solutions form smaller, more stable nanoparticles. The ability to control the precursor concentration allows for customization of the nanoparticle size, which is essential for their use in fields such as catalysis and drug delivery, where the particle size directly affects the efficiency and function of the material.

The contact time between metal ions and biological agents also plays a significant role in nanoparticle synthesis. Longer exposure times usually result in the formation of larger, more stable nanoparticles, as the metal ions have more time to interact with the reducing agents. For example, silver nanoparticles formed by *F. oxysporum* under longer incubation times tend to be larger and more stable, which is useful in antimicrobial applications. Shorter contact times, on the other hand, may lead to the formation of smaller or irregularly shaped nanoparticles, which can be advantageous for applications requiring larger surface areas, such as sensors.

pH is another essential factor affecting nanoparticle synthesis, as it influences both the reduction of metal ions and the surface charge of the nanoparticles. Different pH values result in nanoparticles with varying sizes, surface properties, and stabilities. For example, in the synthesis of copper nanoparticles using *Azadirachta indica* leaf extract, acidic conditions (pH 4-5) produce more stable nanoparticles, whereas alkaline conditions (pH 9-10) result in larger, less stable particles. By adjusting the pH, the size and stability of nanoparticles can be fine-tuned to meet specific requirements, which is particularly useful in applications such as drug delivery, where the particle size and surface charge are critical.

Enzymes are also key factors in the biosynthesis of nanoparticles. These molecules, which can naturally occur in microorganisms or are externally added, catalyze the reduction of metal ions into nanoparticulate forms. Enzymes such as nitrate reductase and hydrogenase help control nanoparticle size, shape, and stability during synthesis. For example, *Pseudomonas fluorescens* utilizes nitrate reductase to reduce silver ions into stable silver nanoparticles, whereas enzymes from *Escherichia coli* can lead to the formation of nanoparticles with distinct shapes, such as rods or spheres, depending on the specific enzymes and conditions used. Compared with chemical methods, the presence of enzymes can lead to more uniform nanoparticles, allowing for precise control over their properties.

The factors of biomass type, temperature, precursor concentration, contact time, pH, and enzyme presence are interdependent and can be optimized for different applications. By adjusting these variables, researchers can tailor nanoparticles to meet the specific needs of various industries, from medical applications such as drug delivery to environmental uses such as water purification.

Green Nanotechnology Applications

Green Nanotechnology in Agriculture

Nano Fertilizers for Sustainable Agriculture

Nanofertilizers have emerged as a significant advancement in sustainable agriculture, addressing the critical issues associated with conventional fertilizers. Traditional fertilizers are often inefficient, with nitrogen uptake accounting for only 20–50%, phosphorus accounting for 10–25%, and potassium accounting for 35–40% of what is applied to crops. The remaining nutrients either evaporate into the atmosphere or leach into the environment, causing a range of environmental issues, such as soil mineral imbalances, eutrophication, and greenhouse gas emissions [158]. These environmental challenges are compounded by the increasing global demand for fertilizers, which is projected to reach 201,663 thousand tonnes by 2020 [159]. This increasing demand threatens irreversible damage to ecosystems.

Green nanotechnology, with the development of smart delivery nanofertilizers, offers a solution to these problems by minimizing nutrient loss. Nanofertilizers, which include nanomaterials such as nanoclays, zeolites, and nanoparticles of essential nutrients such as nitrogen, phosphorus, and zinc, are designed to release nutrients gradually. Compared with traditional fertilizers, this controlled release mechanism ensures more efficient uptake by plants and reduces their environmental impact. The greater surface area of nanoparticles enables better interaction with plants, allowing for more efficient nutrient absorption and a more sustainable agricultural practice [160].

Numerous studies have demonstrated the benefits of nanofertilizers. For example, the foliar application of iron nanoparticles has been shown to increase pod and seed production in black-eyed peas, whereas zinc and magnesium nanoparticles increase the growth and yield of pearl millet. Similarly, foliar applications of boron and phosphorus nanoparticles improved fruit yield and quality in pomegranates, whereas zinc and boron nanoparticles increased potato growth and tuberization. The versatility of nanofertilizers allows for tailored applications across a range of crops, with benefits including improved plant height, chlorophyll content, root growth, and overall yield.

The application of nanoparticles in agriculture is not limited to foliar sprays; fertigation (application through irrigation water) is also a common method for delivering nutrients such as zinc, calcium, and nitrogen. The uptake mechanisms of nanoparticles vary on the basis of their size, shape, and surface charge. Smaller nanoparticles tend to be transported more efficiently through plant tissues, whereas larger particles may face limitations due to their size. For example, positively charged nanoparticles, such as gold (Au) NPs, effectively bind to plant surfaces but poorly internalize, whereas negatively charged nanoparticles exhibit better internalization but weaker binding. These interactions depend on the plant species, and studies have shown that certain species respond more favourably to specific nanoparticle types [161].

In conclusion, the development of nanofertilizers represents a significant leap toward sustainable agriculture. These fertilizers not only improve nutrient use efficiency but also reduce environmental harm by minimizing nutrient runoff. The ability to control the release and uptake of nutrients offers a promising solution to the challenges faced by conventional fertilizers, supporting the transition towards more sustainable and environmentally friendly farming practices (Fig. 7).

Improving Crop Yield and Health

Green nanotechnology has significantly enhanced agricultural practices by improving crop health, increasing yields, and increasing the efficiency of nutrient and pesticide delivery. Nanomaterials such as Carbon Nanotube (CNT) and titanium dioxide (TiO_2) nanoparticles have been utilized to increase seed germination rates, growth rates, and overall plant health. For example, CNTs improve moisture retention, which increases germination rates, whereas TiO_2

nanoparticles have been used effectively to increase wheat productivity [162]. The unique properties of these nanomaterials, such as their large surface area and high reactivity, make them highly efficient at promoting plant growth [163, 164].

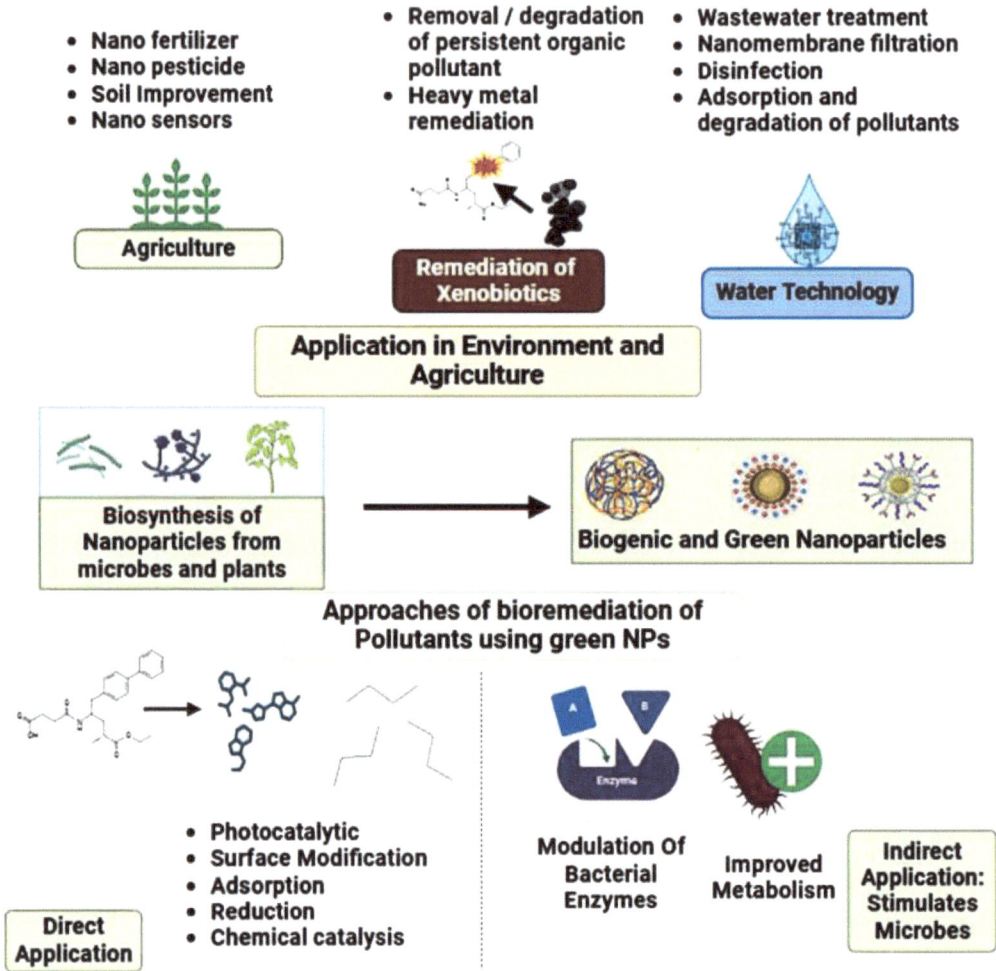

Fig. (7). Applications of green nanoparticles in both direct and indirect ways, including water technology, agriculture, and bioremediation.

Similarly, the development of nanoformulations for pesticides, fungicides, and herbicides—collectively referred to as nanocides—has revolutionized plant protection. These nanoformulations address many of the shortcomings of traditional pesticides, such as leaching, degradation, and ecological toxicity. Nanoformulations improve solubility, bioavailability, and targeted delivery, which not only increases the effectiveness of the Active Ingredients (AIs) but also reduces their environmental impact. These formulations come in various forms, including nanoemulsions, nanosuspensions, nanocapsules, and nanoparticles, each offering specific advantages for targeted pest control.

Compared with conventional pesticides, the small size and large surface area of nanoparticles make them more permeable to pests and plants. For example, silver nanoparticles (Ag NPs) are known for their antifungal properties, effectively controlling fungal infections in plants such as wheat and tomatoes. Furthermore, nanopesticides provide more precise pest control, reducing the number of applications needed and minimizing human exposure to toxic chemicals [165].

Nanoformulations have also shown improved activity in herbicide delivery. By encapsulating herbicides such as paraquat in nanoparticles, their release can be controlled, ensuring greater stability and better adhesion to plant surfaces. This slow-release mechanism not only improves herbicide effectiveness but also reduces herbicide toxicity and the environmental footprint [166].

In conclusion, both nanofertilizers and nanopesticides represent a paradigm shift in sustainable agriculture. These innovations enable more efficient nutrient and pesticide delivery, reduce environmental harm, and improve crop health, laying the foundation for more eco-friendly and productive farming systems.

Nano-Sensors for Monitoring Plant Health

Green nanotechnology is transforming plant health management by enabling the development of nanosensors capable of early disease detection, pest control, and nutrient monitoring. These sensors provide farmers with a powerful tool to identify early signs of plant stress, enabling timely interventions that reduce the need for excessive pesticide and fertilizer use. The ability to detect issues at the molecular level helps maintain healthier crops and protect the environment by minimizing the use of harmful chemicals.

For example, nanosensors have been developed to detect early symptoms of fungal infections in plants. Researchers have utilized silver nanoparticles (Ag NPs), which can interact with biological markers of fungal infections, as key components of these sensors. When fungal pathogens attack plants, they often release specific enzymes or metabolites that can be detected by these sensors [167]. Similarly, researchers have used Carbon Nanotubes (CNTs) to detect oxidative stress, a sign of disease in plants. These sensors offer high sensitivity and the ability to detect disease markers at very low concentrations, enabling farmers to respond rapidly.

In other applications, nanosensors have been used to monitor nutrient deficiencies in soil. For example, titanium dioxide (TiO_2) nanoparticles are employed in sensors designed to detect nitrogen and phosphorus deficiencies, which are common in agricultural soils. These sensors work by binding to specific ions of these nutrients, triggering a detectable signal. By detecting nutrient imbalances early, farmers can apply fertilizers more precisely, reducing both costs and environmental impact [168]. Additionally, these sensors can monitor pesticide levels in the soil, ensuring that toxic chemicals do not accumulate to harmful levels.

The components of nanosensors typically include nanomaterials that increase sensitivity, recognition elements that allow specificity, and signal transduction mechanisms that translate the interaction into a readable signal. Common detection methods include electrical, optical, and mechanical techniques. For example, researchers have developed optical sensors using gold nanoparticles (Au NPs), which change color in the presence of specific pathogen proteins or DNA sequences, providing farmers with visual cues for early detection.

These sensors have applications in soil quality management, pest detection, and even DNA-based diagnostics for plant diseases. For example, nanosensors have been used for the early detection of pathogens such as *Xanthomonas* and Pseudomonas species, which cause significant crop losses [169]. By detecting the presence of these pathogens before visible symptoms appear, farmers can implement treatments that prevent further spread, improving crop health and yield.

The development of these nanosensors represents a leap forward in precision agriculture, offering real-time, targeted solutions that reduce chemical usage and promote sustainable farming practices. With continued advancements in nanotechnology, these sensors will become even more versatile, providing critical data for optimizing agricultural practices while protecting the environment (Fig. 8).

Green Nanotechnology in Medicine

Green nanotechnology, particularly through the use of biomineralized materials, is creating new opportunities for sustainable and effective biomedical applications. Biomineralization, a natural process in which living organisms produce inorganic materials, is being harnessed to develop biocompatible materials for medical use. These biomineralized materials, such as calcium phosphate, silica, and iron oxide nanoparticles, exhibit enhanced properties such as biocompatibility, bioactivity, and controlled release, making them ideal candidates for drug delivery, diagnostic tools, and medical implants.

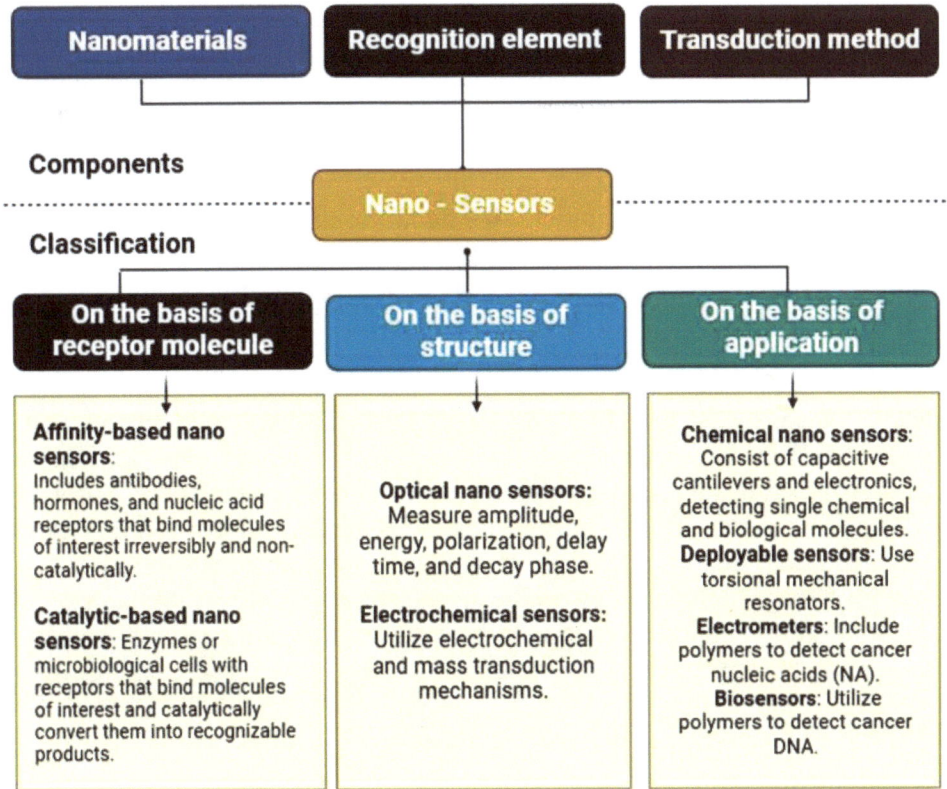

Fig. (8). Components and classification of nanosensors, highlighting their types and functions in various applications.

In drug delivery systems, biomineralized materials offer advantages in terms of biocompatibility and controlled release. For example, calcium phosphate nanoparticles can be used to deliver drugs in a controlled manner, enhancing their therapeutic effect while minimizing side effects [170]. These materials also promote the bioavailability of drugs by facilitating their absorption and providing a stable release profile. Additionally, iron oxide nanoparticles, which are widely used in Magnetic Resonance Imaging (MRI), have shown promise in drug delivery applications, where they can be used to target specific tissues, reducing the systemic side effects of drugs.

In medical diagnostics, biomineralized nanoparticles, such as gold and silica, are particularly useful because of their unique optical properties, which enable high sensitivity and precision in diagnostic tests. Silica nanoparticles, for example, have been functionalized to carry biomolecules for targeted diagnostics, offering a nontoxic and effective method for disease detection. Their biomineralized nature allows them to be seamlessly integrated into biological systems without triggering immune responses, improving the accuracy of diagnostics [171].

The use of biomineralized materials in medical devices also presents significant advantages. Biodegradable and biocompatible materials such as silica-based nanoparticles can be used to develop scaffolds for tissue engineering, where they encourage cellular growth and tissue regeneration. These materials degrade naturally within the body, eliminating the need for surgical removal and reducing the risk of inflammation or rejection. Iron oxide nanoparticles have also been used in magnetic drug targeting, where they are employed to guide drugs to specific sites within the body *via* an external magnetic field, further increasing the efficiency of treatment [172].

Moreover, in cancer therapy, biomineralized materials such as gold nanoparticles have gained attention because of their ability to deliver targeted therapies. Functionalized gold nanoparticles can bind to cancer cell receptors, delivering drugs or genes specifically to tumor sites. This targeted approach enhances the efficacy of the treatment while minimizing damage to healthy tissues. Additionally, these nanoparticles can be employed in photothermal therapy, where they absorb light and convert it into heat to destroy cancer cells selectively, offering a noninvasive treatment option.

Green Nanotechnology in Environmental Protection

Green nanotechnology has made significant strides in various environmental applications, particularly in water purification, waste remediation, and air pollution control. In water purification, nanomaterials such as nanoadsorbents and nanofiltration devices are employed to remove contaminants such as heavy metals, organic pollutants, and pathogens from water. These nanoadsorbents, including carbon-based nanomaterials, possess a large surface area that enhances their ability to adsorb toxic substances efficiently. Nanofiltration devices, on the other hand, use nanostructured membranes with precisely engineered pore sizes that allow for the separation of particles at the molecular level, resulting in a more effective and faster filtration process than traditional methods do. These technologies not only improve removal rates but also offer energy-efficient alternatives to conventional water treatment systems, which are often energy intensive and generate harmful byproducts.

In the field of waste remediation, green nanotechnology plays a crucial role in neutralizing or breaking down hazardous substances into nontoxic compounds, significantly improving environmental cleanup efforts (Fig. **9**). Various nanomaterials, including nanoscale Zero-Valent Iron (nZVI), nanocatalysts, and nanoenzymes, have been used in this regard. The high surface area and reactivity of these nanomaterials allow them to effectively degrade complex pollutants, such as heavy metals and organic chemicals, into safer substances. For example, nZVI reacts with contaminants such as chlorinated solvents and heavy metals, reducing them to harmless forms and thus minimizing the need for harmful chemical reagents in remediation processes. This approach accelerates the degradation process and reduces the environmental footprint of waste treatment technologies, providing an eco-friendly alternative for cleaning contaminated water, soil, and sediments.

For air pollution control, green nanotechnology offers promising solutions through nanoglass-based air filters and nanocatalysts. Nanofilters made from materials such as carbon nanotubes and graphene are highly effective at capturing fine Particulate Matter (PM2.5 and PM10), as well as gases such as carbon dioxide (CO_2), nitrogen oxides (NOx), and Volatile Organic Compounds (VOCs). These filters function by utilizing the porous structure of nanomaterials, which trap pollutants on their surface, improving the overall filtration efficiency compared with that of conventional filters. Additionally, nanocatalysts accelerate the chemical conversion of harmful gases into nontoxic substances, contributing to cleaner industrial emissions. For example, these catalysts can breakdown nitrogen oxides in exhaust gases into nitrogen and oxygen, reducing the harmful effects of industrial pollution. By integrating these nanobased systems into various sectors, air quality can be significantly improved, while industries can lower their carbon footprint and mitigate the impact of climate change [173].

Green Nanotechnology in Renewable Energy

Green nanotechnology has emerged as a key driver of innovation in the development of sustainable energy systems, offering promising solutions to global challenges related to energy efficiency, environmental impact, and climate change. By integrating nanotechnology with renewable energy sources, particularly solar and hydrogen energy, researchers are creating more efficient, cost-effective, and environmentally friendly technologies. These advancements are helping reduce dependency on fossil fuels while improving energy sustainability and mitigating the harmful effects of climate change. Solar and hydrogen energy, two of the most abundant resources on Earth, have vast potential to reshape global energy systems. Solar energy alone delivers more power in one day than the entire world consumes annually, whereas hydrogen offers a clean energy carrier, producing only water as a byproduct when used in fuel cells [164].

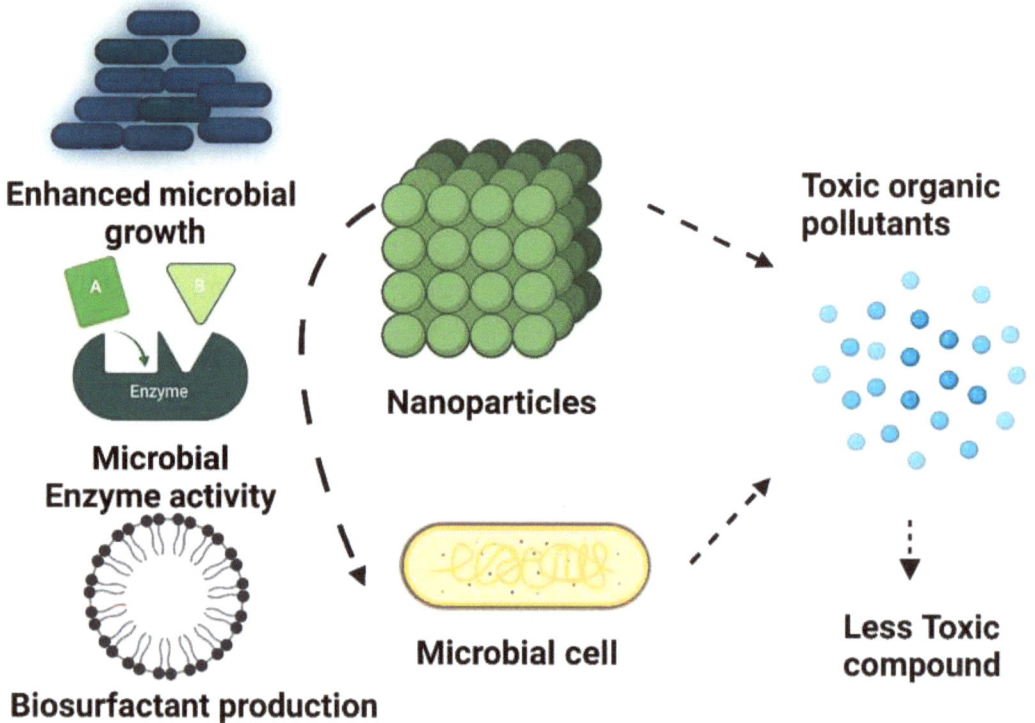

Fig. (9). The role of nanoparticles synthesized by microbial cells in mitigating the toxicity of organic pollutants.

One of the key challenges with solar energy has been the high cost and environmental impact associated with the production of efficient solar cells. Traditional silicon-based solar cells, especially single-crystal silicon, achieve an efficiency of approximately 24.7%, but the manufacturing process is resource intensive and involves thick silicon wafers and expensive photolithography. Green nanotechnology addresses these issues by exploring alternative, sustainable materials and more eco-friendly production techniques. For example, thinner semiconductor wafers, which are less than 200 nm thick, reduce material use and improve production efficiency. Materials such as amorphous silicon, microcrystalline silicon, and thin films such as Cadmium Telluride (CdTe) and Copper Indium Gallium Selenide (CIGS) are gaining attention because of their lower cost, better environmental compatibility, and suitability for a variety of substrates. Furthermore, the use of biosynthesized quantum dots represents another breakthrough, as these tiny particles absorb a broad spectrum of sunlight, increasing the efficiency of solar cells. These quantum dots can be synthesized through eco-friendly methods, reducing the reliance on harmful chemicals while also improving solar energy capture across various wavelengths, from infrared to ultraviolet.

In addition to these innovations, the field of solar thermal energy benefits from green nanofluid sustainable liquids containing bioderived nanoparticles. These nanofluids significantly enhance the heat absorption and thermal transfer capabilities of solar collectors. For example, nanoparticles can be engineered to optimize solar energy absorption while minimizing heat loss through infrared emissions, which directly contributes to improved performance. Additionally, these nanofluids increase the thermal conductivity of the fluid, leading to better heat exchange and a reduction in the need for energy-intensive cooling systems. This makes solar thermal systems more efficient and sustainable by reducing overall energy consumption.

With respect to hydrogen energy, hydrogen serves as a clean, versatile energy carrier, particularly when used in fuel cells. Fuel cells powered by hydrogen are zero-emission solutions that produce only water as a byproduct, which makes them an ideal clean energy option for applications ranging from transportation to stationary power generation. However, traditional hydrogen fuel cells, such as Proton Exchange Membrane Fuel Cells (PEMFCs), rely on platinum-based catalysts, which are costly and environmentally harmful to extract. Green nanotechnology is addressing this challenge by developing bioderived catalysts that are more affordable and sustainable. For example, research has led to the creation of catalysts made from plant-based materials and biologically derived compounds that offer similar or even improved performance compared with that of platinum, thus reducing the environmental impact and cost of fuel cell production. In addition to PEMFCs, Solid Oxide Fuel Cells (SOFCs), which operate at high temperatures (800–1000°C), are also being improved through the application of green nanotechnology. Sustainable ceramic materials are being explored for SOFC electrolytes, offering a more environmentally friendly option for large-scale power applications.

To further enhance hydrogen energy storage and fuel cell performance, green nanotechnology is also advancing the development of more sustainable materials for hydrogen storage. Carbon nanotubes, for example, are produced from renewable resources, offering high-density hydrogen storage (Fig. **10**). These materials are particularly beneficial for fuel cell applications, where quick hydrogen release is essential for maintaining optimal performance. Similarly, core-shell nanoparticles are synthesized *via* green methods to enable fast hydrogen release and improve the efficiency of storage systems. The development of fuel cell membranes made from biocompatible nanocomposites and plant-based conductive polymers has also increased proton conductivity, reducing the need for synthetic materials while improving the overall performance of fuel cells.

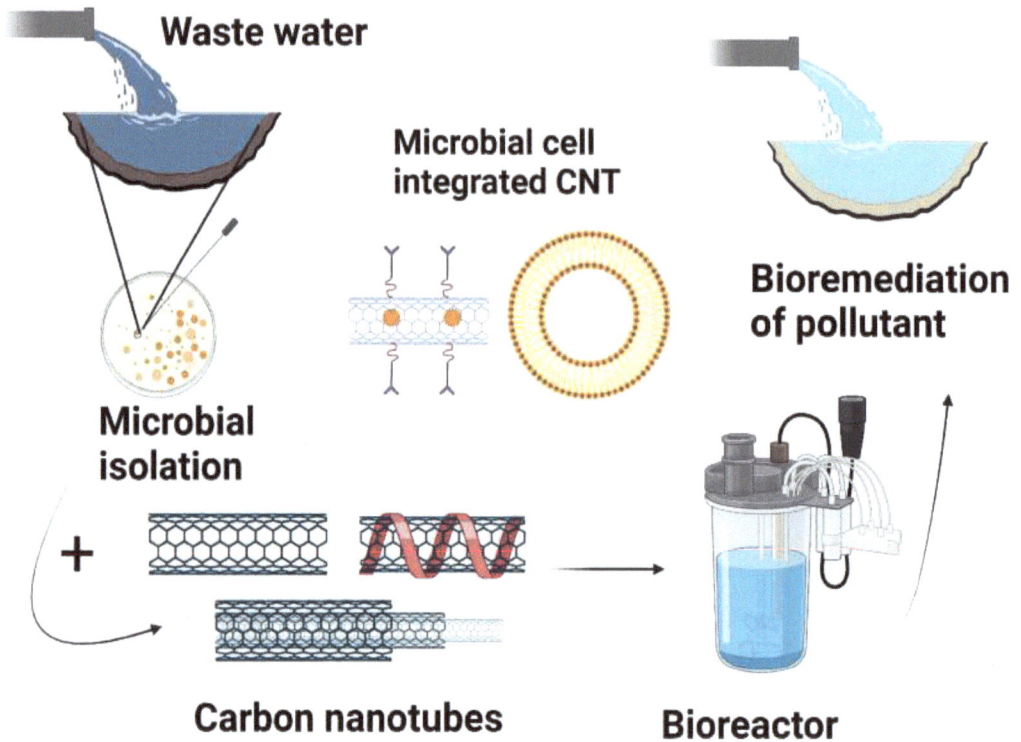

Fig. (10). The production of carbon nanotubes and their application in wastewater treatment.

The miniaturization of fuel cells is another exciting frontier where green nanotechnology is playing a pivotal role. By integrating Microelectromechanical Systems (MEMSs) with sustainable nanotechnology, researchers are designing smaller, more efficient fuel cells. These microfuel cells, which are produced *via* eco-friendly microfabrication techniques, hold great promise for portable electronics and other small-scale applications, offering a clean energy source that is both compact and efficient.

Ultimately, the integration of green nanotechnology with solar and hydrogen energy systems is accelerating the shift toward more sustainable energy solutions. By improving efficiency and reducing the cost and environmental impact of solar cells and fuel cells, these technologies are becoming more accessible, making renewable energy a viable alternative to fossil fuels. For example, multijunction solar cells that combine green-synthesized quantum dots and plant-based nanowires allow for better light capture across a broader spectrum of wavelengths, significantly increasing solar efficiency. Similarly, hybrid fuel cells that combine inorganic semiconductors with biodegradable organic polymers provide flexible, cost-effective solutions for various energy applications. Furthermore, green nanocatalysts for water splitting enable the direct production of hydrogen from water, further supporting the growth of a sustainable hydrogen economy.

Environmental Benefits of Green Microbial Nanotechnology

Green microbial nanotechnology has garnered significant interest for its potential to address pressing environmental challenges in sustainable and innovative ways. By utilizing microorganisms such as bacteria, fungi, and algae to synthesize nanoparticles, this approach offers numerous advantages in terms of waste transformation, minimizing chemical usage, enhancing pollutant removal, and supporting sustainable industrial practices.

Waste Transformation into Valuable Products

One of the most impactful environmental benefits of green microbial nanotechnology is its ability to convert waste materials into valuable nanomaterials. This process contributes to waste management while promoting a circular economy. Agricultural, industrial, and food waste, which can often contribute to environmental pollution, can be transformed by microorganisms into useful nanoparticles. For example, agricultural residues such as rice husks, wheat straw, and sugarcane bagasse have been shown to serve as ideal substrates for microbes to synthesize various nanoparticles. These organic wastes are rich in carbon and other nutrients, which microorganisms utilize to produce valuable nanomaterials, such as metallic nanoparticles, which can be used in catalysis, antimicrobial applications, or environmental remediation.

An example of this is the conversion of corn husks into silver nanoparticles by *Fusarium oxysporum*, a fungal species. These silver nanoparticles are then used in antimicrobial applications, demonstrating the sustainable and effective use of agricultural waste. Similarly, the conversion of orange peel waste by *Aspergillus niger* into gold nanoparticles has been investigated. These biogenic gold nanoparticles have shown promising results in drug delivery systems, making them valuable in the pharmaceutical industry. By utilizing waste to produce high-value products, green microbial nanotechnology reduces the volume of waste ending up in landfills and creates new economic opportunities within a circular economy framework.

Reduction in Chemical Usage in Nanoparticle Synthesis

Green microbial nanotechnology provides an alternative to traditional chemical methods of nanoparticle synthesis, reducing the need for toxic chemicals and harsh reaction conditions. Compared with conventional methods, microorganisms synthesize nanoparticles through natural processes that involve enzymatic reduction, biosorption, and nucleation, which are more environmentally friendly.

For example, *Saccharomyces cerevisiae*, a type of yeast, has been used to synthesize semiconductor nanoparticles, such as Cadmium Sulfide (CdS) and gold-silver alloy nanoparticles. The use of these microorganisms prevents the need for toxic chemicals such as cadmium salts in the nanoparticle synthesis process, which could otherwise pose significant environmental and health risks. Similarly, Penicillium species have been found to synthesize nanoparticles from toxic metals such as copper and zinc, converting them into nontoxic forms through extracellular biosynthesis mechanisms. This ability to detoxify metals by turning them into stable nanomaterials has been explored for applications in wastewater treatment and heavy metal removal from industrial effluents.

Microalgae are also significant contributors to green nanotechnology (Fig. **11**). Diatoms, such as *Navicula* and *Thalassiosira*, are known for their silica-rich frustules, which are naturally occurring nanostructures. These frustules can be functionalized to increase their ability to remove heavy metals from water. In one example, *Phaeodactylum tricornutum*, a type of diatom, was used to produce silica nanoparticles with high adsorption capacity for arsenic removal from contaminated water. The natural biosynthesis of such nanomaterials eliminates the need for toxic chemicals and provides a sustainable method for heavy metal remediation [174].

Enhanced Pollutant Removal from Contaminated Environments

Microbial nanoparticles have shown exceptional capabilities in environmental remediation, especially in removing heavy metals and organic pollutants from wastewater. Owing to the high surface area and reactivity of these nanoparticles, they are effective adsorbents that neutralize and remove contaminants from polluted environments. For example, *Aspergillus tubingensis*, a fungus, is known for producing iron oxide nanoparticles that can effectively remove heavy metals such as lead (Pb), nickel (Ni), and copper (Cu) from industrial effluents. These biogenic nanoparticles are not only highly effective at removing pollutants but can also be regenerated and reused multiple times, reducing the need for fresh materials and supporting continuous pollution control efforts.

Microbial silver nanoparticles are also used in wastewater treatment to break down toxic dyes and pharmaceutical residues. One notable example is the use of silver nanoparticles synthesized from *Trichoderma harzianum*, which has been proven to degrade synthetic dyes in textile wastewater. Furthermore, these nanoparticles have shown significant antimicrobial activity, reducing harmful bacterial populations in wastewater and providing a greener and more efficient alternative to conventional chemical treatments. The use of biogenic nanoparticles in wastewater treatment also eliminates the production of harmful byproducts typically associated with chemical-based treatments.

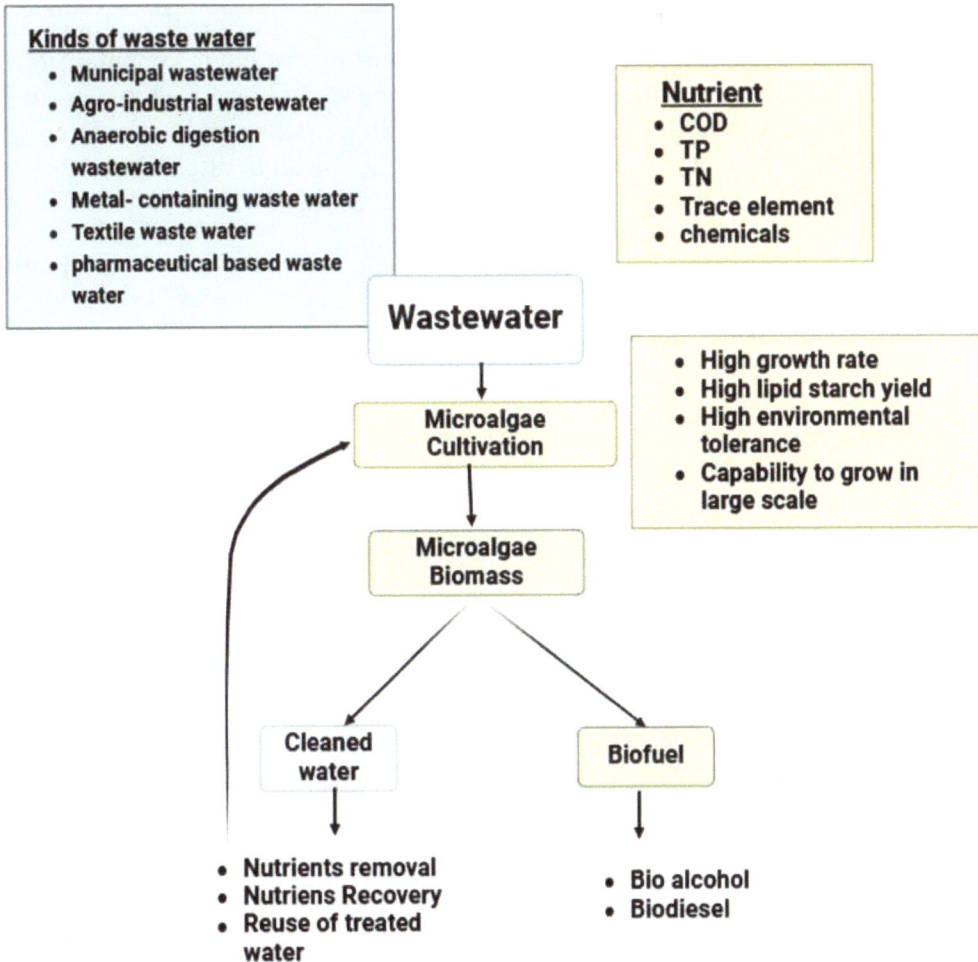

Fig. (11). Classification of wastewater and its components, along with the contribution of microalgae in water treatment and biofuel generation.

Support for Sustainable Industrial Practices

Green microbial nanotechnology offers industries a pathway to adopt more sustainable practices by integrating eco-friendly nanoparticle production and pollution remediation methods. The synthesis of nanoparticles *via* microbial processes typically requires lower energy inputs and mild reaction conditions, which aligns with goals to reduce greenhouse gas emissions and energy consumption.

For example, *Shewanella oneidensis* has been used to produce biogenic palladium nanoparticles, which are effective catalysts for the reduction of toxic aromatic compounds found in pharmaceutical waste. By utilizing these biogenic palladium catalysts, chemical manufacturing processes can significantly reduce the generation of hazardous byproducts, thereby lowering the environmental impact of industrial activities. Similarly, biogenic platinum nanoparticles produced by *Rhodococcus species* have been employed in hydrogenation reactions, offering a green alternative to traditional platinum catalysts, which often require harsh chemical conditions and result in high energy consumption.

In addition, the integration of microbial nanotechnology in bioremediation supports the cleanup of contaminated soils and water sources. For example, *Geobacter sulfurreducens*, a bacterium known for its ability to produce conductive nanoparticles, has been used in the bioremediation of soils contaminated with heavy metals such as chromium and arsenic. These biogenic nanoparticles can facilitate the immobilization of toxic metals, preventing their migration and toxicity in the environment. By incorporating these green nanotechnological solutions, industries can significantly reduce their reliance on toxic chemicals and minimize their environmental footprint.

Biodegradable Nanomaterials for Pollution Control

Microbial nanotechnology offers innovative solutions for addressing pollution in air, water, and soil, with an emphasis on the use of biodegradable nanomaterials for effective pollution control. In water treatment, microbial nanotechnology provides advanced methods for purifying water by using nanomaterials with enhanced adsorption, catalytic, and filtration properties. For example, Pseudomonas aeruginosa has been shown to produce nanoparticles capable of removing a variety of contaminants, including heavy metals and organic pollutants, from contaminated water sources.

Air pollution, a growing environmental concern, can also be mitigated *via* the use of microbial nanotechnology. Biogenic nanomaterials such as those produced by *Azotobacter vinelandii*, which produces polyhydroxyalkanoate (PHA) nanoparticles, have been applied as filters for removing nitrogen oxides (NOx) and Volatile Organic Compounds (VOCs) from industrial air. These biogenic nanoparticles work by adsorbing harmful gases and converting them into less toxic compounds, providing a sustainable alternative to traditional air filtration systems.

Soil contamination, particularly by pesticides, heavy metals, and organic pollutants, can be managed with the help of microbial nanomaterials. For example, *Bacillus subtilis* has been employed to produce biogenic nanoparticles capable of immobilizing heavy metals in contaminated soils, thereby reducing their bioavailability and preventing them from entering the food chain. Similarly, nanocomposites that combine biogenic nanoparticles with plant roots have been developed to increase the removal of pollutants from soils, highlighting the potential for the use of nanotechnology in conjunction with natural remediation strategies [175].

Challenges and Risks in Microbial Nanotechnology

Microbial nanotechnology holds immense promise for addressing a wide range of environmental and technological challenges, yet it also presents significant health, environmental, and technical risks that need to be carefully managed. These challenges must be addressed before this technology can be fully harnessed on a global scale [176].

Health and Environmental Risks

One of the primary concerns surrounding microbial nanotechnology is the potential toxicity of certain nanoparticles and their long-term effects on human health and the environment. Compared with larger particles, nanoparticles are extremely small, which allows them to penetrate biological barriers more easily. This increased reactivity can cause cellular damage, as observed in studies on metal nanoparticles such as silver, gold, and copper, which are known to be toxic to aquatic organisms. For example, silver nanoparticles, often used for their antimicrobial properties, have been shown to accumulate in fish and other aquatic organisms, leading to bioaccumulation and potential disruptions in the food chain. The long-term environmental impact of these nanoparticles remains poorly understood, and their ability to accumulate in ecosystems without being broken down poses significant ecological risks.

The microbial synthesis of nanoparticles, although eco-friendly in comparison with chemical methods, can also lead to the production of byproducts that might be hazardous if not carefully managed. For example, certain strains of bacteria used in nanoparticle production may generate toxic metabolites that need to be neutralized or contained. If these byproducts are released into the environment, they could harm local ecosystems. In one case, a study involving the synthesis of gold nanoparticles using *Bacillus subtilis* revealed that while the process itself was sustainable, the waste products generated by the bacteria were potentially harmful to the environment. This highlights the need for strict containment and proper disposal practices to prevent contamination of water sources and soil.

Regulatory and Safety Concerns

The lack of clear regulatory frameworks for microbial nanotechnology is another significant challenge. Unlike traditional chemical methods, which have well-established safety standards, microbial-based nanoparticle synthesis operates within a much less regulated domain. The absence of universally accepted guidelines for the production, use, and disposal of nanoparticles raises concerns about their safe integration into industries such as healthcare, environmental remediation, and energy. The scaling up of microbial nanotechnology from laboratory settings to industrial applications adds complexity to this issue. For example, industrial-scale production of nanoparticles may lead to higher concentrations of these materials being released into the environment, increasing the potential for human exposure and environmental contamination.

In 2009, the U.S. Environmental Protection Agency (EPA) issued a draft report on nanomaterials, acknowledging the need for clearer regulatory standards but also emphasizing the difficulties in assessing the risks of nanoparticles owing to their size and reactivity. Without comprehensive regulatory frameworks, there is uncertainty regarding how to manage the potential hazards of microbial nanotechnology properly, which could hinder the widespread adoption of these technologies. Some countries, such as Canada, have begun to address these concerns by establishing guidelines for nanomaterial safety, but these regulations remain in progress globally.

Technical Challenges in Particle Control

Another major challenge in microbial nanotechnology lies in the difficulty of precisely controlling the properties of nanoparticles during synthesis. Unlike traditional chemical methods, which allow for the fine-tuning of variables such as temperature, pressure, and reagent concentration to control particle size and shape, microbial processes are much more variable. The microorganisms used in nanoparticle synthesis operate in dynamic environments, with growth conditions, nutrient availability, and microbial strain all influencing the final product. This lack of precision makes it difficult to achieve uniformity in particle size, shape, and distribution, which is crucial for applications such as drug delivery and environmental remediation.

For example, research into the microbial synthesis of iron oxide nanoparticles using Shewanella oneidensis has shown that variations in growth conditions can lead to significant differences in particle morphology. While these nanoparticles have the potential to be used in environmental clean-up efforts, particularly in removing contaminants such as heavy metals, the inconsistency in particle characteristics poses a challenge to their application in real-world settings.

Similarly, the synthesis of gold nanoparticles by *Escherichia coli* has demonstrated variability in size, which can affect the ability of the particles to target specific cells in biomedical applications.

Ensuring consistent and scalable production of high-quality nanoparticles remains a challenge in microbial nanotechnology. Variations in the microbial strain used, growth conditions, and nutrient availability can affect the synthesis process. Researchers are exploring methods to standardize conditions and optimize microbial growth to reduce this variability. For example, some studies have focused on optimizing the growth media and fermentation conditions for microbial nanoparticle production to increase the consistency and yield of nanoparticles. However, achieving scalability while maintaining high quality remains an obstacle to the commercial application of microbial nanotechnology.

Future Directions and Innovations in Green Microbial Nanotechnology

The future of green microbial nanotechnology holds immense promise for addressing critical global challenges related to environmental sustainability, energy efficiency, and pollution reduction. By combining cutting-edge advancements in nanoscience with environmentally friendly practices, this field is paving the way for innovations that have the potential to revolutionize industries while promoting sustainable solutions. Green nanotechnology, particularly microbial nanotechnology, focuses on harnessing the unique properties of microorganisms to produce nanoparticles and nanostructures with applications in waste management, pollution control, resource efficiency, and renewable energy production. These innovations are essential for reducing the environmental impact of industrial processes and promoting more sustainable production methods.

Genetic Engineering for Enhanced Nanoparticle Synthesis

One of the most promising trends in green nanotechnology is the application of genetic engineering to optimize the synthesis of nanoparticles. By modifying the genetic makeup of microorganisms, researchers can increase the production of nanoparticles with specific properties, such as increased stability, reactivity, and functionality. This approach allows for the targeted use of nanoparticles in various fields, including healthcare, environmental remediation, and energy storage. For example, engineered microbes can be used to produce silver nanoparticles, which have antimicrobial properties and can be applied in medical devices or water purification systems. Furthermore, genetic engineering enables the creation of nanomaterials that are more environmentally friendly than traditional chemical synthesis methods, which often involve toxic chemicals. This shift to sustainable microbial processes plays a crucial role in supporting eco-friendly production and reducing the environmental toxicity of nanoparticles [177].

Circular Economy and Waste Reduction

Green microbial nanotechnology is also key to advancing the concept of a circular economy. By utilizing agricultural, industrial, and food waste as substrates for microbial nanoparticle production, this technology turns waste into valuable resources. For example, microbes can be engineered to produce nanoparticles from agricultural residues such as rice husks or fruit peels, reducing the need for raw materials while providing useful byproducts. This not only helps mitigate pollution but also promotes the recovery of resources and fosters more sustainable production practices. Additionally, microorganisms can be used to degrade organic waste into biofuels, bioplastics, and other biochemicals, generating renewable energy and materials while reducing waste volume. The integration of microbial nanotechnology into circular economy models can close the loop on waste, turning it into a resource rather than a burden.

Development of Advanced Nanostructures

The development of complex nanostructures *via* microbial synthesis is another exciting direction for the future of green nanotechnology. Advancements in synthetic biology, coupled with tools such as CRISPR-Cas9, have enabled precise control over the size, shape, and composition of nanostructures. These breakthroughs have allowed the creation of multifunctional nanostructures with applications in various fields, such as theranostic nanoparticles, that can be used to diagnose and treat diseases simultaneously. Furthermore, through biotemplating and self-assembly mechanisms, researchers are producing hierarchical nanostructures with intricate architectures, which can be used in environmental sensors to detect pollutants or toxins. A promising example is the production of nanoparticle-based sensors by microbes to detect heavy metals in contaminated water sources. The versatility of microbial nanotechnology makes it suitable for a range of applications, from healthcare to environmental monitoring.

Green Chemistry and Engineering

At the core of green microbial nanotechnology are the principles of green chemistry and green engineering, which focus on designing processes that minimize environmental impact while improving the efficiency and functionality of nanomaterials. Green chemistry advocates the use of safer chemicals and methods in nanoparticle synthesis, ensuring that the materials produced are nontoxic and sustainable. For example, using plant-based reducing agents or biosynthesized materials instead of hazardous chemicals in nanoparticle production can significantly reduce the environmental footprint. Green engineering, on the other hand, strives to optimize industrial processes by reducing energy consumption, waste, and harmful emissions. A prime example of this is the development of biocatalytic processes, where microorganisms are used to synthesize nanoparticles in a more energy-efficient manner than traditional chemical processes. By combining these principles, green microbial nanotechnology not only promotes sustainability but also advances industrial processes in a more eco-friendly direction.

Pioneering Science for a Sustainable Future

As we look ahead, the future of green microbial nanotechnology is poised to drive significant advancements across multiple sectors, including medicine, energy, and environmental science. By harnessing the natural ability of microorganisms to synthesize nanoparticles and nanostructures, researchers are developing innovative solutions for environmental remediation, resource recovery, and clean energy production. For example, microbial nanotechnology is being used to eliminate environmental pollutants such as oil spills and heavy metals, offering a more sustainable alternative to chemical-based methods. Additionally, engineered microbes can be employed to capture nutrients from wastewater and recycle them as fertilizers, reducing nutrient pollution and promoting sustainable agriculture. The development of eco-friendly nanomaterials for energy storage, such as microbial

CONCLUSION

Microbial nanotechnology stands at the cutting edge of sustainable innovation, merging microbiology and nanoscience to create environmentally friendly nanomaterials. This transformative approach utilizes microorganisms such as bacteria, fungi, algae, and yeasts to synthesize nanoparticles, offering an eco-conscious alternative to traditional nanoparticle production methods, which often rely on toxic chemicals and energy-intensive processes. By harnessing renewable resources and converting industrial, agricultural, and food waste into valuable nanomaterials, microbial nanotechnology not only minimizes the environmental impact but also contributes to the principles of a circular economy. This capability for waste recycling reduces the ecological footprint of industrial processes and supports a more sustainable approach to manufacturing and resource management.

One of the most compelling advantages of microbial nanotechnology lies in its potential for environmental remediation. Microorganism-derived nanoparticles, including silver, gold, and iron oxide, have demonstrated exceptional capabilities in adsorbing and neutralizing a wide range of pollutants from air, water, and soil. Owing to their high surface area, reactivity, and biocompatibility, these materials are particularly effective at addressing environmental contaminants such as heavy metals, dyes, pesticides, and pharmaceutical residues. For example, silver nanoparticles synthesized by fungi such as *Aspergillus tubingensis* have been shown to degrade toxic dyes and remove heavy metals such as lead and copper from wastewater. These biogenic nanoparticles not only provide a cleaner alternative to chemical treatments but also offer enhanced reusability, promoting long-term pollution control. This contributes to cleaner ecosystems, safer water supplies, and reduced contamination in soil and air.

In the agricultural sector, microbial nanotechnology has revolutionized crop management through the development of nanofertilizers and nanopesticides. These innovations help optimize nutrient delivery, improve crop health, and increase productivity while reducing the reliance on harmful chemicals. Nanofertilizers offer controlled nutrient release, enhancing plant uptake and minimizing fertilizer runoff into surrounding ecosystems. This targeted approach not only increases agricultural yields but also supports sustainable farming practices by promoting soil health and reducing the environmental consequences of excessive fertilizer use. Moreover, the development of nanosensors for the early detection of plant diseases and nutrient deficiencies enables precision agriculture, allowing farmers to intervene at the right moment and reducing the need for excessive pesticide and fertilizer applications, which further minimizes environmental impacts.

Microbial nanotechnology also holds vast potential in medicine, where its biodegradable and biocompatible nanoparticles are being explored for drug delivery systems, diagnostic applications, and therapeutic treatments. Gold and silver nanoparticles, which are produced through microbial methods, are particularly beneficial for drug encapsulation, offering controlled release and enhanced bioavailability. These nanoparticles can target specific tissues, improving treatment efficacy while minimizing side effects. In diagnostics, microbial nanotechnology is being used in assays for early disease detection, such as that of cancer and infectious diseases, providing faster, more accurate results. By reducing the reliance on synthetic materials and offering safer, more efficient treatment options, microbial nanotechnology is transforming medical approaches and making healthcare more sustainable.

In the energy sector, microbial nanotechnology is leading renewable energy solutions. Biologically derived nanomaterials are being incorporated into solar cells, hydrogen fuel cells, and energy storage systems to improve their efficiency and reduce production costs. Biosynthesized quantum dots and nanowires, for example, are being used to increase light absorption in solar cells, significantly increasing their energy conversion rates. Additionally, eco-friendly catalysts synthesized by microorganisms, such as *Shewanella oneidensis*, are improving the performance of hydrogen fuel cells, providing a cleaner alternative to fossil fuels.

These advancements have contributed to the reduction of greenhouse gas emissions, supported the transition to a low-carbon economy, and helped mitigate climate change by improving the efficiency of renewable energy systems.

Despite its tremendous potential, the widespread implementation of microbial nanotechnology faces several challenges. Issues surrounding scalability, regulatory standards, and safety concerns remain significant barriers. Owing to differences in microbial strains and environmental conditions, the variability in nanoparticle characteristics, such as size, shape, and composition, poses challenges to the reproducibility and consistency of products, particularly at larger scales. This variability makes it difficult to standardize production processes for commercial use. Furthermore, while microbial nanoparticles offer great promise, there are concerns about their toxicity to humans and ecosystems. To address these challenges, robust safety assessments, comprehensive regulatory frameworks, and the development of standardized protocols are essential to ensure responsible production and use. Ongoing research and innovation will be crucial to overcoming these hurdles, enabling the large-scale commercialization of microbial nanotechnology.

The future of microbial nanotechnology is promising and innovative. Advancements in genetic engineering techniques for optimizing microbial strains will increase the efficiency of nanoparticle synthesis, enabling more precise control over nanoparticle properties. The development of scalable production methods, exploration of new applications, and refinement of regulatory practices are key to unlocking the full potential of this field. With continued investment in research and development, microbial nanotechnology has the potential to address some of the world's most pressing challenges, including pollution, resource depletion, and climate change. By focusing on sustainability, environmental responsibility, and interdisciplinary collaboration, microbial nanotechnology can significantly contribute to a cleaner, greener, and more sustainable future, driving innovation across industries and creating a positive impact on global well-being. Through these advancements, microbial nanotechnology will not only enhance industrial practices but also play an essential role in addressing global sustainability goals, benefiting industries and communities worldwide.

REFERENCES

[1]   Grasso G, Zane D, Dragone R. Microbial nanotechnology: challenges and prospects for green biocatalytic synthesis of nanoscale materials for sensoristic and biomedical applications. Nanomaterials 2020; 10: 11. Available from: https://www.mdpi.com/2079-4991/10/1/11/htm

[2]   Salem SS. A mini review on green nanotechnology and its development in biological effects. Arch Microbiol 2023; 205(4): 128.
[http://dx.doi.org/10.1007/s00203-023-03467-2] [PMID: 36944830]

[3]   Salem SS, Fouda A. Green synthesis of metallic nanoparticles and their prospective biotechnological

applications: An overview. Biol Trace Elem Res 2021; 199(1): 344-70.
[http://dx.doi.org/10.1007/s12011-020-02138-3] [PMID: 32377944]

[4]   Shah MP, Bharadvaja N, Kumar L. editors biogenic nanomaterials for environmental sustainability: Principles, practices, and opportunities 2024. Available from: https://link.springer.com/10.1007/978--031-45956-6

[5]   Khan SH. Green nanotechnology for the environment and sustainable development 2020. Available from: https://link.springer.com/chapter/10.1007/978-3-030-17724-9_2
[http://dx.doi.org/10.1007/978-3-030-17724-9_2]

[6]   Garg D, Sridhar K, Stephen Inbaraj B, Chawla P, Tripathi M, Sharma M. Nano-biofertilizer formulations for agriculture: a systematic review on recent advances and prospective applications. Bioengineering 2023; 10(9): 1010-0. Available from: https://www.mdpi.com/2306-5354/10/9/1010/htm
[http://dx.doi.org/10.3390/bioengineering10091010]

[7]   Jain H. From pollution to progress: Groundbreaking advances in clean technology unveiled. Innovation and Green Development 2024; 3(2): 100143.
[http://dx.doi.org/10.1016/j.igd.2024.100143]

[8]   Kanchi S, Ahmed S. Green metal nanoparticles: synthesis, characterization and their applications. The Macabresque 2018; 1-694. Available from: https://onlinelibrary.wiley.com/doi/10.1002/9781119418900

[9]   John MS, Nagoth JA, Ramasamy KP, *et al.* Synthesis of bioactive silver nanoparticles by a pseudomonas strain associated with the antarctic psychrophilic protozoon *euplotes focardii.* Mar Drugs 2020; 18(1): 38.
[http://dx.doi.org/10.3390/md18010038] [PMID: 31947807]

[10]   Alves GB, Melo FL, Oliveira EE, *et al.* Comparative genomic analysis and mosquito larvicidal activity of four *Bacillus thuringiensis serovar israelensis* strains. Sci Rep 2020; 10(1): 5518.
[http://dx.doi.org/10.1038/s41598-020-60670-7] [PMID: 32218451]

[11]   Wang Z, Li N, Zhou X, *et al.* Optimization of fermentation parameters to improve the biosynthesis of selenium nanoparticles by *Bacillus licheniformis* F1 and its comprehensive application. BMC Microbiol 2024; 24(1): 271.
[http://dx.doi.org/10.1186/s12866-024-03410-5] [PMID: 39033096]

[12]   Bhardwaj LK, *et al.* COVID-19 and the interplay with antibacterial drug resistance Available from: https://services.igi-global.com/resolvedoi/resolve.aspx?doi=10.4018/979-8-3693-4139-1.ch010
[http://dx.doi.org/10.4018/979-8-3693-4139-1.ch010]

[13]   Hallaj-Nezhadi S, Hamdipour R, Shahrvirani M. Antimicrobial activity of *bacillus sp.* isolated strains of wild honey. BMC Complement Med Ther 2022; 22(1)

[14]   LaGier MJ, McDaniel M, Ragner A, Castillo A. Identification and characterization of a potential antibiotic producing strain of *Pantoea ananatis.* J Genomics 2022; 10: 26-32.
[http://dx.doi.org/10.7150/jgen.70066] [PMID: 35145564]

[15]   Saravanan M, Gopinath V, Chaurasia MK, Syed A, Ameen F, Purushothaman N. Green synthesis of anisotropic zinc oxide nanoparticles with antibacterial and cytofriendly properties. Microb Pathog 2018; 115: 57-63.
[http://dx.doi.org/10.1016/j.micpath.2017.12.039] [PMID: 29248514]

[16]   Abd Alamer IS, Tomah AA, Ahmed T, Li B, Zhang J. Biosynthesis of silver chloride nanoparticles by rhizospheric bacteria and their antibacterial activity against phytopathogenic bacterium *Ralstonia solanacearum.* Molecules 2021; 27(1): 224.
[http://dx.doi.org/10.3390/molecules27010224] [PMID: 35011455]

[17]   Ivankovic T, Turk H, Hrenovic J, Schauperl Z, Ivankovic M, Ressler A. Antibacterial activity of silver doped hydroxyapatite toward multidrug-resistant clinical isolates of *Acinetobacter baumannii.* J

Hazard Mater 2023; 458: 131867.
[http://dx.doi.org/10.1016/j.jhazmat.2023.131867] [PMID: 37331061]

[18]   Maťátková O, Michailidu J, Miškovská A, Kolouchová I, Masák J, Čejková A. Antimicrobial properties and applications of metal nanoparticles biosynthesized by green methods. Biotechnol Adv. Elsevier Inc. 2022; 58.

[19]   Shah MP, Shah N, Eds. Environmental approach to remediate refractory pollutants from industrial wastewater treatment plant. Elsevier 2024.

[20]   Zayed M, El-Garawani IM, El-Sabbagh SM, *et al.* Structural diversity, LC-MS-MS analysis and potential biological activities of *Brevibacillus laterosporus* extract. Metabolites 2022; 12(11): 1102.
[http://dx.doi.org/10.3390/metabo12111102] [PMID: 36422242]

[21]   Shleeva MO, Kondratieva DA, Kaprelyants AS. *Bacillus licheniformis*: a producer of antimicrobial substances, including antimycobacterials, which are feasible for medical applications. Pharmaceutics. 2023; 15.

[22]   Murhekar S, Wright MH, Greene AC, Brownlie JC, Cock ie. Inhibition of *Shewanella* spp. growth by *Syzygium australe* and *Syzygium luehmannii* extracts: natural methods for the prevention of fish spoilage. J Food Sci Technol 2017; 54(10): 3314-26.
[http://dx.doi.org/10.1007/s13197-017-2782-6] [PMID: 28974817]

[23]   Malhotra A, Dolma K, Kaur N, *et al.* Biosynthesis of gold and silver nanoparticles using a novel marine strain of *Stenotrophomonas*. Bioresour Technol 2013; 142: 727-31.
[http://dx.doi.org/10.1016/j.biortech.2013.05.109] [PMID: 23791020]

[24]   Najjar A, Hassan EA, Zabermawi N, *et al.* Optimizing the catalytic activities of methanol and thermotolerant *Kocuria flava* lipases for biodiesel production from cooking oil wastes. Sci Rep 2021; 11(1): 13659.
[http://dx.doi.org/10.1038/s41598-021-93023-z] [PMID: 34211018]

[25]   Lv Q, Zhang B, Xing X, *et al.* Biosynthesis of copper nanoparticles using *Shewanella loihica* PV-4 with antibacterial activity: Novel approach and mechanisms investigation. J Hazard Mater 2018; 347: 141-9.
[http://dx.doi.org/10.1016/j.jhazmat.2017.12.070] [PMID: 29304452]

[26]   Zou L. On-going applications of shewanella species in microbial electrochemical system for bioenergy, bioremediation and biosensing. World J Microbiol Biotechnol. 2019; 35.

[27]   Zhang Yunfei, Zhao Qiang, Chen Baoliang. Reduction and removal of Cr(VI) in water using biosynthesized palladium nanoparticles loaded *Shewanella oneidensis* MR-1. Sci Total Environ 2022; 805: 150336.
[http://dx.doi.org/10.1016/j.scitotenv.2021.150336]

[28]   Ahmed E, Kalathil S, Shi L, Alharbi O, Wang P. Synthesis of ultra-small platinum, palladium and gold nanoparticles by *Shewanella loihica* PV-4 electrochemically active biofilms and their enhanced catalytic activities. J Saudi Chem Soc 2018; 22(8): 919-29.
[http://dx.doi.org/10.1016/j.jscs.2018.02.002]

[29]   Wang W, Zhang B, He Z. Bioelectrochemical deposition of palladium nanoparticles as catalysts by *Shewanella oneidensis* MR-1 towards enhanced hydrogen production in microbial electrolysis cells. Electrochim Acta 2019; 318: 794-800.
[http://dx.doi.org/10.1016/j.electacta.2019.06.038]

[30]   Yang Z, Liu Z, Dabrowska M, *et al.* Biostimulation of sulfate-reducing bacteria used for treatment of hydrometallurgical waste by secondary metabolites of urea decomposition by *Ochrobactrum sp.* POC9: From genome to microbiome analysis. Chemosphere 2021; 282: 131064.
[http://dx.doi.org/10.1016/j.chemosphere.2021.131064] [PMID: 34118631]

[31]   Camas M, Sazak Camas A, Kyeremeh K. Extracellular synthesis and characterization of gold nanoparticles using *Mycobacterium* sp. BRS2A-AR2 isolated from the aerial roots of the ghanaian

mangrove plant, *Rhizophora racemosa.*. Indian J Microbiol 2018; 58(2): 214-21.
[http://dx.doi.org/10.1007/s12088-018-0710-8] [PMID: 29651181]

[32]   Shah R, Oza G, Pandey S, Sharon M. Biogenic fabrication of gold nanoparticles using *Halomonas salina*. J Microbiol Biotech Res 2012; 2012

[33]   Tuo Y, Liu G, Dong B, *et al.* Microbial synthesis of Pd/Fe$_3$O$_4$, Au/Fe$_3$O$_4$ and PdAu/Fe$_3$O$_4$ nanocomposites for catalytic reduction of nitroaromatic compounds. Sci Rep 2015; 5(1): 13515.
[http://dx.doi.org/10.1038/srep13515] [PMID: 26310728]

[34]   Ikram M, Naeem M, Zahoor M, *et al. Bacillus subtilis*: As an efficient bacterial strain for the reclamation of water loaded with textile azo dye, orange II. Int J Mol Sci 2022; 23(18): 10637.
[http://dx.doi.org/10.3390/ijms231810637] [PMID: 36142543]

[35]   Raj R, Dalei K, Chakraborty J, Das S. Extracellular polymeric substances of a marine bacterium mediated synthesis of CdS nanoparticles for removal of cadmium from aqueous solution. J Colloid Interface Sci 2016; 462: 166-75.
[http://dx.doi.org/10.1016/j.jcis.2015.10.004] [PMID: 26454375]

[36]   Ranjan R, Jadeja V. Isolation, characterization and chromatography based purification of antibacterial compound isolated from rare endophytic actinomycetes *Micrococcus yunnanensis.*. J Pharm Anal 2017; 7(5): 343-7.
[http://dx.doi.org/10.1016/j.jpha.2017.05.001] [PMID: 29404059]

[37]   Lee SY, Yang SB, Choi YM, *et al.* Heat-killed *Mycobacterium paragordonae* therapy exerts an anti-cancer immune response *via* enhanced immune cell mediated oncolytic activity in xenograft mice model. Cancer Lett 2020; 472: 142-50.
[http://dx.doi.org/10.1016/j.canlet.2019.12.028] [PMID: 31874244]

[38]   Chen WC, Hsu CC, Lan JCW, Chang YK, Wang LF, Wei YH. Production and characterization of ectoine using a moderately halophilic strain *Halomonas salina* BCRC17875. J Biosci Bioeng 2018; 125(5): 578-84.
[http://dx.doi.org/10.1016/j.jbiosc.2017.12.011] [PMID: 29331525]

[39]   Kong WQ, Lin JY, He X, *et al.* Reduction pathway and mechanism of chloronitrobenzenes synergistically catalyzed by bioPd and *Shewanella oneidensis* MR-1 assisted by calculation. Chemosphere 2017; 187: 62-9.
[http://dx.doi.org/10.1016/j.chemosphere.2017.07.155] [PMID: 28841432]

[40]   Valenzuela-Ibaceta F, Torres-Olea N, Ramos-Zúñiga J, Dietz-Vargas C, Navarro CA, Pérez-Donoso JM. Minicells as an *Escherichia coli* mechanism for the accumulation and disposal of fluorescent cadmium sulphide nanoparticles. J Nanobiotechnology 2024; 22(1): 78.
[http://dx.doi.org/10.1186/s12951-024-02348-0] [PMID: 38414055]

[41]   Chellaiah ER. Cadmium (heavy metals) bioremediation by *Pseudomonas aeruginosa*: a minireview. Appl Water Sci. Springer Verlag 2018; 8.

[42]   Órdenes-Aenishanslins NA, Saona LA, Durán-Toro VM, Monrás JP, Bravo DM, Pérez-Donoso JM. Use of titanium dioxide nanoparticles biosynthesized by *Bacillus mycoides* in quantum dot sensitized solar cells. Microb Cell Fact 2014; 13(1): 90.
[http://dx.doi.org/10.1186/s12934-014-0090-7] [PMID: 25027643]

[43]   Eid AM, Hassan SED, Hamza MF, *et al.* Photocatalytic, antimicrobial, and cytotoxic efficacy of biogenic silver nanoparticles fabricated by *Bacillus amyloliquefaciens*. Catalysts 2024; 14(7): 419.
[http://dx.doi.org/10.3390/catal14070419]

[44]   Al Laham SA, Al Fadel FM. Antibacterial activity of various plants extracts against antibiotic-resistant *Aeromonas hydrophila*. Jundishapur J Microbiol 2014; 7(7): e11370.
[http://dx.doi.org/10.5812/jjm.11370] [PMID: 25368797]

[45]   Abdel-Maksoud G, Abdel-Nasser M, Hassan SED, Eid AM, Abdel-Nasser A, Fouda A. Biosynthesis of titanium dioxide nanoparticles using probiotic bacterial strain, *Lactobacillus rhamnosus*, and

evaluate of their biocompatibility and antifungal activity. Biomass Convers Biorefin 2023.

[46]   Vaigankar DC, Dubey SK, Mujawar SY, D'Costa A, S K S. Tellurite biotransformation and detoxification by *Shewanella baltica* with simultaneous synthesis of tellurium nanorods exhibiting photo-catalytic and anti-biofilm activity. Ecotoxicol Environ Saf 2018; 165: 516-26.
[http://dx.doi.org/10.1016/j.ecoenv.2018.08.111] [PMID: 30223164]

[47]   Anele A, Obare S, Wei J. Recent Trends and Advances of Co_3O_4 Nanoparticles in Environmental Remediation of Bacteria in Wastewater.. Nanomaterials. MDPI 2022; 12.

[48]   Che L, Dong Y, Wu M, Zhao Y, Liu L, Zhou H. Characterization of selenite reduction by *Lysinibacillus* sp. ZYM-1 and photocatalytic performance of biogenic selenium nanospheres. ACS Sustain Chem& Eng 2017; 5(3): 2535-43.
[http://dx.doi.org/10.1021/acssuschemeng.6b02889]

[49]   Musin EV, Kim AL, Dubrovskii AV, Kudryashova EB, Tikhonenko SA. Decapsulation of dextran by destruction of polyelectrolyte microcapsule nanoscale shell by Bacillus subtilis bacteria. Nanomaterials (Basel) 2019; 10(1): 12.
[http://dx.doi.org/10.3390/nano10010012] [PMID: 31861482]

[50]   Taran M, Rad M, Alavi M. Biosynthesis of TiO_2 and ZnO nanoparticles by *Halomonas elongata* IBRC-M 10214 in different conditions of medium. Bioimpacts 2017; 8(2): 81-9.
[http://dx.doi.org/10.15171/bi.2018.10] [PMID: 29977829]

[51]   Springer Proceedings in Physics. Available from: http://www.springer.com/series/361

[52]   Rauf MA, Owais M, Rajpoot R, Ahmad F, Khan N, Zubair S. Biomimetically synthesized ZnO nanoparticles attain potent antibacterial activity against less susceptible *S. aureus* skin infection in experimental animals. RSC Advances 2017; 7(58): 36361-73.
[http://dx.doi.org/10.1039/C7RA05040B]

[53]   Soliman H, Elsayed A, Dyaa A. Antimicrobial activity of silver nanoparticles biosynthesised by *Rhodotorula* sp. strain ATL72. Egyptian Journal of Basic and Applied Sciences 2018; 5(3): 228-33.
[http://dx.doi.org/10.1016/j.ejbas.2018.05.005]

[54]   Iravani S, Korbekandi H, Mirmohammadi SV, Zolfaghari B. Synthesis of silver nanoparticles: chemical, physical and biological methods. Vol. 9. Res Pharm Sci 2014.

[55]   Xia S, Wu H, Ma D, *et al.* Enhancing biocontrol efficacy of *Cryptococcus laurentii* induced by carboxymethyl cellulose: Metabolic pathways and enzyme activities insights. Postharvest Biol Technol 2025; 219: 113239.
[http://dx.doi.org/10.1016/j.postharvbio.2024.113239]

[56]   Cunha FA, Cunha MCSO, da Frota SM, *et al.* Biogenic synthesis of multifunctional silver nanoparticles from *Rhodotorula glutinis* and *Rhodotorula mucilaginosa*: antifungal, catalytic and cytotoxicity activities. World J Microbiol Biotechnol 2018; 34(9): 127.
[http://dx.doi.org/10.1007/s11274-018-2514-8] [PMID: 30084085]

[57]   Ruas FAD, Amorim SS, Leão VA, Guerra-Sá R. *Rhodotorula mucilaginosa* isolated from the manganese mine water in minas gerais, brazil: potential employment for bioremediation of contaminated water. Water Air Soil Pollut. Springer Science and Business Media Deutschland GmbH 2020; 231.

[58]   Adami A, Ress C, Collini C, Pedrotti S, Lorenzelli L. Development of an integrated electrochemical system for *in vitro* yeast viability testing. Biosens Bioelectron 2013; 40(1): 315-22.
[http://dx.doi.org/10.1016/j.bios.2012.07.070] [PMID: 22944021]

[59]   Mussagy CU, dos Santos A. *Phaffia rhodozyma* biorefinery: A sustainable pathway to obtain natural pigments and production of methane biogas as renewable fuel. Chem Eng J 2023; Oct 1: 473.
[http://dx.doi.org/10.1016/j.cej.2023.145350]

[60]   Manzano M, Vizzini P, Jia K, Adam PM, Ionescu RE. Development of localized surface plasmon resonance biosensors for the detection of *Brettanomyces bruxellensis* in wine. Sens Actuators B Chem

2016; 223: 295-300.
[http://dx.doi.org/10.1016/j.snb.2015.09.099]

[61]   Qu Y, You S, Zhang X, *et al.* Biosynthesis of gold nanoparticles using cell-free extracts of *Magnusiomyces ingens* LH-F1 for nitrophenols reduction. Bioprocess Biosyst Eng 2018; 41(3): 359-67.
[http://dx.doi.org/10.1007/s00449-017-1869-9] [PMID: 29188359]

[62]   Kiayi Z, Lotfabad TB, Heidarinasab A, Shahcheraghi F. Microbial degradation of azo dye carmoisine in aqueous medium using *Saccharomyces cerevisiae* ATCC 9763. J Hazard Mater 2019; 373: 608-19.
[http://dx.doi.org/10.1016/j.jhazmat.2019.03.111] [PMID: 30953978]

[63]   Moghaddam AB, Moniri M, Azizi S, *et al.* Biosynthesis of ZnO nanoparticles by a new *Pichia kudriavzevii* yeast strain and evaluation of their antimicrobial and antioxidant activities. Molecules 2017; 22(6): 872.
[http://dx.doi.org/10.3390/molecules22060872] [PMID: 28538674]

[64]   Lian S, Diko CS, Yan Y, Li Z, Zhang H, Ma Q, *et al.* Characterization of biogenic selenium nanoparticles derived from cell-free extracts of a novel yeast *Magnusiomyces ingens*. 3 Biotech 2019; 9(6)

[65]   Liu J, Zheng D, Zhong L, Gong A, Wu S, Xie Z. Biosynthesis of biocompatibility Ag$_2$Se quantum dots in *Saccharomyces cerevisiae* and its application. Biochem Biophys Res Commun 2021; 544: 60-4.
[http://dx.doi.org/10.1016/j.bbrc.2021.01.071] [PMID: 33516883]

[66]   AbdelRahim K, Mahmoud SY, Ali AM, Almaary KS, Mustafa AEZMA, Husseiny SM. Extracellular biosynthesis of silver nanoparticles using *Rhizopus stolonifer.*. Saudi J Biol Sci 2017; 24(1): 208-16.
[http://dx.doi.org/10.1016/j.sjbs.2016.02.025] [PMID: 28053592]

[67]   Zhang S, Zhao Q, Xue W, *et al.* The isolation and identification of *Candida glabrata* from avian species and a study of the antibacterial activities of Chinese herbal medicine *in vitro*. Poult Sci 2021; 100(4): 101003.
[http://dx.doi.org/10.1016/j.psj.2021.01.026] [PMID: 33676095]

[68]   Kobashigawa JM, Robles CA, Martínez Ricci ML, Carmarán CC. Influence of strong bases on the synthesis of silver nanoparticles (AgNPs) using the ligninolytic fungi *Trametes trogii.*. Saudi J Biol Sci 2019; 26(7): 1331-7.
[http://dx.doi.org/10.1016/j.sjbs.2018.09.006] [PMID: 31762592]

[69]   Díaz-García E, Valenzuela-Quintanar AI, Sánchez-Estrada A, *et al.* Phenolic compounds synthesized by *Trichoderma longibrachiatum* native to semi-arid areas show antifungal activity against phytopathogenic fungi of horticultural interest. Microbiol Res (Pavia) 2024; 15(3): 1425-40.
[http://dx.doi.org/10.3390/microbiolres15030096]

[70]   Husseiny SM, Salah TA, Anter HA. Biosynthesis of size controlled silver nanoparticles by *Fusarium oxysporum*, their antibacterial and antitumor activities. Beni Suef Univ J Basic Appl Sci 2015; 4(3): 225-31.
[http://dx.doi.org/10.1016/j.bjbas.2015.07.004]

[71]   Uras IS, Karsli B, Konuklugil B, Ocsoy I, Demirbas A. Organic–inorganic nanocomposites of *Aspergillus terreus* extract and its compounds with antimicrobial properties. Sustainability (Basel) 2023; 15(5): 4638.
[http://dx.doi.org/10.3390/su15054638]

[72]   Mohanta YK, Nayak D, Biswas K, *et al.* Silver nanoparticles synthesized using wild mushroom show potential antimicrobial activities against food borne pathogens. Molecules 2018; 23(3): 655.
[http://dx.doi.org/10.3390/molecules23030655] [PMID: 29538308]

[73]   Popli D, Anil V, Subramanyam AB. Endophyte fungi, cladosporium species-mediated synthesis of silver nanoparticles possessing *in vitro* antioxidant, anti-diabetic and anti-alzheimer activity. Artif Cells Nanomed Biotechnol 2018; 46(sup1): 676-83.

[74]   Rashid Tavga Sulaiman, Galali Yaseen, Awla Hayman Kakakhan, Sajadi S Mohammad. Recent advances in applications, antimicrobial, cytotoxic activities and their associated mechanism of green silver nanoparticles: a review. Results Chem 2024; 11: 101849.
[http://dx.doi.org/10.1016/j.rechem.2024.101849]

[75]   Mohamed AA, Fouda A, Abdel-Rahman MA, *et al.* Fungal strain impacts the shape, bioactivity and multifunctional properties of green synthesized zinc oxide nanoparticles. Biocatal Agric Biotechnol 2019; 19: 101103.
[http://dx.doi.org/10.1016/j.bcab.2019.101103]

[76]   Siddiqi KS, Husen A. Fabrication of metal nanoparticles from fungi and metal salts: scope and application. Vol. 11, nanoscale research letters. Springer New York 2016; LLC: 1-15.

[77]   Rajput S, Werezuk R, Lange RM, McDermott MT. Fungal isolate optimized for biogenesis of silver nanoparticles with enhanced colloidal stability. Langmuir 2016; 32(34): 8688-97.
[http://dx.doi.org/10.1021/acs.langmuir.6b01813] [PMID: 27466012]

[78]   Mohamed GA, Ibrahim SRM, Alhakamy NA, Aljohani OS. Fusaroxazin, a novel cytotoxic and antimicrobial xanthone derivative from *Fusarium oxysporum*. Nat Prod Res 2022; 36(4): 952-60.
[http://dx.doi.org/10.1080/14786419.2020.1855165] [PMID: 33930988]

[79]   Vivekanandhan P, Swathy K, Murugan AC, Krutmuang P. Insecticidal efficacy of *Metarhizium anisopliae* derived chemical constituents against disease-vector mosquitoes. J Fungi (Basel) 2022; 8(3): 300.
[http://dx.doi.org/10.3390/jof8030300] [PMID: 35330302]

[80]   Leelavathi MS, Vani L, Reena P. Antimicrobial activity of *Trichoderma harzianum* against bacteria and fungi. Int J Curr Microbiol App Sci 2014; 3

[81]   Bonilla JJA, Guerrero DJP, Sáez RGT, *et al.* Green synthesis of silver nanoparticles using maltose and cysteine and their effect on cell wall envelope shapes and microbial growth of *Candida spp.* J Nanosci Nanotechnol 2017; 17(3): 1729-39.
[http://dx.doi.org/10.1166/jnn.2017.12822]

[82]   Chowdhury S, Basu A, Kundu S. Green synthesis of protein capped silver nanoparticles from phytopathogenic fungus *Macrophomina phaseolina* (Tassi) Goid with antimicrobial properties against multidrug-resistant bacteria. Nanoscale Res Lett 2014; 9(1): 365.
[http://dx.doi.org/10.1186/1556-276X-9-365] [PMID: 25114655]

[83]   Elkady FM, Hashem AH, Salem SS, *et al.* Unveiling biological activities of biosynthesized starch/silver-selenium nanocomposite using *Cladosporium cladosporioides* CBS 174.62. BMC Microbiol 2024; 24(1): 78.
[http://dx.doi.org/10.1186/s12866-024-03228-1] [PMID: 38459502]

[84]   Rahman M, Borah SM, Borah PK, *et al.* Deciphering the antimicrobial activity of multifaceted rhizospheric biocontrol agents of solanaceous crops *viz., Trichoderma harzianum* MC2, and *Trichoderma harzianum* NBG. Front Plant Sci 2023; 14: 1141506.
[http://dx.doi.org/10.3389/fpls.2023.1141506] [PMID: 36938007]

[85]   Baazeem A, Almanea A, Manikandan P, Alorabi M, Vijayaraghavan P, Abdel-Hadi A. *In vitro* antibacterial, antifungal, nematocidal and growth promoting activities of *Trichoderma hamatum* fb10 and its secondary metabolites. J Fungi (Basel) 2021; 7(5): 331.
[http://dx.doi.org/10.3390/jof7050331] [PMID: 33923354]

[86]   El Domany EB, Essam TM, Ahmed AE, Farghali AA. Biosynthesis physico-chemical optimization of gold nanoparticles as anti-cancer and synergetic antimicrobial activity using *Pleurotus ostreatus* fungus. J Appl Pharm Sci 2018; 8(5): 119-28.

[87]   Shen W, Qu Y, Pei X, *et al.* Catalytic reduction of 4-nitrophenol using gold nanoparticles biosynthesized by cell-free extracts of *Aspergillus sp*. WL-Au. J Hazard Mater 2017; 321: 299-306.
[http://dx.doi.org/10.1016/j.jhazmat.2016.07.051] [PMID: 27637096]

[88]   El-Zawawy NA, Ali SS, Nouh HS. Exploring the potential of *Rhizopus oryzae* AUMC14899 as a novel endophytic fungus for the production of l-tyrosine and its biomedical applications. Microb Cell Fact 2023; 22(1): 31.
[http://dx.doi.org/10.1186/s12934-023-02041-1] [PMID: 36804031]

[89]   Huang S, Chen H, Li W, Zhu X, Ding W, Li C. Bioactive chaetoglobosins from the mangrove endophytic fungus *Penicillium chrysogenum*. Mar Drugs 2016; 14(10): 172.
[http://dx.doi.org/10.3390/md14100172] [PMID: 27690061]

[90]   Salem SS, Mohamed AA, Gl-Gamal MS, Talat M, Fouda A. Biological decolorization and degradation of azo dyes from textile wastewater effluent by *Aspergillus niger*. Egypt J Chem 2019; 62(10): 1799-813.

[91]   Shamsuzzaman MA, Mashrai A, Khanam H, Aljawfi RN. Biological synthesis of ZnO nanoparticles using *C. albicans* and studying their catalytic performance in the synthesis of steroidal pyrazolines. Arab J Chem 2017; 10: S1530-6.
[http://dx.doi.org/10.1016/j.arabjc.2013.05.004]

[92]   Fouda A, EL-Din Hassan S, Salem SS, Shaheen TI. *In-vitro* cytotoxicity, antibacterial, and UV protection properties of the biosynthesized Zinc oxide nanoparticles for medical textile applications. Microb Pathog 2018; 125: 252-61.
[http://dx.doi.org/10.1016/j.micpath.2018.09.030] [PMID: 30240818]

[93]   Abo Elsoud MM, Al-Hagar OEA, Abdelkhalek ES, Sidkey NM. Synthesis and investigations on tellurium myconanoparticles. Biotechnol Rep (Amst) 2018; 18: e00247.
[http://dx.doi.org/10.1016/j.btre.2018.e00247] [PMID: 29876299]

[94]   Sarkar J, Mollick MMR, Chattopadhyay D, Acharya K. An eco-friendly route of γ-Fe$_2$O$_3$ nanoparticles formation and investigation of the mechanical properties of the HPMC-γ-Fe$_2$O$_3$ nanocomposites. Bioprocess Biosyst Eng 2017; 40(3): 351-9.
[http://dx.doi.org/10.1007/s00449-016-1702-x] [PMID: 27826743]

[95]   Intra B, Mungsuntisuk I, Nihira T, Igarashi Y, Panbangred W. Identification of actinomycetes from plant rhizospheric soils with inhibitory activity against *Colletotrichum spp.*, the causative agent of anthracnose disease. BMC Res Notes 2011; 4(1): 98.
[http://dx.doi.org/10.1186/1756-0500-4-98] [PMID: 21457542]

[96]   Vijayanandan AS, Balakrishnan RM. Biosynthesis of cobalt oxide nanoparticles using endophytic fungus *Aspergillus nidulans*. J Environ Manage 2018; 218: 442-50.
[http://dx.doi.org/10.1016/j.jenvman.2018.04.032] [PMID: 29709813]

[97]   Manivasagan P, Venkatesan J, Senthilkumar K, Sivakumar K, Kim SK. Biosynthesis, antimicrobial and cytotoxic effect of silver nanoparticles using a novel *Nnocardiopsis sp.* mbrc-1. Biomed Res Int 2013; 2013

[98]   Buszewski B, Railean-Plugaru V, Pomastowski P, *et al.* Antimicrobial activity of biosilver nanoparticles produced by a novel *Streptacidiphilus durhamensis* strain. J Microbiol Immunol Infect 2018; 51(1): 45-54.
[http://dx.doi.org/10.1016/j.jmii.2016.03.002] [PMID: 27103501]

[99]   Abd-Elnaby HM, Abo-Elala GM, Abdel-Raouf UM, Hamed MM. Antibacterial and anticancer activity of extracellular synthesized silver nanoparticles from marine *Streptomyces rochei* MHM13. Egypt J Aquat Res 2016; 42(3): 301-12.
[http://dx.doi.org/10.1016/j.ejar.2016.05.004]

[100] Alani F, Moo-Young M, Anderson W. Biosynthesis of silver nanoparticles by a new strain of *Streptomyces sp.* compared with *Aspergillus fumigatus*. World J Microbiol Biotechnol 2012; 28(3): 1081-6.
[http://dx.doi.org/10.1007/s11274-011-0906-0] [PMID: 22805829]

[101] Sholkamy EN, Ahamd MS, Yasser MM, Eslam N. Anti-microbiological activities of bio-synthesized

silver Nano-stars by *Saccharopolyspora hirsuta.*. Saudi J Biol Sci 2019; 26(1): 195-200.
[http://dx.doi.org/10.1016/j.sjbs.2018.02.020] [PMID: 30627051]

[102] El Awady M, Hamed A, El Abacy D, Battah M, Hassan M. Bioactive secondary metabolite from marine *Streptomyces parvulus* strain mda with potential antibacterial activity. Egypt J Chem 2023; 0(0): 0-0.

[103] Shin J, Park S, Lee JS, Lee EJ, Youn HD. Complete genome sequence and comparative analysis of *Streptomyces seoulensis*, a pioneer strain of nickel superoxide dismutase. Genes Genomics 2020; 42(3): 273-81.
[http://dx.doi.org/10.1007/s13258-019-00878-8] [PMID: 31797314]

[104] Bennur T, Ravi Kumar A, Zinjarde SS, Javdekar V. *Nocardiopsis* species: a potential source of bioactive compounds. J Appl Microbiol 2016; 120(1): 1-16.
[http://dx.doi.org/10.1111/jam.12950] [PMID: 26369300]

[105] Mohammed FA, Abu-Hussien SH, Dougdoug NKE, Koutb N, Korayem AS. *Streptomyces fradiae* mitigates the impact of potato virus y by inducing systemic resistance in two egyptian potato (*Solanum tuberosum* L.) cultivars. Microb Ecol 2024; 87(1): 131.
[http://dx.doi.org/10.1007/s00248-024-02437-5] [PMID: 39419884]

[106] Otari S V, Patil RM, Ghosh SJ, Thorat ND, Pawar SH. Intracellular synthesis of silver nanoparticle by actinobacteria and its antimicrobial activity. Spectrochim Acta A Mol Biomol Spectrosc 2015; 136(PB): 1175-80.
[http://dx.doi.org/10.1016/j.saa.2014.10.003]

[107] Al-Dhabi NA, Mohammed Ghilan AK, Arasu MV. Characterization of silver nanomaterials derived from marine *Streptomyces sp.* Al-Dhabi-87 and its *in vitro* application against multidrug resistant and extended-spectrum beta-lactamase clinical pathogens. Nanomaterials (Basel) 2018; 8(5): 279.
[http://dx.doi.org/10.3390/nano8050279] [PMID: 29701657]

[108] Składanowski M, Wypij M, Laskowski D, Golińska P, Dahm H, Rai M. Silver and gold nanoparticles synthesized from *Streptomyces sp.* isolated from acid forest soil with special reference to its antibacterial activity against pathogens. J Cluster Sci 2017; 28(1): 59-79.
[http://dx.doi.org/10.1007/s10876-016-1043-6]

[109] Sowani H, Mohite P, Munot H, *et al.* Green synthesis of gold and silver nanoparticles by an actinomycete *Gordonia amicalis* HS-11: Mechanistic aspects and biological application. Process Biochem 2016; 51(3): 374-83.
[http://dx.doi.org/10.1016/j.procbio.2015.12.013]

[110] Hassan SELD, Salem SS, Fouda A, Awad MA, El-Gamal MS, Abdo AM. New approach for antimicrobial activity and bio-control of various pathogens by biosynthesized copper nanoparticles using endophytic actinomycetes. Journal of Radiation Research and Applied Sciences 2018; 11(3): 262-70.
[http://dx.doi.org/10.1016/j.jrras.2018.05.003]

[111] Hassan SED, Fouda A, Radwan AA, *et al.* Endophytic actinomycetes *Streptomyces spp* mediated biosynthesis of copper oxide nanoparticles as a promising tool for biotechnological applications. J Biol Inorg Chem 2019; 24(3): 377-93.
[http://dx.doi.org/10.1007/s00775-019-01654-5] [PMID: 30915551]

[112] Balagurunathan R, Radhakrishnan M, Rajendran RB, Velmurugan D. Biosynthesis of gold nanoparticles by actinomycete *Streptomyces viridogens* strain HM10. Indian J Biochem Biophys 2011; 48(5): 331-5.
[PMID: 22165291]

[113] Khalil MA, El-Shanshoury AERR, Alghamdi MA, Sun J, Ali SS. *Streptomyces catenulae* as a novel marine actinobacterium mediated silver nanoparticles: characterization, biological activities, and proposed mechanism of antibacterial action. Front Microbiol 2022; 13: 833154.
[http://dx.doi.org/10.3389/fmicb.2022.833154] [PMID: 35572675]

[114] Manivasagan P, Alam MS, Kang KH, Kwak M, Kim SK. Extracellular synthesis of gold bionanoparticles by *Nocardiopsis sp.* and evaluation of its antimicrobial, antioxidant and cytotoxic activities. Bioprocess Biosyst Eng 2015; 38(6): 1167-77.
[http://dx.doi.org/10.1007/s00449-015-1358-y] [PMID: 25645365]

[115] El-Naggar NEA, El-Sawah AA, Elmansy MF, *et al.* Process optimization for gold nanoparticles biosynthesis by *Streptomyces albogriseolus* using artificial neural network, characterization and antitumor activities. Sci Rep 2024; 14(1): 4581.
[http://dx.doi.org/10.1038/s41598-024-54698-2] [PMID: 38403677]

[116] Presentato A, Piacenza E, Anikovskiy M, Cappelletti M, Zannoni D, Turner RJ. Biosynthesis of selenium-nanoparticles and -nanorods as a product of selenite bioconversion by the aerobic bacterium *Rhodococcus aetherivorans* BCP1. N Biotechnol 2018; 41: 1-8.
[http://dx.doi.org/10.1016/j.nbt.2017.11.002] [PMID: 29174512]

[117] El-Mahdy A, Shaaban M. Antifungal activity of biologically synthesized metallic nanoparticles from streptomyces species SS7. Egypt J Med Microbiol 2024; 33(2): 0.
[http://dx.doi.org/10.21608/ejmm.2024.274035.1229]

[118] Karthik L, Kumar G, Keswani T, Bhattacharyya A, Reddy BP, Rao KVB. Marine actinobacterial mediated gold nanoparticles synthesis and their antimalarial activity. Nanomedicine 2013; 9(7): 951-60.
[http://dx.doi.org/10.1016/j.nano.2013.02.002] [PMID: 23434675]

[119] Ramya S, Shanmugasundaram T, Balagurunathan R. Biomedical potential of actinobacterially synthesized selenium nanoparticles with special reference to anti-biofilm, anti-oxidant, wound healing, cytotoxic and anti-viral activities. J Trace Elem Med Biol 2015; 32: 30-9.
[http://dx.doi.org/10.1016/j.jtemb.2015.05.005] [PMID: 26302909]

[120] Ahmad MS, Yasser MM, Sholkamy EN, Ali AM, Mehanni MM. Anticancer activity of biostabilized selenium nanorods synthesized by *Streptomyces bikiniensis* strain Ess_amA-1. Int J Nanomedicine 2015; 10: 3389-401.
[PMID: 26005349]

[121] Balraj B, Senthilkumar N, Siva C, *et al.* Synthesis and characterization of Zinc Oxide nanoparticles using marine *Streptomyces sp.* with its investigations on anticancer and antibacterial activity. Res Chem Intermed 2017; 43(4): 2367-76.
[http://dx.doi.org/10.1007/s11164-016-2766-6]

[122] Fatima R, Priya M, Indurthi L, Radhakrishnan V, Sudhakaran R. Biosynthesis of silver nanoparticles using red algae *Portieria hornemannii* and its antibacterial activity against fish pathogens. Microb Pathog 2020; 138: 103780.
[http://dx.doi.org/10.1016/j.micpath.2019.103780] [PMID: 31622663]

[123] Bhuyar P, Rahim MHA, Sundararaju S, Ramaraj R, Maniam GP, Govindan N. Synthesis of silver nanoparticles using marine macroalgae *Padina sp.* and its antibacterial activity towards pathogenic bacteria. Beni Suef Univ J Basic Appl Sci 2020; 9(1)

[124] Husain S, Afreen S, Hemlata , Yasin D, Afzal B, Fatma T. Cyanobacteria as a bioreactor for synthesis of silver nanoparticles-an effect of different reaction conditions on the size of nanoparticles and their dye decolorization ability. J Microbiol Methods 2019; 162: 77-82.
[http://dx.doi.org/10.1016/j.mimet.2019.05.011] [PMID: 31132377]

[125] González-Ballesteros N, Rodríguez-Argüelles MC, Prado-López S, *et al.* Macroalgae to nanoparticles: Study of *Ulva lactuca* L. role in biosynthesis of gold and silver nanoparticles and of their cytotoxicity on colon cancer cell lines. Mater Sci Eng C 2019; 97: 498-509.
[http://dx.doi.org/10.1016/j.msec.2018.12.066] [PMID: 30678937]

[126] Abdel-Raouf N, Al-Enazi NM, Ibraheem IBM, Alharbi RM, Alkhulaifi MM. Biosynthesis of silver nanoparticles by using of the marine brown alga *Padina pavonia* and their characterization. Saudi J Biol Sci 2019; 26(6): 1207-15.

[http://dx.doi.org/10.1016/j.sjbs.2018.01.007] [PMID: 31516350]

[127] Pugazhendhi A, Prabakar D, Jacob JM, Karuppusamy I, Saratale RG. Synthesis and characterization of silver nanoparticles using *Gelidium amansii* and its antimicrobial property against various pathogenic bacteria. Microb Pathog 2018; 114: 41-5.
[http://dx.doi.org/10.1016/j.micpath.2017.11.013] [PMID: 29146498]

[128] Aboelfetoh EF, El-Shenody RA, Ghobara MM. Eco-friendly synthesis of silver nanoparticles using green algae (*Caulerpa serrulata*): reaction optimization, catalytic and antibacterial activities. Environ Monit Assess 2017; 189(7): 349.
[http://dx.doi.org/10.1007/s10661-017-6033-0] [PMID: 28646435]

[129] WA F. Green Biosynthesis of silver nanoparticles using marine red algae *Acanthophora specifera* and its antibacterial activity. J Nanomed Nanotechnol 2016; 07(06)
[http://dx.doi.org/10.4172/2157-7439.1000409]

[130] de Aragão AP, de Oliveira TM, Quelemes PV, *et al.* Green synthesis of silver nanoparticles using the seaweed *Gracilaria birdiae* and their antibacterial activity. Arab J Chem 2019; 12(8): 4182-8.
[http://dx.doi.org/10.1016/j.arabjc.2016.04.014]

[131] Madhiyazhagan P, Murugan K, Kumar AN, *et al.* S argassum *muticum*-synthesized silver nanoparticles: an effective control tool against mosquito vectors and bacterial pathogens. Parasitol Res 2015; 114(11): 4305-17.
[http://dx.doi.org/10.1007/s00436-015-4671-0] [PMID: 26281786]

[132] Ebrahimzadeh Z, Salehzadeh A, Naeemi AS, Jalali A. Silver nanoparticles biosynthesized by Anabaena flos-aquae enhance the apoptosis in breast cancer cell line. Bull Mater Sci 2020; 43(1): 92.
[http://dx.doi.org/10.1007/s12034-020-2064-1]

[133] Moshfegh A, Jalali A, Salehzadeh A, Sadeghi Jozani A. Biological synthesis of silver nanoparticles by cell-free extract of *Polysiphonia* algae and their anticancer activity against breast cancer MCF-7 cell lines. Micro & Nano Lett 2019; 14(5): 581-4.
[http://dx.doi.org/10.1049/mnl.2018.5260]

[134] Senthilkumar P, Surendran L, Sudhagar B, Ranjith Santhosh Kumar DS, Kumar Ranjith Santhosh DS. Facile green synthesis of gold nanoparticles from marine algae *Gelidiella acerosa* and evaluation of its biological potential. SN Appl Sci 2019; 1(4)

[135] Colin JA, Pech-Pech ie, Oviedo M, Águila SA, Romo-Herrera JM, Contreras OE. Gold nanoparticles synthesis assisted by marine algae extract: Biomolecules shells from a green chemistry approach. Chem Phys Lett 2018; 708: 210-5.
[http://dx.doi.org/10.1016/j.cplett.2018.08.022]

[136] González-Ballesteros N, Prado-López S, Rodríguez-González JB, Lastra M, Rodríguez-Argüelles MC. Green synthesis of gold nanoparticles using brown algae *Cystoseira baccata*: Its activity in colon cancer cells. Colloids Surf B Biointerfaces 2017; 153: 190-8.
[http://dx.doi.org/10.1016/j.colsurfb.2017.02.020] [PMID: 28242372]

[137] Li L, Zhang Z. Biosynthesis of gold nanoparticles using green alga *Pithophora oedogonia* with their electrochemical performance for determining carbendazim in soil. Int J Electrochem Sci 2016; 11(6): 4550-9.
[http://dx.doi.org/10.20964/2016.06.13]

[138] Ramakrishna M, Rajesh Babu D, Gengan RM, Chandra S, Nageswara Rao G. Green synthesis of gold nanoparticles using marine algae and evaluation of their catalytic activity. J Nanostructure Chem 2016; 6(1): 1-13.
[http://dx.doi.org/10.1007/s40097-015-0173-y]

[139] Uma Suganya KS, Govindaraju K, Ganesh Kumar V, *et al.* Blue green alga mediated synthesis of gold nanoparticles and its antibacterial efficacy against Gram positive organisms. Mater Sci Eng C 2015; 47: 351-6.
[http://dx.doi.org/10.1016/j.msec.2014.11.043] [PMID: 25492207]

[140] Mishra V, Arya A, Chundawat TS. High catalytic activity of Pd nanoparticles synthesized from green alga *Chlorella vulgaris* in buchwald-hartwig synthesis of N-Aryl piperazines. Curr Organocatal 2019; 7(1): 23-33.
[http://dx.doi.org/10.2174/2213337206666190515091945]

[141] Prasad BSN, Padmesh T, Kumar VG, Govindaraju K. Seaweed (*Sargassum wightii* Greville) assisted green synthesis of palladium nanoparticles. Research Journal of Pharmacy and Technology 2015; 8(4): 392-4.
[http://dx.doi.org/10.5958/0974-360X.2015.00066.9]

[142] Gu H, Chen X, Chen F, Zhou X, Parsaee Z. Ultrasound-assisted biosynthesis of CuO-NPs using brown alga *Cystoseira trinodis*: Characterization, photocatalytic AOP, DPPH scavenging and antibacterial investigations. Ultrason Sonochem 2018; 41: 109-19.
[http://dx.doi.org/10.1016/j.ultsonch.2017.09.006] [PMID: 29137732]

[143] Arya A, Gupta K, Chundawat TS, Vaya D. Biogenic synthesis of copper and silver nanoparticles using green alga *Botryococcus braunii* and its antimicrobial activity. Bioinorg Chem Appl 2018; 2018
[http://dx.doi.org/10.1155/2018/7879403]

[144] Vishnu S, Ramaswamy P, Narendhran S, Sivaraj R. Potentiating effect of ecofriendly synthesis of copper oxide nanoparticles using brown alga: antimicrobial and anticancer activities. Vol. 39. Bull Mater Sci 2016.

[145] Abboud Y, Saffaj T, Chagraoui A, *et al.* Biosynthesis, characterization and antimicrobial activity of Copper Oxide Nanoparticles (CONPs) produced using brown alga extract (*Bifurcaria bifurcata*). Appl Nanosci 2014; 4(5): 571-6.
[http://dx.doi.org/10.1007/s13204-013-0233-x]

[146] Salem DMSA, Ismail MM, Aly-Eldeen MA. Biogenic synthesis and antimicrobial potency of Iron Oxide (Fe$_3$O$_4$) nanoparticles using algae harvested from the Mediterranean Sea, Egypt. Egypt J Aquat Res 2019; 45(3): 197-204.
[http://dx.doi.org/10.1016/j.ejar.2019.07.002]

[147] El-Kassas HY, Aly-Eldeen MA, Gharib SM. Green synthesis of Iron Oxide (Fe$_3$O$_4$) nanoparticles using two selected brown seaweeds: Characterization and application for lead bioremediation. Acta Oceanol Sin 2016; 35(8): 89-98.
[http://dx.doi.org/10.1007/s13131-016-0880-3]

[148] Mahdavi M, Namvar F, Ahmad M, Mohamad R. Green biosynthesis and characterization of magnetic iron oxide (Fe$_3$O$_4$) nanoparticles using seaweed (*Sargassum muticum*) aqueous extract. Molecules 2013; 18(5): 5954-64.
[http://dx.doi.org/10.3390/molecules18055954] [PMID: 23698048]

[149] Kahzad N, Salehzadeh A. Green synthesis of CuFe$_2$O$_4$@Ag nanocomposite using the *Chlorella vulgaris* and evaluation of its effect on the expression of nora efflux pump gene among *Staphylococcus aureus* strains. Biol Trace Elem Res 2020; 198(1): 359-70.
[http://dx.doi.org/10.1007/s12011-020-02055-5] [PMID: 32067154]

[150] Shokoofeh N, Moradi-Shoeili Z, Naeemi AS, Jalali A, Hedayati M, Salehzadeh A. Biosynthesis of Fe$_3$O$_4$@Ag nanocomposite and evaluation of its performance on expression of norA and norB efflux pump genes in ciprofloxacin-resistant *Staphylococcus aureus*. Biol Trace Elem Res 2019; 191(2): 522-30.
[http://dx.doi.org/10.1007/s12011-019-1632-y] [PMID: 30788722]

[151] Hussein AK. Applications of nanotechnology in renewable energies—A comprehensive overview and understanding. Renew Sustain Energy Rev 2015; 42: 460-76.
[http://dx.doi.org/10.1016/j.rser.2014.10.027]

[152] Malik S, Dhasmana A, Preetam S, Mishra YK, Chaudhary V, Bera SP, *et al.* Exploring microbial-based green nanobiotechnology for wastewater remediation: A sustainable strategy. Nanomaterials. MDPI 2022; 12.

[http://dx.doi.org/10.3390/nano12234187]

[153]   Kumari S, Tehri N, Gahlaut A, Hooda V. Actinomycetes mediated synthesis, characterization, and applications of metallic nanoparticles. Inorg Nano-Met Chem. Taylor and Francis Ltd 2020; 51: pp. 1386-95.
[http://dx.doi.org/10.1080/24701556.2020.1835978]

[154]   Nasrollahzadeh M, Sajadi SM, Issaabadi Z, Sajjadi M. Biological sources used in green nanotechnology. Interface Science and Technology. Elsevier B.V. 2019; pp. 81-111.
[http://dx.doi.org/10.1016/B978-0-12-813586-0.00003-1]

[155]   Nath D, Banerjee P. Green nanotechnology – A new hope for medical biology. Environ Toxicol Pharmacol 2013; 36(3): 997-1014.
[http://dx.doi.org/10.1016/j.etap.2013.09.002] [PMID: 24095717]

[156]   Nasrollahzadeh M, Sajadi SM, Sajjadi M, Issaabadi Z. Applications of nanotechnology in daily life. Interface Science and Technology. Elsevier 2019; pp. 113-43. Internet Available from: https://linkinghub.elsevier.com/retrieve/pii/B9780128135860000043

[157]   Yadav A, Rai M. Phytosynthesis of silver nanoparticles and their role as antimicrobials. Synthesis and Applications of Nanoparticles 2022; 357-69. Available from: https://link.springer.com/chapter/10.1007/978-981-16-6819-7_16
[http://dx.doi.org/10.1007/978-981-16-6819-7_16]

[158]   Bahrulolum H, Nooraei S, Javanshir N, Tarrahimofrad H, Mirbagheri VS, Easton AJ, *et al.* Green synthesis of metal nanoparticles using microorganisms and their application in the agrifood sector. J Nanobiotechnol 2021; 19
[http://dx.doi.org/10.1186/s12951-021-00834-3]

[159]   Comer Benjamin M, Fuentes Porfirio, Dimkpa Christian O, *et al.* Prospects and challenges for solar fertilizers. Joule 2019; 3(7): 1578-605.
[http://dx.doi.org/10.1016/j.joule.2019.05.001]

[160]   Nazir R, Ayub Y, Tahir L. Green-nanotechnology for precision and sustainable agriculture. Biogenic Nano-Particles and their Use in Agro-ecosystems 2020; 317-57. Available from: https://link.springer.com/chapter/10.1007/978-981-15-2985-6_18
[http://dx.doi.org/10.1007/978-981-15-2985-6_18]

[161]   Nanotechnology research directions for societal needs in. 2020 Available from: http://www.springer.com/series/8882

[162]   Malik S, Muhammad K, Waheed Y. Nanotechnology: a revolution in modern industry. Molecules 2023; 28(2): 661-1.
[http://dx.doi.org/10.3390/molecules28020661]

[163]   Hou J, Wang L, Wang C, *et al.* Toxicity and mechanisms of action of titanium dioxide nanoparticles in living organisms. J Environ Sci (China) 2019; 75: 40-53.
[http://dx.doi.org/10.1016/j.jes.2018.06.010] [PMID: 30473306]

[164]   Apollon W. An overview of microbial fuel cell technology for sustainable electricity production. Membranes. Multidisciplinary Digital Publishing Institute (MDPI) 2023; 13.
[http://dx.doi.org/10.3390/membranes13110884]

[165]   Grasso G, Zane D, Dragone R. Microbial nanotechnology: Challenges and prospects for green biocatalytic synthesis of nanoscale materials for sensoristic and biomedical applications. Nanomaterials. MDPI AG 2020; 10.
[http://dx.doi.org/10.3390/nano10010011] [PMID: 31861471] [PMCID: PMC7023511]

[166]   Kumar A, Tyagi PK, Tyagi S, Ghorbanpour M. Integrating green nanotechnology with sustainable development goals: a pathway to sustainable innovation. Discover Sustainability. Springer Nature 2024; 5.
[http://dx.doi.org/10.1007/s43621-024-00610-x]

[167] Nizamani Mir Muhammad, Hughes Alice C, Zhang Hai-Li, Wang Yong. Revolutionizing agriculture with nanotechnology: innovative approaches in fungal disease management and plant health monitoring. Sci Total Environ 2024; 928: 172473.
[http://dx.doi.org/10.1016/j.scitotenv.2024.172473]

[168] Haydar Md Salman, Ghosh Dibakar, Roy Swarnendu. Slow and controlled release nanofertilizers as an efficient tool for sustainable agriculture: recent understanding and concerns. Plant Nano Biol 2024; 7: 100058.
[http://dx.doi.org/10.1016/j.plana.2024.100058]

[169] Guo KW. Green nanotechnology of trends in future energy: a review. Int J Energy Res 2012; 36(1): 1-17. Available from: https://onlinelibrary.wiley.com/doi/full/10.1002/er.1928 [Internet].
[http://dx.doi.org/10.1002/er.1928]

[170] Qiu Chong, Wu Yanyan, Guo Qiuyan, *et al.* Preparation and application of calcium phosphate nanocarriers in drug delivery. Mater Today Bio 2022; 17: 100501.
[http://dx.doi.org/10.1016/j.mtbio.2022.100501]

[171] Kour D, Khan SS, Kumari S, *et al.* Microbial nanotechnology for agriculture, food, and environmental sustainability: current status and future perspective. Folia Microbiol 2024; 69(3): 491-520.
[http://dx.doi.org/10.1007/s12223-024-01147-2]

[172] Abd-Elsalam KA. Special issue: microbial nanotechnology. Microorganisms. Multidisciplinary Digital Publishing Institute (MDPI) 2024; 12.
[http://dx.doi.org/10.3390/microorganisms12020352]

[173] Zikalala NE, Azizi S, Zikalala SA, Kamika I, Maaza M, Zinatizadeh AA, *et al.* An evaluation of the biocatalyst for the synthesis and application of zinc oxide nanoparticles for water remediation—a review. Catalysts 2022; 12(11): 1442-2. Available from: https://www.mdpi.com/2073-4344/12/11/1442/htm

[174] Vinayak V, Joshi K B, Gordon R, Schoefs B. Diatom nanotechnology: progress and emerging applications. 2017; 55-78.
[http://dx.doi.org/10.3390/bioengineering10091010]

[175] Dhanker R, Hussain T, Tyagi P, Singh KJ, Kamble SS. The emerging trend of bio-engineering approaches for microbial nanomaterial synthesis and its applications. Frontiers in Microbiology. Frontiers Media S.A. 2021; 12.
[http://dx.doi.org/10.3389/fmicb.2021.638003]

[176] Salem SS, Badawy MSEM, Al-Askar AA, Arishi AA, Elkady FM, Hashem AH. Green biosynthesis of selenium nanoparticles using orange peel waste: characterization, antibacterial and antibiofilm activities against multidrug-resistant bacteria. Life 2022; 12(6): 893-3. Available from: https://www.mdpi.com/2075-1729/12/6/893/htm

[177] Govindasamy R, Gayathiri E, Sankar S, Venkidasamy B, Prakash P, Rekha K, *et al.* Emerging trends of nanotechnology and genetic engineering in cyanobacteria to optimize production for future applications. Life. Life. MDPI 2022; 12.
[http://dx.doi.org/10.3390/life12122013]

CHAPTER 6

Microbial Innovations for Sustainable Biocontrol and Bioremediation: Exploring Strategies in Wastewater Treatment and Environmental Restoration

Shanvi Rana[1] and **Geetansh Sharma**[1,*]

[1] *School of Bioengineering and Food Technology, Shoolini University, Solan, Himachal Pradesh-173229, India*

Abstract: Microbial-based approaches have emerged as effective solutions for sustainable wastewater treatment and environmental restoration. Bioremediation, utilising diverse microbial agents, is pivotal in mitigating pollution by degrading organic contaminants, reducing heavy metals, and treating industrial and agricultural effluents. This chapter explores innovative microbial strategies, such as bio-stimulation, bioaugmentation, and biosurfactants produced by strains like *Bacillus thuringiensis* and *Bacillus toyonensis*. These biosurfactants demonstrate high stability across varying pH, temperature, and salinity, making them suitable for oil residue and pathogen remediation applications. Additionally, this chapter delves into the symbiotic potential of endophytic microbes, which not only enhance plant resilience to pests but also contribute to bioremediation through the degradation of pollutants in the rhizosphere. Rhizoremediation, a key focus area, emphasises the synergistic interactions between plant roots and microbial communities for contaminant removal. This chapter highlights sustainable wastewater treatment and environmental conservation approaches by examining these microbial insights, promoting a shift towards eco-friendly and biologically-driven solutions.

Keywords: Bioremediation, Biosurfactants, Endophytic microbes, Microbial agents, Rhizoremediation, Sustainable wastewater treatment.

INTRODUCTION

The rapid growth of industrialisation, urbanisation, and agricultural expansion has contributed significantly to environmental pollution, mainly water contamination, which remains one of the most challenging issues globally. The introduction of hazardous pollutants such as pharmaceutical compounds, heavy metals, synthetic

---

* **Corresponding author Geetansh Sharma:** School of Bioengineering and Food Technology, Shoolini University, Solan, Himachal Pradesh-173229, India; E-mail: geetanshsharma.gs@gmail.com

Harshita Jain & Maulin P. Shah (Eds.)

dyes, and pathogenic microorganisms in water bodies has been through the release of untreated or inadequately treated wastewater from various sectors, including municipal, industrial, and agricultural sources. Even at trace levels, these pollutants pose significant threats to aquatic ecosystems and human health, making effective wastewater treatment methods essential [1, 2].

Traditional methods of wastewater treatment, such as coagulation, ion exchange, membrane filtration, and reverse osmosis, are widely used, yet these techniques often suffer from significant energy consumption, low cost-effectiveness, and the production of hazardous by-products, which makes an eco-friendly and sustainable solution indispensable [3]. Bioremediation, which has shown promise in using microorganisms for the degradation or immobilisation of pollutants, thus is a promising solution. This approach is particularly appealing because it is cost-effective, versatile, and environmentally sustainable [4]. Microbial bioremediation utilises the metabolic potential of Indigenous microbes, which can oxidise, transform, or immobilise contaminants, effectively restoring polluted environments [5]. Moreover, innovative approaches like genetically modified organisms and microbial consortia have improved the effectiveness of bioremediation [6].

Out of the vast number of microbes with potential applications in treating wastewater, fungi, such as *Trichoderma sp.*, have gained special consideration because of their strong enzyme production and metabolic diversity. Fungal species are regarded to be efficient in the depolymerization of complex pollutants, such as dyestuffs and pharmaceutical components, making them good biological remediation agents in industry-specific wastewater [7, 8]. These experiments establish that *T. harzianum* can effectively remove persistent pollutants, including crystal violet and acetaminophen, from industrial effluents [9, 10].

This chapter discusses the importance of *Trichoderma species* in the bioremediation of industrial wastewater, focusing on their ecological resilience, enzymatic functions, and feasibility for large-scale implementation. Integrating morphological and molecular methodologies underscores the precise identification and taxonomic categorisation of *Trichoderma species* isolated from industrial environments. The aim is to improve their understanding of potential ecological functions in wastewater management, which will present a sustainable approach to the increasing global issue of water contamination.

BIOSORBENTS FOR WASTEWATER BIOREMEDIATION

Typically, biosorbents are divided into three categories based on their source, including natural, biological, and waste-based sources (Table **1**). Many biomaterials that are derived from natural sources have been used as biosorbents for the removal of pollutants. Biomass derived from microbes, plants, animals, and their by-products has been widely studied due to their effectiveness in pollutant removal [1, 2]. Recently, the focus has shifted towards the use of agricultural waste materials, polysaccharides, and industrial process by-products [3]. Among the materials in this category, chitosan, a naturally occurring amino polysaccharide, has received considerable attention due to its higher amino and hydroxyl functional groups, making it particularly effective in the removal of various aquatic pollutants. Furthermore, biological compounds such as bacteria, cyanobacteria, and algae, which cover microalgae and macroalgae, yeasts, fungi, and lichens, have been recognized for their ability to adsorb and recover heavy metal ions. This is because these compounds exhibit superior efficiency, cost-effectiveness, and large availability. They consist of a high concentration of chelating functional groups, which greatly enhance their affinity toward the metal ions [4, 11].

Table 1. Classification, categories, and important examples of bio-adsorbents.

S.NO.	Microbial Technologies	Definition	Pollutants	Polluted Sites	Advantages	Disadvantages	References
1	Microbial Bio-remediation	This is an eco-friendly approach to waste management in which the natural capabilities of algae, fungi, and bacteria are utilized to remove organic and inorganic types of pollutants from industrial waste.	Organic [phenols, chlorophenols, azo dyes, endocrine-disrupting chemicals, polyaromatic hydrocarbons, pesticides, persistent organic pollutants, polychlorinated biphenyls] and inorganic pollutants [cadmium, chromium, lead, arsenic, mercury]; radionuclides, industrial effluents, solid waste treatment.	Contaminated soil, water, and wastewater.	Eco-friendly and makes use of natural processes for effective degradation of pollutants.	It is sometimes time-consuming and usually dependent on environmental conditions or the nature of the pollutant.	[112]
2	Bio-stimulation	Nutrients or changes in environmental conditions induce bioremediation while enhancing microbial activity.	Various pollutants requiring degradation *via* stimulated native microorganisms.	Aquifers	Makes effective use of naturally adapted in-situ microorganisms to perform remediation.	Its efficiency is governed by the right delivery of nutrients or oxidizing agents.	[113]

(Table 1) cont.....

S.NO.	Microbial Technologies	Definition	Pollutants	Polluted Sites	Advantages	Disadvantages	References
3	Bio-augmentation	Incorporation of specific microorganisms to increase the degradation rate of pollutants.	Tetrachloroethylene, and Trichloroethylene.	Soil and groundwaters.	Microbes introduce good microbial strains, enhancing pollutant degradation.	Needs proper choice and maintenance of the microbes to be introduced.	[114]
4	Electro-bioremediation	This widely practised hybrid technology combines bioremediation and electro-kinetics to treat environmental pollution. Electrokinetic phenomena are utilized to improve the transport as well as the orientation of microorganisms and pollutants, resulting in a faster biodegradation process.	Organic pollutants, inorganic pollutants, and industrial effluents.	Contaminated soil and wastewater.	Improves the bioavailability of pollutants and microbial activity, thereby accelerating the remediation process.	They require careful control of electrical parameters and are quite energy-intensive.	[116]
5	Bio-sparging	A type of remediation that injects air at a pressure below the water table to increase the oxygen levels in the aquifer to enhance biodegradation by native bacteria.	Kerosene, Diesel, BTEX compounds, and Naphthalene.	Groundwater contamination.	It accelerates the natural biodegradation of hydrocarbons effectively.	Limited by soil permeability and pollutant biodegradability.	[115]
6	Bio-venting	A remediation technique for stimulating in-situ aerobic biodegradation through enhancement of oxygen flow and maintenance of suitable conditions for the soil microbial community.	Oil products.	Contaminated soils.	Facilitates efficient, natural pollutant degradation without significant disturbance.	Limited to aerobically degradable pollutants, which requires reliance on soil permeability.	[117]

Besides the aforementioned natural biosorbents, many other biomaterials have been widely studied in academic literature. Some of these include rice husks [1], coconut shells [2], plant barks [11], leaves [12], sawdust [13], sugarcane bagasse [14], and peat moss [9]. One example is fly ash, which is a combustion by-product from coal. There has been research into the potential for using it as a sorbent [7, 10]. Fly ash is highly alkaline and develops a negative surface charge at pH values above neutral so that it can strip metal ions from aqueous solutions by mechanisms such as precipitation, electrostatic attraction, and ion exchange [8, 12]. A study shows that ashes obtained from biological materials, such as wood and bone ash, can be effective alternatives to conventional adsorbents used in wastewater treatment [12].

TYPES OF BIOSORBENTS

Natural Biosorbents

Several naturally occurring materials with natural adsorbent characteristics have been readily available in the environment. Although many naturally occurring adsorbents exist, clay, zeolite, and siliceous materials have been reported to be effective for removing varied pollutants from wastewater or water, such as dyes, heavy metals, pharmaceutical compounds, and other organic contaminants.

Clay

Clay is the most widely available bio-sorbent in nature. The qualities, including low price, extreme adsorption capabilities, high porosity, an expansive surface area, ease of chemical alteration, and usability for treatment, collectively make it an indispensable adsorbent [13, 14]. It has a stratified structure. These, based on variation in their configurations, come within one of the following classes: [i] Kaolinite (including serpentine, pyrophyllite, vermiculite, and sepiolite), [ii] Micas, notably illite, and [iii] smectites, including, for example, montmorillonite and saponite [18]. Among the three types, montmorillonite has the smallest size of crystallites; thus, it has the highest surface area and displays the maximum cation exchange capacity. Studies have claimed that montmorillonite is 20 times cheaper than activated carbon, with the greatest cation exchange features and surface area [15, 16]. Clay has a net negative charge on its surface, due to which contaminants can be neutralised by adsorption. It has been shown that clay is adsorbable by anions after some modification on its surface. Pure clay minerals (according to Ahmaruzzaman M. A) have not been found to exhibit any attraction towards the oxyanions of chromate; however, after exposure to a cationic surfactant, HDTMA bromide, it was found that a large amount of chromate got adsorbed to clay minerals including kaolinite, montmorillonite, and pillared montmorillonite [17].

Siliceous Materials

Naturally occurring siliceous materials include dolomite, several types of glass, perlite silica beads, and alunite [18]. The presence of silanol groups on the surface of the silica beads enhances the chemical reactivity of their hydrophilic surfaces and enables irreversible non-specific adsorption [14]. Their large availability, accessibility, economic feasibility, mechanical strength, large surface area, and porous nature make them good adsorbents [19]. Since they are reported to resist alkaline solutions, siliceous materials can only be employed in acidic environments with pH below 8, which is observed by a study [20]. Alunite is one of the primary siliceous adsorbents widely used for dye adsorption [21]. This mineral is classified under the jarosite group and has 50% SiO_2. Non-layered alunite fails to exhibit satisfactory adsorption properties; hence, alunite-layered compounds have been utilised for the removal of dyes from wastewater and water. Other silica-based materials, like dolomite glass and perlite, are applied to remove dye from water and wastewater [22]. The removal of Rhodamine-B-a cationic dye from water by natural perlite has been reported in a study [23]. Perlites, due to low density, can spread oil and float on the surface; thus, it is capable of oil sorption [24].

Zeolites

Zeolites are crystalline hydrated aluminosilicates with a highly porous three-D structure and a negatively charged lattice [11, 25]. The structural framework consists of pores occupied by alkali, alkaline earth cations, and water [26]. The exchangeable cations in the interstitial spaces act to counterbalance the negative charge. Large surface areas, coupled with an excellent ion exchange capacity and ease of economic feasibility, make zeolite a popular adsorbent [27]. It is known that there are more than 40 natural zeolite species, and the most common one is clinoptilolite [28]. Chemical modifications, including surface functionalization, ion exchange, and acid treatment, improve the adsorption capability of zeolites [29]. The raw Chinese zeolite has been known to be very effective in removing ammonium ions from aqueous solutions [30]. Furthermore, MnOx-coated zeolites have been reported to be highly efficient in removing Pb [II] and Cu [II] ions from water by adsorption mechanism [31].

Biological and Synthetic Hybrid Materials

Biological adsorbents mainly come from biological sources and microorganisms, such as fungi, algae, yeasts, bacteria, and even shells of higher organisms. These materials are strongly associated with various pollutants in aquatic and terrestrial ecosystems. The following sections, divided into three parts, will elaborate on the different categories of biological biosorbents along with their characteristics.

Chitin and Chitosan

Chitin is one of the most significant natural biopolymers and holds a position as the second-most abundant polymer after cellulose. It is a tough, white, inelastic, nitrogenous polysaccharide that forms the skeleton of crustaceans, insects, and similar living creatures [32]. The classification of the biopolymer as chitin or chitosan depends on the percentage of N-acetyl-glucosamine units it contains, where less than 50% of these units are chitin and more than 50% are chitosan [33]. The main sources of chitosan and chitin are crustaceans, such as crayfish, krill, and crab [34]. These types of quantities are abundant and available easily as food-processing industry by-products [Assam, Silchar]. A copolymer of chitin is made up of units of N-glucosamine and N-acetyl-glucosamine. Several research studies showed that chitosan-based and chitin-based biosorbents are excellent media for adsorption. For instance, Nair and Madhavan in 1984 reported that the adsorption of mercury ions from solutions in an aqueous form was done employing chitosan. A study showed the feasibility of chitosan powder for the removal of cadmium ions within a concentration range of 1–10 ppm by the use of different particle sizes [35]. Chitosan is found to be very efficient for the removal of reactive dyes completely. As chitosan is soluble in an acidic medium, it cannot be used as an insoluble sorbent. The chemical modification of chitosan overcomes the solubility problem but simultaneously controls diffusion characteristics, resulting in increased removal efficiency and reducing vulnerability to environmental interference [36].

Alginate-Based Polymers

These polymers mainly originate from brown algae, consisting of linear polysaccharides. Their structure makes them highly soluble in water [37]. Water solubility decreases the efficiency of alginate-based polymers in removing radionuclides and heavy metals. After being processed through ion-exchange reactions involving multivalent metal ions, these materials can be used as adsorbents or for the removal of different contaminants. Furthermore, they can be converted into hydrogels in the presence of calcium. These have gained vast applications in removing several heavy metals [38]. The sorption ability of alginate compounds varies between 70% and 91%, and adsorption generally occurs in an endothermic process [39].

Peat is an inexpensive, readily available, and abundant bio-sorbent. The raw material peat is composed of cellulose, lignin, fulvic, and humic acids. It is described as a porous composite of soil that contains a large amount of organic matter (humic substances) in various stages of decomposition [40]. Lignin and humic acid can form chemical linkages and contain polar functional groups, which are a combination of ketones, aldehydes, alcohols, carboxylic acids, phenolic hydroxides, and ethers. Peat comes in four different forms depending upon the nature of the source: moss peat, woody peat, sedimentary peat, and herbaceous peat. Several workers could show that peat successfully absorbs dyes from a solution. Nevertheless, raw peat has disadvantages that include poor chemical stability, high affinity to water, shrinkage tendency, low mechanical strength, and tendency to leach out humic compounds like fulvic acid. Chemical structure modifications of peat improve its adsorption capacity and selectivity.

Biomass

Bio-adsorbents are the white-rot fungi as well as several microbial cultures and dead biomass, including living biomass, and thus a potential source for bio-adsorption [41 - 43]. Though microbial biomass is produced in significant quantities through fermentation processes designed to produce antibiotics and enzymes, a large amount of by-products is produced (Table **2**). Such by-products can be utilised for the bio-adsorption of pollutants [44 - 47]. Diverse microbial biomasses that primarily have been employed to recover hazardous heavy metal ions throughout various aquatic systems have developed potential biosorption capabilities. Among the functional groups such as phosphate, amino, carboxyl, and thiol groups existing in the cell wall of fungi facilitate the interaction of the dye with the fungi [48 - 50].

Table 2. Types of Bioadsorbents

Nature	Category of biosorbents	Examples	References
Natural	Clay	Serpentine, pylophyllite, vermiculite, sepiolite, illite, montmorillonite, and saponite.	[117]
-	Siliceous materials	Glasses, silica beads, alunite, dolomite, and perlite.	[118]
-	Zeolites	Clinoptilolite, Natrolite, Phillipsite, and Stilvite.	[119]
Biological	Chitin and Chitosan	3,4-dimethoxybenzaldehyde chitosan derivative [Chi/DMB], xanthate-modified magnetic cross-linked chitosan [XMCS], and chitosan beads.	[120]
-	Alginate-Based Polymers	Radionuclides and Heavy metals [*e.g.*, lead, mercury, cadmium].	[121]
-	Biomass	Dyes and Organic pollutants.	[122]

MECHANISMS OF ADSORPTION

Physical Adsorption

Reversible and energy-efficient, physical adsorption, also known as physisorption, is a surface phenomenon controlled by weak van der Waals forces between the adsorbent and adsorbate [51]. With negative enthalpy [Δ-ΔH] and Gibbs free energy [Δ-ΔG] values demonstrating its spontaneous nature, this exothermic process is thermodynamically favourable [52]. Several variables, including temperature, pressure, surface area, porosity, and the functional groups on the adsorbent surface, affect how effective physical adsorption is. Adsorbents with high surface area and well-developed pore structures improve contaminant removal, while lower temperatures and higher pressures increase adsorption capacity [53, 54]. By improving interactions with contaminants, functional groups like hydroxyl [-OH] and carboxyl [-COOH] contribute indirectly to adsorption [55]. For instance, biochar with a high porosity and functional groups has shown a strong ability to adsorb carbon dioxide and heavy metals [56, 57]. Because physical adsorption is reversible and uses less energy than chemisorption processes, it is especially appealing for applications where adsorbent regeneration is crucial [58]. Physical adsorption has been used in environmental remediation and wastewater treatment to remove a wide range of contaminants, such as organic pollutants, heavy metals, and gases like carbon dioxide [59, 60]. For example, because of its large surface area and the presence of functional groups that improve metal binding, biochar made from agricultural waste can efficiently adsorb lead [Pb] and chromium [Cr] ions [61]. Furthermore, by utilising their high porosity and selective binding capabilities, cutting-edge materials such as Metal-Organic Frameworks [MOFs] have demonstrated promise in enhancing CO_2 adsorption [62]. Physical adsorption can be limited in its standalone application due to its drawbacks, including low selectivity for certain pollutants and a reliance on external circumstances despite its benefits [63]. Research into enhancing adsorbent materials through surface modifications, pore structure enhancements, and combining physical adsorption with additional treatment mechanisms like chemisorption and catalytic processes has been spurred by these difficulties [64]. Physical adsorption remains a fundamental component of sustainable environmental solutions by tackling these problems and providing flexible and affordable methods of pollution reduction [65, 66].

Chemical Adsorption

Chemical adsorption, or chemisorption, is a process that involves the adhesion of molecules of a substance to another material, usually a solid, through the formation of bonds that are strong in a chemical sense. Fig. (**1**). Illustrates graphically the basic difference between physical and chemical adsorption. In a way, it is contrasted with physical adsorption, whereby the process is governed by weak Van der Waals forces [67]. The process involved in the formation of ionic and covalent bonds between the adsorbate and the adsorbent makes for a more stable and stronger attachment [68]. This is a very specific process as it depends upon the chemical compatibility between the adsorbate and the surface material. Chemisorption is typically observed in heterogeneous catalysis, where reaction species are adsorbed onto a solid catalyst surface, allowing chemical reactions to be carried out [69]. For instance, in the process of hydrogenation of alkenes, both molecules of hydrogen and alkene are chemisorbed upon the catalyst surface, where they form bonds with atoms on the surface, which therefore leads to the desired reaction coupled with desorption of the final product [70].

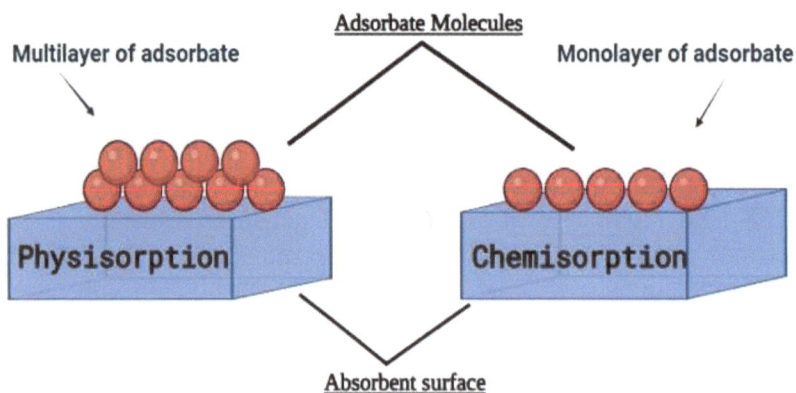

Fig. (1). Difference in physical and chemical adsorption.

SAMs, a very common example of chemisorption, are self-assembled monolayers that result from the adsorption of reactive reagents like thiols [RS-H] on metal surfaces such as gold [71]. Strong bonds such as Au-SR form while H_2 is released. The monolayer is very densely packed and may protect the underlying surface. The chemisorption phenomenon is different from physisorption because it alters the chemical structure of the adsorbate and surface [72]. The energy associated with chemisorption is much more significant, often more than 0.5 eV per adsorbed species, as opposed to the weak intermolecular forces of physisorption. The Kinetics of chemisorption also have multiple stages. There is first an adsorbate contact with the surface, which eventually gets locked in a weakly bound precursor state that more or less mimics physisorption. With time, the adsorbate transits into a chemisorbed state through the formation of strong chemical bonds to the surface. The process is normally exothermic; it's activated by a negative change in the Gibbs free energy, that is ΔG [67].

Applications of chemisorption include environmental engineering because it has a huge input in removing contaminants [68]. More so, chemisorption is crucial in the application that leads to the invention of gas sensors and catalysts, hence effective chemical transformations [69]. In some cases, dissociative chemisorption occurs, where diatomic molecules such as hydrogen, oxygen, or nitrogen dissociate into individual atoms upon adsorption. This phenomenon is usually observed on metallic surfaces such as copper and is driven by the energy provided through translational or vibrational excitation of the gas molecules [70]. The changes occurring in the surface structure of an adsorbent may even be drastic, like relaxation or reconstruction, that could eventually enhance the adsorption [72]. Advanced experimental techniques, such as atomic force microscopy, have even demonstrated the transition between the two and thus opened wider windows into the nature of these surface interactions [71].

MICROBIAL TECHNOLOGIES FOR ENVIRONMENTAL REMEDIATION

Bio-remediation Techniques

Human activities and poor waste management are among the significant problems worldwide, particularly in regions with lenient policies on the environment. Many developing countries have realised this is a gold mine for farmers since it is nutrient-rich. However, the long-term application of wastewater in irrigation deteriorates the soil by altering its physical, chemical, and biological properties [72 - 75]. In addition, there is the possibility of an accumulation of heavy metals [71, 74] and dyes.

Untreated sewage sludge application in agricultural soils harbours pathogenic microorganisms leading to significant health concerns [76 - 79]. Sludge is a good source of nutrients such as N and P, as well as organic matter; however, it can harbour hazardous substances like heavy metals and chlorinated hydrocarbons [80]. Another environmental threat is crude oil and other petroleum products. They cause soil and water pollution through leakage during transportation, storage, or extraction. The product includes benzene, which is a carcinogen that causes leukaemia [81]. It creates a complex organic mixture that poses a risk to ecosystems [82]. The amount of heavy metal contamination in the soil, air, and water is on the rise globally as industrial and urban activities are flourishing. These pollutants render resources unproductive for agricultural usage and are harmful to humans and animals [83]. Hence, proper treatment of industrial effluents is a requirement to remove contaminants from the effluent before releasing it into the environment [84]. Physicochemical methods have been applied in industrial waste treatment, but such methods are costly and unsustainable from an environmental perspective, and they produce a secondary pollutant in most cases, such as sludge, which requires additional disposal [85]. The microbiological approach is coming as one of the alternatives to diminish pollutants. Microorganisms and their metabolites are naturally or artificially able to degrade, remove, and fix environmental pollutants [86, 87]. PGPR was applied with great success in the enhancement of plant growth while remediating polluted soils and water bodies, eliminating environmental toxins [88]. Microbial enzyme systems can mineralize industrial effluents under various environmental conditions [89]. Bioremediation utilizes microbial metabolic processes to convert toxic compounds into less harmful or even harmless forms [90]. This method has some advantages over physicochemical treatments, including the generation of less sludge, shorter treatment periods, wider temperature tolerance, and ease of use [91]. Bioremediation has a cost-effective and environmentally friendly approach by utilizing microorganisms with the ability to degrade persistent and xenobiotic pollutants in industrial wastewater. Recent studies have established the role of fungi in pesticide and pollutant degradation. Fungi are known for their effectiveness in overcoming some challenges in pollutant degradation, such as the recalcitrant nature of contaminants, enzyme functionality, and survival of microorganisms under field conditions [92, 93]. Synthetic microbial communities are promising to support sustainable bioremediation processes. Advancements in molecular methods and genetic engineering have facilitated the ability to enhance bioremediation further through the generation of transgenic microbes designed to degrade certain pollutants [94]. However, the environmental risks and regulations placed by the U.S. and Europe have made the extensive application of genetically modified organisms infeasible. Future research should concentrate on identifying efficient microbial degraders, studying catabolic enzymes and genes for the detoxification of pollutants, and engineering robust microbial strains for practical applications in the field. Addressing the challenges of cost-efficiency and scalability will be essential to fully leverage bioremediation technologies for sustainable environmental restoration [95].

Bio-stimulation and Bioaugmentation

Bioaugmentation has been shown to treat soils and groundwater contaminated with compounds like tetrachloroethylene and trichloroethylene through the introduction of selected indigenous or genetically engineered microbial species, which ensures their breakdown into non-toxic byproducts like ethylene and chlorides [96]. Likewise, bio-stimulation through supplementation of nutrients, including nitrogen or oxygen, stimulates native microorganisms that are effective in activating microbial degradation of contaminants [97]. All these approaches have been proven effective not only *in-situ* but also *ex-situ* approaches and are hence environmentally and economically friendly towards environmental rehabilitation [98]. These technologies depend on microbial activity in removing harmful compounds in saturated soils and groundwater, with the availability of nutrients and electron donors or acceptors [99, 100].

Electro-bioremediation

Electro-bioremediation is an emerging hybrid technology that combines bioremediation and electrokinetics for the efficient treatment of environmental pollutants, including wastewater [101]. This approach addresses the limitations of in-situ bioremediation, such as insufficient mass transfer of nutrients and electron acceptors, limited bioavailability of contaminants, and the adaptability of indigenous microbes to specific pollutants [102]. In this method, a low direct current [0.2–2 V/cm] is passed through subsurface media to improve the transport of pollutants and microbes by using mechanisms such as electroosmosis, electrophoresis, and electromigration [101, 102]. Electro-bioremediation has shown promise in treating contaminated soils and wastewater by accelerating biodegradation processes and enhancing the efficiency of pollutant removal [103, 104]. It has been proven that this technology can degrade various contaminants. Such studies have shown that, for example, degradation of phenol is achievable in microbial cultures such as *Cupriavidus basilensis* and *Shewanella oneidensis* [105]. Another example is the electrooxidation and immobilized Bacillus strains, where up to 91.5% COD removal in diluted wastewater was achieved through a significant reduction of chemical oxygen demand in treated tannery wastewater [106]. The success of electro-bioremediation depends on several environmental factors, such as soil moisture, groundwater chemistry, geological strata, contaminant heterogeneity, and electrode material properties. Suitable electrodes must be cost-effective, durable, non-toxic, and efficient in hydrocarbon removal [102]. Despite its promise, the large-scale field applications of electro-bioremediation are still under investigation. Further research should optimize the electrode materials, refine operational parameters, and explore hybrid technologies that combine electro-bioremediation with chemical or phytoremediation methods. It is also crucial to undertake field trials to evaluate the cost-effectiveness and environmental impacts of the technology before full-scale implementation [102].

Bio-sparging

Bio-sparging is a technique that introduces air into the soil, similar to bioventing, to stimulate microbial activity and facilitate the removal of contaminants from contaminated sites. Unlike conventional biodegradation techniques, bioventing injects air into saturated zones, promoting the upward migration of flammable organic chemicals to adjacent unsaturated zones. The success of sparging largely depends on the porosity of the soil [107]. SVE and IAS mainly depend upon the higher airflow rates to facilitate the volatilisation of contaminants, though bio-sparging focuses on the enhanced degradation through microorganisms. This technique is applied extensively for the extraction of diesel and kerosene from water sources through the provision of oxygen to the microorganisms during increased bioremediation for the acceleration of degradation mechanisms in that region [108]. Other technologies that can be applied in the removal of organic pollutants include BTEX compounds like adsorption, microbial degradation, bio-sparging, PRBs, and application of modified or synthesized zeolites. However, there is limited research regarding inexpensive, readily available materials like natural zeolites for BTEX adsorption [109]. Bio-sparging injects compressed air below the water table, increasing oxygen levels in the groundwater, which improves the rate of biodegradation by the naturally occurring bacteria [110]. This has been particularly successful in remediating kerosene- and diesel-contaminated aquifers while promoting the biodegradation of BTEX compounds and naphthalene [111]. This technology is susceptible to soil permeability and pollutant biodegradability [112].

Bio-venting

Bioventing is one promising *in situ* bioremediation technique that supports natural, aerobic degradation of compounds using microorganisms native to soil [110]. The process requires the air supply to be managed; this is because, without sufficient oxygen, it becomes difficult to activate microbes to accelerate pollutant degradation [111]. Bio-venting ensures adequate levels of nutrients and soil moisture for optimal transformation of contaminants. Fig. (**2**) represents the different microbial technologies. This technology has been successfully applied to remediate soils contaminated with petroleum-based products [109]. Bioventing facilitates the delivery of oxygen to the unsaturated zone, which promotes the activity of native microbes and the transformation of pollutants into nontoxic byproducts [108]. Moreover, nutrient and moisture supplementation promote microbial activity in the process. Bioventing has recently been highlighted as one of the key *in situ* bioremediation strategies that have effectively stimulated native microbial communities through aeration and facilitated the precipitation of heavy metal pollutants [112].

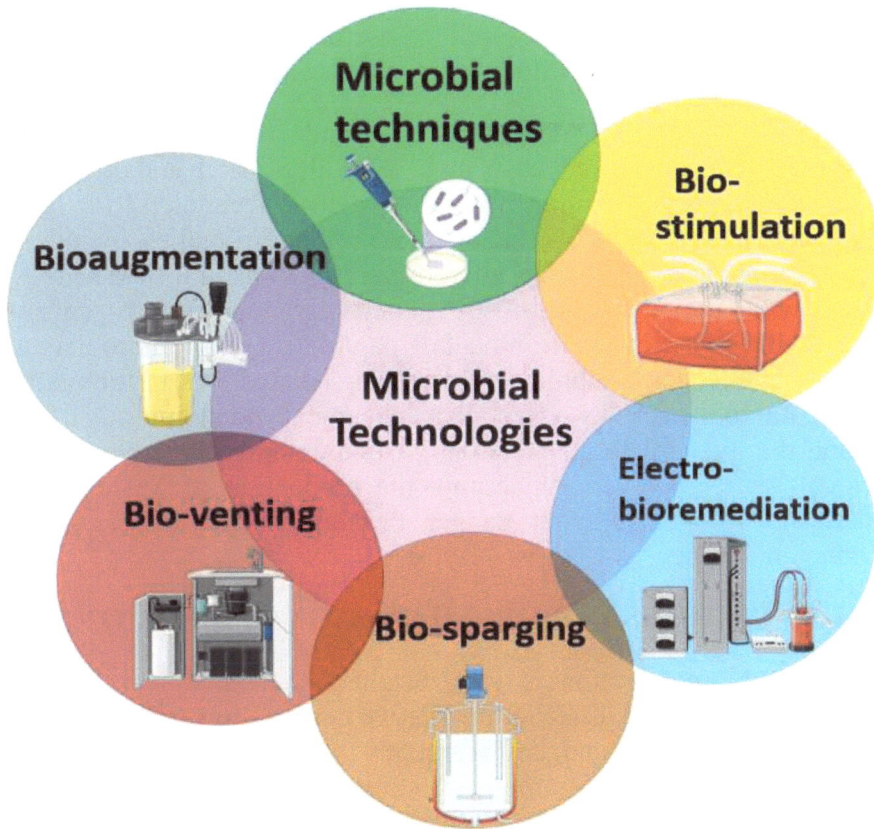

Fig. (2). Different microbial technologies for bioremediation.

CHALLENGES AND LIMITATIONS

Bioremediation is a very promising technique for the restoration of environmental materials, but there are some significant constraints to the effectiveness of this process. The main constraint is that not all compounds are biodegradable, which generally restricts the overall success of the bioremediation process. In the cases where the materials are biodegradable, the downstream processing of these materials can sometimes produce toxic by-products. Additionally, microbial strains effective at one site may not perform well at other sites due to specific site-related factors, such as differing environmental conditions. The complexity of bioremediation is further compounded by the intricate nature of microbial metabolic processes, the types of pollutants involved, and the necessity for adequate nutrient levels. The process of bioremediation is usually time-consuming and labour-intensive because it involves the excavation of soil, construction, and customizing special site layouts. Moreover, bioremediation is often conducted underground or in isolated areas far from residential zones, but the utilization of heavy machinery and pumps will create noise and disturbances. In particular, ethical issues also raise questions about the use of some bacterial strains because they could affect the local microflora in ways that cannot be predicted.

Challenges and limitations. Major aspects may include technological and economic ones, such as the high costs involved with research and development, in addition to limitations to scaling up bioremediation technology for large-scale field applications. Environmental factors also present difficulties, including a lack of information on the behaviour of microorganisms in natural field settings, risks associated with the introduction of genetically modified organisms, and dependence on specific environmental parameters for microbial activity. Operational challenges include problems associated with stimulating and maintaining microbial activity, effective management and monitoring of microbes *in situ*, and the incomplete understanding of contaminant-microbe interactions. One critical issue, however, concerning the production of waste products, in particular, the second pollutant danger through sludge, lingers during this bioremediation activity.

FUTURE DIRECTIONS IN BIOREMEDIATION AND BIOCONTROL

The future of bioremediation is very bright as technological developments continue to move forward, offering innovative answers toward the enhancement of efficiency, scalability, and environmental compatibility of bioremediation processes. The main focus areas for future development would involve innovations in biosorbent materials, advanced genetic engineering of microbial strains, integration into emerging technologies, and hybrid approaches for enhancing pollutant removal. These innovations would help overcome the present limitations and promote sustainable environmental management practices.

Innovations in Biosorbent Materials

The development of advanced biosorbents represents a significant leap forward in bioremediation technology. Researchers are exploring the use of agricultural by-products, such as crop residues, and industrial waste materials like fly ash as cost-effective and environmentally sustainable sources for biosorbents. These natural materials are being engineered to absorb a wide array of pollutants, including heavy metals, organic contaminants, and dyes. Hybrid biosorbents, composed of natural organic materials with synthetic polymers or nanoparticles, are gaining attention because of their enhanced adsorption capacity, improved regeneration ability, and increased stability under changing environmental conditions. Harnessing the potential of these novel materials may make bioremediation processes more efficient and applicable in various environmental contexts.

Advanced Genetic Engineering

The capability of genetic engineering in the realm of bioremediation can revolutionize it because new microbial strains can be created that are enhanced in their capability to degrade pollutants. Synthetic biology and gene editing techniques have been developed through which microorganisms can be engineered to target specific contaminants more efficiently. This includes improvements in the breakdown of complex pollutants, such as petrochemical by-products, pesticides, and heavy metals. Optimization of the functionality of microbial enzymes *via* genetic modification would considerably enhance the rate and stability of such biodegradation processes even under different fluctuations of environmental conditions. In addition, bioremediation carried out with GMOs shall provide strains degrading in several pollutants simultaneously, an overall enhancement to the performance of bioremediation.

Integration with Emerging Tools and Technologies

The possibility of integrating bioremediation with emerging tools or technologies, most particularly nanotechnology, presents really promising opportunities in enhancing the degradation rates of pollutants. Nanomaterials, such as magnetic nanoparticles or carbon-based nanomaterials, are being explored for their ability to interact with and enhance the bioavailability of contaminants, thereby making them more accessible to microbial degradation. Nanoparticles may also be used as delivery systems for microorganisms or enzymes, ensuring targeted and efficient remediation at contaminated sites. This integration may result in a significant increase in the pace and precision of bioremediation efforts. The Smart bioreactors represent another promising development, using the convergence of Artificial Intelligence and the Internet of Things. These are equipped with sensors monitoring environmental parameters and microbial activity and providing data in real-time to make adaptive management of the bioremediation processes possible. The AI algorithms may analyse these data to adjust conditions like temperature, pH, and nutrient levels, ensuring maximum efficiency in the bioremediation process. This level of automation reduces the need for manual intervention and provides more precise control over the microbial environment, which leads to more successful outcomes.

Hybrid Technologies for Enhanced Remediation

Hybrid technologies are becoming one of the effective solutions to promote the bioavailability of pollutants while increasing the efficiency of bioremediation processes. One is electro-bioremediation, wherein electrochemical processes are integrated into bioremediation to accelerate pollutant degradation. Low-voltage electric fields can be used in electro-bioremediation to enhance the mobilization of contaminants, for example, heavy metals towards microbial cells, thereby fastening the degradation process. It has been found suitable for both soil and water remediation. Bio-sparging and bio-venting, which are *in situ* techniques for the remediation of groundwater and soil, are also being enhanced by hybrid approaches. Bio-sparging injects oxygen into contaminated groundwater to stimulate naturally occurring microbes. Bio-venting enhances VOC degradation in soil by allowing airflow through contaminated areas. When combined with other forms of bioremediation, like bioaugmentation with genetically engineered microorganisms or supplementing the addition of electron donors, these hybrid technologies can effectively remediate an expanded range of pollutants while functioning in multiple environmental scenarios.

CONCLUSION

Industrial waste contributes enormously to environmental pollution by releasing highly toxic organic and inorganic pollutants, which pose threats to ecosystems and public health. Although the conventional physicochemical remediation methods have been successful, they are environmentally destructive and expensive. Microbial remediation has been developed as a sustainable, eco-friendly, and low-cost option, but no method can be effective. Each remediation technology, therefore, has its pros and cons, calling for ongoing research to enhance microbial solution optimization and bring about field-scale economic viability. Bioremediation has great potential in treating contaminated environments using the pollutant-degrading capabilities of certain microbes. Its mass application is, however, restricted due to limitations such as scarce knowledge about the behaviour of microbes in field conditions, difficulties in stimulating and handling microbes and ensuring effective interaction between microbes and pollutants. Innovations in bioaugmentation, genetic engineering, and biotransformation have enhanced the efficiency of microbes, degrading hitherto resistant pollutants like polychlorinated biphenyls and chlorinated solvents. Further improvement in the accessibility and efficacy of microbes has been achieved through techniques like gas sparging, surfactant addition, and heat-based solubilization. Standardized protocols for monitoring contaminant reduction, assessing microbial biodegradation potential, and validating field-level performance are crucial to advancing bioremediation. Techniques emerging from molecular biology-based approaches and *in situ* physicochemical assessments offer transformative potential for evaluating and implementing microbial remediation strategies. Despite its advantages, bioremediation requires careful microbial handling, monitoring, and optimization of environmental conditions, which limits its widespread adoption. Future research should focus on overcoming these barriers through advanced microbial engineering and microbiome strategies, such as cultivating acid-tolerant species like Pseudomonas using *Aspergillus niger* culture filtrates. This approach could facilitate wastewater treatment and biofertilizer production, further enhancing its applicability. In conclusion, microbial remediation is a powerful tool by which environmental sustainability can be achieved. With further enhancements in genetic engineering, optimization of enzymes, and modification in microbial communities, scalable and environment-friendly bioremediation technologies would be developed for field testing and societal acceptance. They will consequently lead to a cleaner, greener, and healthier future.

REFERENCES

[1]   Mandal S, Dutta T, Chakraborty A, Saha S, Ghosh S. Environmental pollution: causes, effects, and control measures. Environ Sci Pollut Res Int 2021; 28: 23074-88.
        [http://dx.doi.org/10.1007/s11356-021-12515-2]

[2]   Rahman MM, Saha S, Ghosh S, Mandal S. Heavy metal contamination in water: a global issue. J Hazard Mater 2022; 423: 127051.
        [http://dx.doi.org/10.1016/j.jhazmat.2021.127051]

[3]   Dasgupta D, Bandyopadhyay M, Bhattacharya S. Wastewater treatment: conventional and advanced techniques. Environ Eng Sci 2020; 37: 5-17.
        [http://dx.doi.org/10.1089/ees.2019.0270]

[4]   Jin S, Hu H, Li Y, Liu G, Zhang Y. Bioremediation of contaminated water using microorganisms. J Appl Microbiol 2022; 132: 2773-89.
        [http://dx.doi.org/10.1111/jam.15689]

[5]   Chen X, Zhang Y, Zhang L, Wu X, Wang J. Microbial bioremediation: current strategies and future directions. Biodegradation 2020; 31: 213-27.
        [http://dx.doi.org/10.1007/s10532-020-09946-1]

[6]   Kumar A, Singh R, Sahu N, Rani S, Sharma M. Genetically modified microorganisms in wastewater treatment. Biotechnol Lett 2021; 43: 1187-96.
        [http://dx.doi.org/10.1007/s10529-021-03115-7]

[7]   Zaidi A, Amri S, Khan MS, Zaidi MA, Figueiredo MB. Fungi in bioremediation of industrial effluents. Biol Fertil Soils 2009; 46: 495-508.
        [http://dx.doi.org/10.1007/s00374-009-0382-6]

[8]   Timmusk S, Aro T, Hiltunen K, *et al.* Fungal applications in bioremediation of industrial pollutants. Appl Microbiol Biotechnol 2017; 101: 4855-64.
        [http://dx.doi.org/10.1007/s00253-017-8285-0]

[9]   Bai J, Zhu J, Li L, Xu S, Wang Y. Fungal bioremediation of pharmaceutical compounds in industrial effluents. Microbiol Res 2008; 163: 187-95.
        [http://dx.doi.org/10.1016/j.micres.2007.07.009]

[10]   Finley A, Yang Y, Zhang Z, Deng X, Yu Z. *Trichoderma species* in bioremediation of wastewater containing synthetic dyes and pharmaceuticals. Biotechnol Adv 2010; 28: 727-34.
        [http://dx.doi.org/10.1016/j.biotechadv.2010.07.003]

[11]   Singh SK, Singh RP, Kumar R. Use of microbial and algal biosorbents for heavy metal removal from industrial effluents: a review. Int J Environ Sci Technol 2022; 19: 467-81.
        [http://dx.doi.org/10.1007/s13762-021-03544-z]

[12]   Sinha S, Tiwari RP, Nandi R. Heavy metal removal by bioremediation using plant materials. J Environ Biol 2010; 31: 745-51.

[13]   Cheah WL, Tan CM, Yusoff MS, Tan LH. Removal of heavy metals from aqueous solution using

sawdust. J Hazard Mater 2016; 307: 144-51.
[http://dx.doi.org/10.1016/j.jhazmat.2015.11.008]

[14]   Zheng L, Zhang X, Lin H, Wang M, Wei W. Biosorption of heavy metals from industrial wastewater using sugarcane bagasse. Water Sci Technol 2013; 68: 78-85.
[http://dx.doi.org/10.2166/wst.2013.290]

[15]   Chojnacka K, Michalak I. Waste biomass as an alternative to conventional adsorbents in water treatment. Bioresour Technol 2009; 100: 6351-7.
[http://dx.doi.org/10.1016/j.biortech.2009.07.014]

[16]   Olu-Owolabi BI, Akinmoladun FO, Adeyemi FA. Biosorption of heavy metals using clay: a review. J Environ Chem Eng 2016; 4: 2154-62.
[http://dx.doi.org/10.1016/j.jece.2016.04.012]

[17]   Ahmaruzzaman M. A review on the adsorption of heavy metals by clay minerals, with special focus on the removal of cationic dye. J Hazard Mater 2008; 154: 184-91.
[http://dx.doi.org/10.1016/j.jhazmat.2007.10.015]

[18]   Shichi K, Takagi S. Adsorption of heavy metals onto clay minerals: kaolinite, micas, and smectites. Clay Sci 2000; 11: 1-7.
[http://dx.doi.org/10.52943/claysci.11.1_1]

[19]   Uddin MT. Adsorption of heavy metal ions by montmorillonite and its modifications: a review. J Environ Chem Eng 2017; 5: 585-601.
[http://dx.doi.org/10.1016/j.jece.2017.01.015]

[20]   Babel S, Kurniawan TA. Low-cost adsorbents for heavy metals removal: a review. J Hazard Mater 2003; 97: 219-43.
[http://dx.doi.org/10.1016/S0304-3894(02)00263-7] [PMID: 12573840]

[21]   Agarwal S, Jain H. Investigating the potential of floral waste as a vermicompost and dual-functional biosorbent for sustainable environmental management. Water Air Soil Pollut 2024; 235(5): 322-2.
[http://dx.doi.org/10.1007/s11270-024-07144-y]

[22]   Siddique R, Manjula S, Reddy G, Kadiyala P, Rao AB. Siliceous materials in environmental pollution control. Environ Sci Pollut Res Int 2017; 24: 10732-47.
[http://dx.doi.org/10.1007/s11356-017-9045-4]

[23]   Crini G, Lichtfouse E, Morin-Crini N. Applications of chitosan-based biosorbents for water treatment. Environ Sci Pollut Res Int 2006; 13: 277-82.

[24]   Ahmed A, Ram S. Removal of metal ions from water using silica-based adsorbents. J Chem Eng Data 1992; 37: 1063-6.

[25]   Akar T, Iscan G, Demir A, Bektas S, Okutucu B. Adsorption of dyes from aqueous solutions using alunite-based materials. Desalination 2016; 383: 19-25.

[26]   Gaur A, Sharma R. Synthesis and application of silica-based adsorbents for the removal of pollutants. J Environ Chem Eng 2017; 5: 4207-14.

[27]   Vijayakumar S, Murugesan K, Karthikeyan S, Rajendran S, Kumar PV. Adsorption of rhodamine-B dye from aqueous solution by natural perlite. J Environ Chem Eng 2012; 2: 269-78.

[28]   Bastani P, Farhadian M, Mirbagheri SA, Ghaffari H. Removal of oil from aqueous solutions using perlite: an experimental study. J Hazard Mater 2006; 137: 1105-11.

[29]   Jha M, Singh D. Use of zeolites in adsorption of heavy metals: a review. J Chem Eng Data 2016; 61: 2253-63.

[30]   Wang L, Xu X, Li G, Zhang Z, Liu Y. Removal of heavy metal ions using zeolites: a review. Environ Sci Pollut Res Int 2010; 17: 1-9.

[31]   Hor H, Foo L, Lee K, Lim C. Ion exchange and surface modifications of zeolites: advances in

adsorption of heavy metals from aqueous solutions. J Hazard Mater 2016; 302: 68-76.
[http://dx.doi.org/10.1016/j.jhazmat.2015.09.024]

[32] Barbosa P, Rodrigues A, Sousa R, Martins F. Zeolites in wastewater treatment: current advances and challenges. Environ Sci Pollut Res Int 2016; 23: 14226-41.
[http://dx.doi.org/10.1007/s11356-016-6505-3]

[33] Ennaert T, Sels B, De Vos D, De Smet J, Van Der Voort P. Surface modification of zeolites for environmental remediation applications. J Environ Chem Eng 2016; 4: 1813-23.
[http://dx.doi.org/10.1016/j.jece.2016.05.015]

[34] Jain H. From pollution to progress: groundbreaking advances in clean technology unveiled. Innov Green Dev 2024; 3(2): 100143-3.
[http://dx.doi.org/10.1016/j.igd.2024.100143]

[35] Lee Y, Hsu S, Hsueh C, Huang T. Removal of heavy metal ions by MnOx-coated zeolites. J Hazard Mater 2015; 283: 272-80.
[http://dx.doi.org/10.1016/j.jhazmat.2014.09.039]

[36] Elieh-Ali-Komi D, Hamblin MR. Chitin and chitosan: properties and applications. Biopolymers 2016; 106: 431-9.
[http://dx.doi.org/10.1002/bip.22701]

[37] Khor E, Lim L, Zulfakar M. Chitosan: a biopolymer with wide applications. Biotechnol Bioprocess Eng; BBE 2003; 8: 138-44.
[http://dx.doi.org/10.1007/BF02930225]

[38] de Queiroz Antonino A, de Oliveira P, da Silva G, de Lima M, Rodrigues E. Chitosan-based adsorbents for environmental remediation: a review. Carbohydr Polym 2017; 157: 1-12.
[http://dx.doi.org/10.1016/j.carbpol.2016.09.061]

[39] Nair P, Madhavan A. Studies on chitosan-based adsorbents for mercury ion removal from aqueous solutions. J Hazard Mater 1984; 8: 225-38.
[http://dx.doi.org/10.1016/0304-3894[84]90006-0]

[40] Jha M, Singhal R, Sharma R. Adsorption of cadmium ions from aqueous solutions using chitosan. Bioresour Technol 1988; 25: 41-6.
[http://dx.doi.org/10.1016/0960-8524[88]90145-2]

[41] Ahmaruzzaman M, Rashid S, Hossain A. Chitosan for water treatment: a review of adsorption characteristics and applications. J Hazard Mater 2008; 153: 1217-25.
[http://dx.doi.org/10.1016/j.jhazmat.2007.09.011]

[42] Yu S, Zhang W, Guo X, Zheng Y. Alginate-based polymers for adsorption and environmental applications. J Environ Sci (China) 2017; 53: 7-16.
[http://dx.doi.org/10.1080/10643389.2018.1547621]

[43] Gok O, Aytas S. A new approach for removal of heavy metals from aqueous solutions: alginate-based adsorbents. Desalination 2009; 249: 1119-25.
[http://dx.doi.org/10.1016/j.desal.2008.08.020]

[44] Pang Y, Liu Y, Zhang J, *et al.* Adsorption of heavy metals by alginate/chitosan hydrogel: preparation, characterization, and application. Bioresour Technol 2010; 101: 4795-802.
[http://dx.doi.org/10.1016/j.biortech.2010.01.001]

[45] Jain H. Exploring the emergence of sustainable practices in healthcare research and application as a path to a healthier future. In: Prabhakar PK, Leal Filho W, Eds. in Preserving Health, Preserving Earth: The Path to Sustainable Healthcare. Cham: Springer Nature Switzerland 2024; pp. 121-37.
[http://dx.doi.org/10.1007/978-3-031-60545-1_7]

[46] Gazso A, Barta Z, Ujhelyi Z, Fenyvesi E, Horvath G. Bioremediation of heavy metals by microbial biomass. Sci Total Environ 2001; 272: 109-15.
[http://dx.doi.org/10.1016/S0048-9697[00]00892-2]

[47]   Gupta VK, Sharma S, Ali I, Saini V. Bioremediation of hazardous pollutants by microorganisms: a review. Biotechnol Adv 2016; 34: 663-85.
[http://dx.doi.org/10.1016/j.biotechadv.2016.04.003]

[48]   Abdi M, Kazemi M. Biosorption of heavy metals by fungi and their potential applications: a review. Biochem Eng J 2015; 98: 45-58.
[http://dx.doi.org/10.1016/j.bej.2015.02.010]

[49]   Kaushik P, Yadav K, Awasthi M, Chatterjee S, Verma R. Biomass from white-rot fungi as a potential adsorbent for the removal of heavy metals. Environ Sci Pollut Res Int 2009; 16: 527-36.
[http://dx.doi.org/10.1007/s11356-009-0136-3]

[50]   Ngah WSW, Hanafiah MAKM. Removal of heavy metal ions from aqueous solutions by adsorption onto chitosan-based adsorbents: a review. Bioresour Technol 2008; 99: 3935-48.
[http://dx.doi.org/10.1016/j.biortech.2007.06.011] [PMID: 17681755]

[51]   Anber AR, Matouq M. Adsorption of heavy metals from aqueous solutions using new adsorbents. J Hazard Mater 2008; 150: 207-15.
[http://dx.doi.org/10.1016/j.jhazmat.2007.09.040]

[52]   Chen M, Li L, Zhou L, Wu L, Gao B, Liu J. Adsorption of heavy metal ions on chitosan and its derivatives: a review. Chemosphere 2018; 192: 37-47.
[http://dx.doi.org/10.1016/j.chemosphere.2017.10.153]

[53]   Li H, Wei S, Yu S, Xie H, Xu Z, Zhang Z. A review on adsorption of heavy metals from water by chitosan and its derivatives. Carbohydr Polym 2017; 164: 264-74.
[http://dx.doi.org/10.1016/j.carbpol.2017.02.080]

[54]   Harvey J, Lowry T, Lory M, Upton M, Ward G. Enhancing adsorption properties of biopolymers for water treatment applications. Water Res 2011; 45: 2114-24.
[http://dx.doi.org/10.1016/j.watres.2010.11.035]

[55]   Jain H. Groundwater vulnerability and risk mitigation: a comprehensive review of the techniques and applications. Groundw Sustain Dev 2023; 22: 100968-8.
[http://dx.doi.org/10.1016/j.gsd.2023.100968]

[56]   Yuan Z, Li W, Li J, Li Y, Wu L, Yang X. Surface modification of biochar to enhance its sorption capacity for heavy metals. J Environ Manage 2011; 92(5): 1370-7.
[http://dx.doi.org/10.1016/j.jenvman.2010.10.002]

[57]   Cantrell KB, Hunt PG, Uchimiya M, Novak JM, Schlegel VA. Impact of biochar and activated carbon on soil nitrogen dynamics and N_2O emissions. J Environ Qual 2012; 41(4): 1124-31.
[http://dx.doi.org/10.2134/jeq2011.0123]

[58]   Kong S, Li S, Li X, Wu D. A review of biochar's adsorption capacity for removal of environmental pollutants. Environ Sci Pollut Res Int 2011; 18(1): 1-15.
[http://dx.doi.org/10.1007/s11356-010-0373-4] [PMID: 20614196]

[59]   Dong X, Zhang L, Zeng G, *et al.* Advances in biochar adsorbents for the removal of pollutants from wastewater: a review. J Environ Manage 2013; 118: 212-20.
[http://dx.doi.org/10.1016/j.jenvman.2013.01.048]

[60]   Lu Y, Zhong Y, Yao Z, Wu H, Zhang B. Adsorption of heavy metals by biochar: a review. Bioresour Technol 2012; 114: 6-14.
[http://dx.doi.org/10.1016/j.biortech.2012.03.072]

[61]   Graham A, Hanan J, Leclerc A, Johnson S. Metal-organic frameworks for enhanced CO_2 adsorption. Nat Commun 2012; 3: 1027.
[http://dx.doi.org/10.1038/ncomms2011]

[62]   Ahmad M, Lee SH, Bhatnagar A, Sillanpää M. A critical review on heavy metal ions and toxic organics removal using adsorbents and sorbents. J Chem Eng Process Technol 2012; 3: 2-10.

[http://dx.doi.org/10.4172/2157-7048.1000131]

[63]   Tan Y, Wang X, Zhang Q, Zhang S, Zhu L. Surface modification of adsorbents for pollutants removal from aqueous solution. J Chem Technol Biotechnol 2015; 90(2): 383-400.
[http://dx.doi.org/10.1002/jctb.4533]

[64]   Pulido-Novicio EE, Martinez-Aguirre A, Gonzalez F, De la Cruz-Malave A, Vera JL. Application of adsorption in environmental treatment systems. J Hazard Mater 2001; 87(3): 169-78.
[http://dx.doi.org/10.1016/S0304-3894[01]00207-6]

[65]   Oura M, Lifshits MA, Saranin AA, Zotov AV, Katayama M. Surface science: an introduction. 2nd ed., Berlin: Springer 2003.
[http://dx.doi.org/10.1007/978-3-662-05179-5]

[66]   Rettner CT, Auerbach SM. Adsorption of small molecules on metal surfaces: a study of surface chemistry. J Phys Chem 1996; 100(11): 4599-605.
[http://dx.doi.org/10.1021/jp953108u]

[67]   Gasser RS. Surface reactions and catalysis: fundamental principles and applications. Chem Rev 1985; 85(2): 29-52.
[http://dx.doi.org/10.1021/cr00067a002]

[68]   Norskov JK. Hydrogenation of alkenes and chemisorption on metal surfaces. J Catal 1990; 123(1): 1-16.
[http://dx.doi.org/10.1016/0021-9517[90]90281-J]

[69]   Clark RS. Self-assembled monolayers of thiols on gold surfaces. J Chem Phys 1974; 60(1): 179-81.
[http://dx.doi.org/10.1063/1.1680833]

[70]   Huber SJ, Lin Y, Zhao X, Wadhwa P. Understanding the mechanisms of chemisorption and physisorption in surface chemistry. Langmuir 2019; 35(23): 7408-18.
[http://dx.doi.org/10.1021/acs.langmuir.9b00562]

[71]   Narwal RP, Joshi M, Yadav KN, Antil RS. Effect of wastewater irrigation on soil properties and crop production. J Agric Sci 1988; 30(2): 139-46.

[72]   Joshi M, Yadav KN. Impact of wastewater on agricultural soils. Environ Pollut 2005; 137(1): 47-53.
[PMID: 15809107]

[73]   Antil RS, Yadav KN, Narwal RP. Effect of irrigation with sewage water on soil quality. Environ Sci Pollut Res Int 2007; 14(1): 61-5.

[74]   Kharche AD, Agarwal MK, Rao KK. Effect of sewage water irrigation on soil and crop productivity. J Environ Sci (China) 2011; 23(2): 187-92.

[75]   Chambers RM, Lapen DR, Edwards M, Gottschall N. Soil contamination and health risks due to sewage sludge application. Soil Sci Soc Am J 2001; 65(3): 687-95.

[76]   Lapen DR, Edwards M, Gottschall N, Chambers RM. Effects of sludge application on agricultural soils and crops. J Agric Food Chem 2008; 56(16): 6229-36.

[77]   Edwards M, Gottschall N, Chambers RM, Lapen DR. Hazardous waste materials in sewage sludge. Environ Toxicol Chem 2009; 28(9): 2035-43.

[78]   Gottschall N, Lapen DR, Edwards M, Chambers RM. Sewage sludge contamination in agricultural systems: potential human health risks. Sci Total Environ 2009; 407(15): 4586-96.

[79]   Jang M, Lee S, Kim J, Lee M, Shin K, Hong S, *et al.* Application of sewage sludge for soil improvement. Waste Manag 2010; 30(3): 498-503.

[80]   Kirkeleit J, Tryggestad A, Heggelund L, *et al.* Health effects of exposure to crude oil and benzene: a review. Environ Health Perspect 2006; 114(12): 1845-50.

[81]   Adelana SM, Falade OT, Odukoya OO, Olufunmilayo OP, Ogundele AF. Environmental and ecotoxicological impact of petroleum products: a review. J Environ Manage 2011; 92(4): 1080-8.

[82] Wahid SR, Khan A, Hussain A, Malik MA, Bashir M, Mehmood A. Heavy metal contamination in agricultural soils and water: a global challenge. Environ Toxicol Chem 2000; 19(3): 701-10.
[http://dx.doi.org/10.21203/rs.3.rs-5319547/v1]

[83] Ali H. Industrial effluent treatment for the removal of heavy metals: a review. J Environ Manage 2010; 91(8): 1924-34.

[84] Zhang X, Wei X, Zhang S, Zhang H. Physicochemical methods for industrial wastewater treatment: challenges and solutions. J Hazard Mater 2004; 117(1-2): 179-88.
[http://dx.doi.org/10.1515/revce-2021-0094]

[85] Uqab S, Khan MI, Irshad A, Rehman A, Anwar S, Riaz M. Biodegradation of pollutants by microorganisms. Environ Sci Pollut Res Int 2016; 23(1): 51-68.

[86] U.S. Environmental Protection Agency. Bioremediation of contaminated soil and groundwater: principles and applications. Washington, DC: EPA 2016.

[87] Gouda S, Das G, Sen SK, Shin HS, Patra JK. PGPR: a promising approach for enhancing the growth of plants in polluted soil and water. Biol Control 2018; 46(3): 413-26.

[88] Pandey G, Chauhan RS, Singh R. Biodegradation of industrial effluents by microorganisms. J Environ Sci (China) 2012; 24(3): 1185-95.

[89] Xu L, Zhou Q. Bioremediation and its potential in soil and groundwater pollution cleanup. Environ Sci Pollut Res Int 2017; 24(9): 8285-98.

[90] Kulshrestha S, Husain T. Bioremediation technologies: innovations and applications. Environ Chem Lett 2007; 5(3): 143-8.

[91] Spina F, Nannipieri P, Franchi M, Ricci S, Ciampolini F. Fungal degradation of pollutants in contaminated sites. J Bioremediat Biodegrad 2018; 9(4): 1-15.

[92] Daghio M, Spina F, Nannipieri P, *et al.* Bioremediation of pollutants using fungal enzymes: a review. Environ Int 2017; 105: 87-95.
[http://dx.doi.org/10.1016/j.envint.2017.05.007]

[93] Bhargava R, Bhatt A, Anwar R, Shrivastava R, Jain S. Microbial degradation of pollutants: current status and future prospects. Appl Microbiol Biotechnol 2020; 104(6): 2357-72.
[http://dx.doi.org/10.1007/s00253-020-10367-7]

[94] Bharagava RN, Ghosh A, Tiwari D, Shukla P. Microbial degradation of xenobiotic compounds for environmental bioremediation. Appl Microbiol Biotechnol 2020; 104(7): 2505-17.
[http://dx.doi.org/10.1007/s00253-020-10401-8]

[95] Niu J, Zhang Y, Zhang Z, Jin R. Bioaugmentation for the biodegradation of tetrachloroethylene and trichloroethylene contaminants in groundwater. Environ Sci Technol 2009; 43(7): 2488-94.
[http://dx.doi.org/10.1021/es803155m]

[96] Zeneli A, Elliott DW, O'Carroll DM, Sleep BE, Liu Y. Enhancement of microbial degradation of contaminants through bio-stimulation with nitrogen and oxygen. J Hazard Mater 2019; 365: 34-41.
[http://dx.doi.org/10.1016/j.jhazmat.2018.10.077]

[97] Bodor A, Bounedjoum N, Vincze GE, *et al.* Bioaugmentation and biostimulation for environmental remediation. J Environ Manage 2020; 260: 110080.
[http://dx.doi.org/10.1016/j.jenvman.2020.110080]

[98] Jørgensen BB. Microbial activity and processes in saturated soils: a key factor in environmental bioremediation. FEMS Microbiol Ecol 2011; 77(3): 397-403.
[http://dx.doi.org/10.1111/j.1574-6941.2011.01150.x]

[99] Girma S. Microbial remediation of contaminated soils and groundwater: a review. Environ Sci Pollut Res Int 2015; 22(11): 8462-75.
[http://dx.doi.org/10.1007/s11356-015-4470-1]

[100] Maszenan AM, Tay JH, Ng WJ, Liu Y. Bioremediation of wastewater using electrokinetics: a review. Crit Rev Environ Sci Technol 2011; 41(14): 1251-76.
[http://dx.doi.org/10.1080/10643380903190771]

[101] Daghio M, Maphosa F, Haggblom MM, *et al.* Electro-bioremediation of contaminated soils: current knowledge on technological applicability and environmental impacts. Appl Microbiol Biotechnol 2017; 101(7): 2575-86.
[http://dx.doi.org/10.1007/s00253-017-8173-7]

[102] Wick LY, Shi L, Harms H. Electrokinetic transport of PAH-degrading bacteria in model aquifers and the effect of surfactants. Environ Sci Technol 2007; 41(16): 5469-75.
[http://dx.doi.org/10.1021/es0700815]

[103] Martínez-Prado MA, Lopez J, Sierra-Alvarez R. Electrokinetic treatment of hydrocarbon-contaminated soils: current developments and applications. J Hazard Mater 2014; 276: 1-10.
[http://dx.doi.org/10.1016/j.jhazmat.2014.04.060]

[104] Friman H, Schechter A, Nitzan Y, Cahan R. Phenol degradation by *Cupriavidus basilensis* under controlled electro-bioremediation conditions. J Hazard Mater 2013; 244-245: 426-33.
[http://dx.doi.org/10.1016/j.jhazmat.2012.11.054]

[105] Kanagasabi R, Rajagopal K, Ramalingam S, Pitchai A, Natarajan N. Electro-bioremediation of tannery effluent using immobilized *Bacillus* strains. Int J Environ Sci Technol 2013; 10(2): 423-32.
[http://dx.doi.org/10.1007/s13762-012-0121-2]

[106] Fosso-Kankeu E, Mulaba-Bafubiandi A. Characterization of porosity and airflow in soil for the optimization of bio-sparging techniques. J Clean Energy Technol 2014; 2(1): 7-12.
[http://dx.doi.org/10.7763/JOCET.2014.V2.86]

[107] Wang L, Chen X. Bioremediation of diesel and kerosene contaminated aquifers using bio-sparging. Int Biodeterior Biodegradation 2006; 58(2): 78-83.
[http://dx.doi.org/10.1016/j.ibiod.2006.07.003]

[108] Chettri D, Verma A K, Verma A K. Bioaugmentation: an approach to biological treatment of pollutants. Biodegradation 2024; 35(2): 117-35.
[http://dx.doi.org/10.1007/s10532-023-10050-5]

[109] Romantschuk M, Lahti-Leikas K, Kontro M, *et al.* Bioremediation of contaminated soil and groundwater by in situ biostimulation. Front Microbiol 2023; 14: 1258148.
[http://dx.doi.org/10.3389/fmicb.2023.1258148]

[110] Maglione G, Zinno P, Tropea A, *et al.* Microbes' role in environmental pollution and remediation: a bioeconomy focus approach. AIMS Microbiol 2024; 10(3): 723.
[http://dx.doi.org/10.3934/microbiol.2024033]

[111] Cartwright B, Higgins T, Kelley B, Ogden S. Soil permeability and bioventing efficiency: an evaluation of kerosene contaminated sites. J Contam Hydrol 2000; 45(1–2): 35-53.

[112] Rabbani A. Microbe-mediated remediation of environmental contaminants. Microbe-mediated remediation of environmental contaminants. Elsevier 2021.
[http://dx.doi.org/10.1016/B978-0-12-821199-1.00022-5]

[113] Zeneli A, Kastanaki E, Simantiraki F, Gidarakos E. Monitoring the biodegradation of TPH and PAHs in refinery solid waste by biostimulation and bioaugmentation. J Environ Chem Eng 2019; 7(3): 103054.
[http://dx.doi.org/10.1016/j.jece.2019.103054]

[114] Bodor A, Petrovszki P, Erdeiné Kis Á, *et al.* Intensification of *ex situ* bioremediation of soils polluted with used lubricant oils: A comparison of biostimulation and bioaugmentation with a special focus on the type and size of the inoculum. Int J Environ Res Public Health 2020; 17(11): 4106.
[http://dx.doi.org/10.3390/ijerph17114106] [PMID: 32526873]

[115] Godheja J, Modi DR, Kolla V, Pereira AM, Bajpai R, Mishra M, *et al.* Environmental remediation: Microbial and nonmicrobial prospects. In: Singh DP, Gupta VK, Prabha R, Eds. Microbial Interventions in Agriculture and Environment: Rhizosphere, Microbiome and Agro-ecology. Singapore: Springer 2019; 2: pp. 379-409.
[http://dx.doi.org/10.1007/978-981-13-8383-0_13]

[116] Shichi and Takagi. 2000; Krishna et al, 2001. Ahmaruzzaman 2008.

[117] Ahmaruzzaman et al. 2008; Crini et al. 2006; Bastani et al. 2006.

[118] Wang et al. 2010, Huang et al. 2010.

[119] Arvand , *et al.* 2013; Chen and Wang, 2012; Gandhi et al 2012; Kyzas and Bikiaris, 2015. Anastopoulos et al 2017.

[120] Pang S, *et al.* Sorption properties of alginate for heavy metal ions. Water Res 2010; 44(19): 5191-200.

[121] Elwakeel KZ, Ahmed MM, Akhdhar A, Sulaiman MGM, Khan ZA. Recent advances in alginate-based adsorbents for heavy metal retention from water: a review. Desalination Water Treat 2022; 272: 50-74.
[http://dx.doi.org/10.5004/dwt.2022.28834]

[122] Adegoke KA, Akinnawo SO, Adebusuyi TA, *et al.* Modified biomass adsorbents for removal of organic pollutants: a review of batch and optimization studies. Int J Environ Sci Technol 2023; 20(10): 11615-44.
[http://dx.doi.org/10.1007/s13762-023-04872-2]

CHAPTER 7

Microbial Interactions with Plants and Environmental Resilience

Mohamed Ebrahim[1], Ziad Samy[2], Mohammad H. El-Zmrany[2], Kinzey M. Abohussein[2], Maria A. Farag[2] and Samah H. Abu-Hussien[3,*]

[1] *Department of Plant Pathology, Faculty of Agriculture, Ain Shams University, Cairo, Egypt*

[2] *Biotechnology Program, New Programs Administration, Faculty of Agriculture, Ain Shams University, Cairo, Egypt*

[3] *Department of Agricultural Microbiology, Faculty of Agriculture, Ain Shams University, Cairo, Egypt*

Abstract: Compared with conventional chemical fertilizers, biofertilizers, which are composed of living microorganisms that promote plant growth by increasing nutrient availability and improving soil health, present a sustainable and eco-friendly alternative. This chapter explores the fundamental principles of biofertilizers, focusing on key symbiotic relationships such as *mycorrhizal* associations and rhizobial symbiosis, both of which play crucial roles in improving nutrient uptake, particularly nitrogen, and phosphorus, which are essential for plant growth and productivity. This chapter also delves into the role of PGPRs, examining their direct effects, such as nitrogen fixation and phosphate solubilization, as well as indirect contributions, including pathogen resistance and induced systemic resistance. Despite the benefits biofertilizers provide, they face significant challenges, including issues with consistency, storage, and effectiveness under field conditions. However, advancements in biotechnology, such as the development of new microbial strains, more effective formulations, and precision application techniques, offer promising solutions to these limitations. This chapter also highlights the growing research needs in terms of understanding the complexity of microbial-plan interactions, improving biofertilizer efficiency, and expanding the applicability of biofertilizers to a broader range of crops. As global agriculture increasingly embraces sustainable practices, biofertilizers have become a key component in achieving high crop yields with reduced environmental impact.

Keywords: Biofertilizers, Mycorrhizal associations, PGPR, Plant-microbe interactions, Sustainable agriculture.

* **Corresponding author Samah H. Abu-Hussien:** Department of Agricultural Microbiology, Faculty of Agriculture, Ain Shams University, Cairo, Egypt; E-mail: samahhashemm@gmail.com

Harshita Jain & Maulin P. Shah (Eds.)

INTRODUCTION

The intricate and dynamic relationships between plants and microorganisms have captivated scientists and agriculturalists for a long time, offering insights into the fundamental processes that underpin ecosystem health and agricultural productivity [1]. At the heart of this biological interplay is the rhizosphere—a microbially rich zone around plant roots where complex and mutually beneficial exchanges occur. Within this environment, plants release organic compounds known as root exudates, which in turn attract an array of microorganisms, including bacteria, fungi, archaea, and protozoa. These microorganisms perform critical functions, such as nutrient cycling, organic matter decomposition, and plant growth regulation, all of which contribute to the health and resilience of both plants and the soil [2]. These symbiotic relationships have evolved over millions of years, leading to sophisticated communication systems between plants and their microbial partners. As global agriculture faces increasing demands for increased productivity while simultaneously confronting environmental challenges, the importance of understanding and harnessing plant-microbe interactions has never been more pressing. Sustainable agricultural systems that leverage these natural processes offer a promising pathway to address the dual challenges of food security and environmental degradation.

Symbiotic relationships, such as those involving mycorrhizal fungi and nitrogen-fixing bacteria, exemplify the potential of plant-microbe partnerships to revolutionize modern farming. *Mycorrhizal* fungi, for example, form intricate networks with plant roots, effectively extending the ability of the root system to absorb water and nutrients from the soil. In exchange, plants provide these fungi with essential carbohydrates produced through photosynthesis. This nutrient exchange not only improves plant health but also enhances soil structure and fertility, promoting sustainable crop production [3]. Similarly, nitrogen-fixing bacteria, particularly those associated with legumes, convert atmospheric nitrogen into plant-available forms, reducing the need for synthetic nitrogen fertilizers. These biological processes offer natural solutions to one of agriculture's most pressing problems: maintaining high crop yields while minimizing environmental damage. As we delve deeper into the mechanisms of these symbiotic interactions, the potential to develop biological alternatives to chemical fertilizers and pesticides becomes increasingly clear. By optimizing these relationships, farmers can reduce their reliance on agrochemicals, thereby decreasing pollution, preserving soil health, and fostering biodiversity [4].

In recent years, PGPRs have garnered significant attention for their ability to support plant growth through both direct and indirect mechanisms. PGPR can improve plant growth by facilitating nutrient acquisition, such as by solubilizing phosphate and fixing nitrogen, as well as by producing phytohormones that enhance root and shoot development. PGPR can indirectly protect plants from pathogens by producing antibiotics, competing for nutrients, and inducing systemic resistance, making plants more resilient to diseases. The use of PGPR in agriculture, along with the integration of other microbial-based technologies, represents a growing area of research with vast potential for practical applications. These microbial solutions, when integrated into traditional farming practices, offer a sustainable approach to agriculture that not only increases productivity but also mitigates the environmental impact of chemical inputs. This chapter explores the various microbial interactions in the rhizosphere, their benefits to plant health and productivity, and the emerging technologies that harness these relationships for more sustainable farming. As the global agricultural landscape continues to evolve, the optimization of plant-microbe interactions holds the key to a future where high-yielding crops and environmental stewardship go hand in hand [5 - 7].

MICROBIAL INTERACTION TYPES

Soil microorganisms play crucial roles in the soil environment through various interactions, including those with plant roots in the rhizosphere, soil components, and other microbial communities. These interactions are essential for maintaining sustainable agroecosystems, promoting plant growth, and ensuring plant health. The rhizosphere of the narrow zone of soil surrounding plant roots is distinct from that of bulk soil because of the presence of root exudates. These exudates increase nutrient availability and microbial biomass, altering the rhizosphere's environmental conditions [8]. As a result, microbial interactions in this zone are shaped not only by the microorganisms themselves but also by their interactions with plants, animals, and other soil constituents (Fig. **1**).

Microbial interactions within the rhizosphere can be broadly categorized into two types: intraspecific interactions and interspecific interactions. Intraspecific interactions involve microorganisms of the same species, whereas interspecific interactions occur between organisms of different species, such as microbial populations that interact with plants or animals.

Intraspecific Interactions

They occur exclusively among individuals within a single microbial population and can be either positive or negative.

Positive Interactions

They are also known as cooperation and contribute to the growth and success of the microbial population. For example, when a small inoculum is used (less than

10%), an extended lag phase ensures successful growth. Similarly, microbial cells may adhere to their normal habitats through minimum infectious doses or coordinate synchronized movements to form colonies. During biofilm formation, cells attach to a protective matrix, which enhances their survival. Positive interactions also facilitate cooperation in the breakdown of insoluble substrates such as lignin and cellulose through enzyme production. Moreover, genetic exchange among microbial members *via* transformation, transduction, and conjugation helps populations develop resistance to environmental stresses [9].

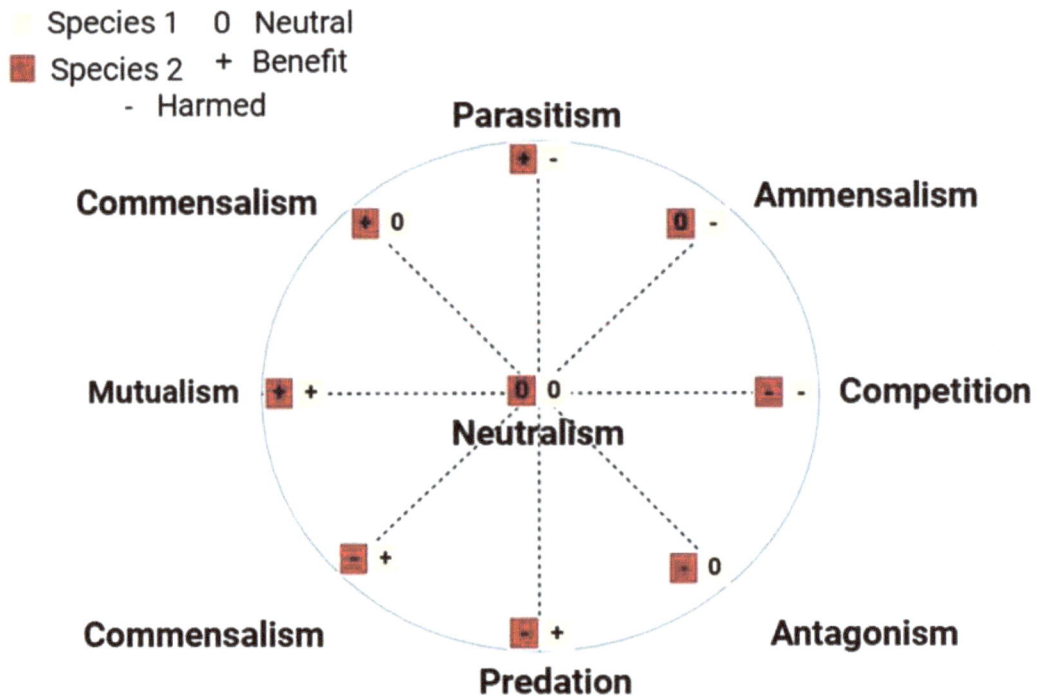

Fig. (1). Types of microbial interactions.

Negative Intraspecific Interactions

They occur through competition among individuals of the same population. Such competition plays a key role in regulating population size and density, thereby ensuring the even distribution of individuals within an ecosystem. Negative interactions often arise owing to limited nutrient availability, where all cells compete for the same substrates in shared ecological niches. Additionally, high microbial density can lead to the accumulation of toxic metabolic byproducts, further inhibiting growth and survival.

Interspecific Interactions

Neutral Interactions (Neutralism)

Soil microorganisms engage in various interactions within diverse microbial populations, which can be classified on the basis of their effects on growth and survival. When these interactions promote the growth of a microbial population, they are termed positive interactions. Conversely, when they inhibit growth or survival, they are known as negative interactions. Some interactions, however, have no significant effect and are considered neutral or indifferent.

Neutral interactions, or naturalistic interactions, occur when two distinct microbial populations share the same environment but do not influence each other in any measurable way. In such cases, neither the population benefits nor the population experiences harm, resulting in a balanced coexistence where no obvious effect is observed on their growth, survival, or metabolic activities. Neutralism is relatively uncommon in natural ecosystems because microorganisms are typically in constant competition or cooperation [10]. However, it can occur under specific conditions where resources are abundant and populations do not overlap in their ecological roles. For example, two microbial populations that consume completely different nutrients may coexist in the same habitat without interacting. Similarly, spore-forming bacteria such as *Bacillus* species might remain dormant in the soil while actively growing fungi decompose organic matter nearby, yet neither affects the other. Another example of neutralism can be observed in environments with physical barriers, such as soil particles, where microorganisms occupy isolated microenvironments. For example, aerobic bacteria residing near the surface of the soil and anaerobic bacteria deeper within the soil layers may coexist without influencing each other's activities, as they thrive under entirely different oxygen levels. In laboratory cultures, neutralism may occur when two microbial species are placed together in a medium but grow independently without interfering with each other's metabolic pathways. For example, one species might utilize a carbon source, whereas the other relies on nitrogen sources that are not limiting. Neutralism is often transient because changes in environmental conditions can quickly alter dynamics. Factors such as nutrient depletion, changes in pH, or the introduction of new species can shift a previously neutral relationship into one of competition or cooperation. For example, two microbial species that initially coexist neutrally may begin competing for a common nutrient if its availability decreases. While neutral interactions are rare compared with positive or negative interactions, they highlight the diversity and complexity of microbial relationships [11]. Microorganisms are highly adaptable, and even neutral relationships can evolve over time as environmental conditions change, leading to new interactions that shape ecosystem stability and functionality.

Positive Microbial Interactions

They play crucial roles in maintaining ecological balance and driving numerous biological processes essential for life on Earth. These interactions, where at least one organism benefits without harming others, demonstrate the cooperative nature of microbial ecosystems. Mutualism exemplifies these beneficial associations, such as the symbiotic relationship between nitrogen-fixing bacteria and leguminous plants, where bacteria provide essential nutrients in exchange for carbon compounds. Commensalism, another form of positive interaction, highlights cases where one microbe benefits from the presence of another without affecting it, as observed in the skin microbiota [12]. Synergism and protocooperation illustrate no obligatory but mutually advantageous collaborations, which are often observed in mixed microbial communities and facilitate processes such as biogas production. Together, these interactions underpin critical functions in nutrient cycling, biodegradation, and even host health, revealing the indispensable role of microbial cooperation in both natural and engineered systems.

Mutualism

Mutualism is a type of positive microbial interaction in which both organisms derive significant benefits (Fig. **2**). This interaction often involves a highly specific or obligatory relationship where the survival or optimal functioning of one or both partners depends on the association. Mutualism can occur in two forms on the basis of interaction patterns: obligate mutualism and facultative mutualism. In obligate mutualism, both partners live together in close proximity and cannot survive independently, as they rely on the interaction for survival or optimal growth. In contrast, facultative mutualism occurs when one partner can survive without the other under certain conditions, although the association remains beneficial. For various purposes, mutualism can be further divided into trophic mutualism, defensive mutualism, and service-service mutualism [13].

Trophic mutualism, also known as resource–resource interaction, involves the exchange of nutrients between two species. This form of mutualism is sometimes called syntrophism, where both partners support each other metabolically. An example of trophic mutualism is the symbiotic relationship between nitrogen-fixing bacteria (*Rhizobium*) and leguminous plants. Bacteria fix atmospheric nitrogen into ammonia, providing essential nutrients for plants, while plants supply carbohydrates and a protective environment for bacteria. Defensive mutualism occurs when one partner provides shelter or protection against predators or pathogens while the other supplies food or other resources. For example, mycorrhizal fungi associate with plant roots, offering protection from soilborne pathogens while obtaining carbohydrates from the plant. This mutualism also enhances plant nutrient uptake, tolerance to environmental stresses, and overall growth. Service–service mutualism, on the other hand, occurs when both partners provide services to each other without exchanging nutrients. While this form of mutualism is less common in soil ecosystems, it still plays a role in certain environments.

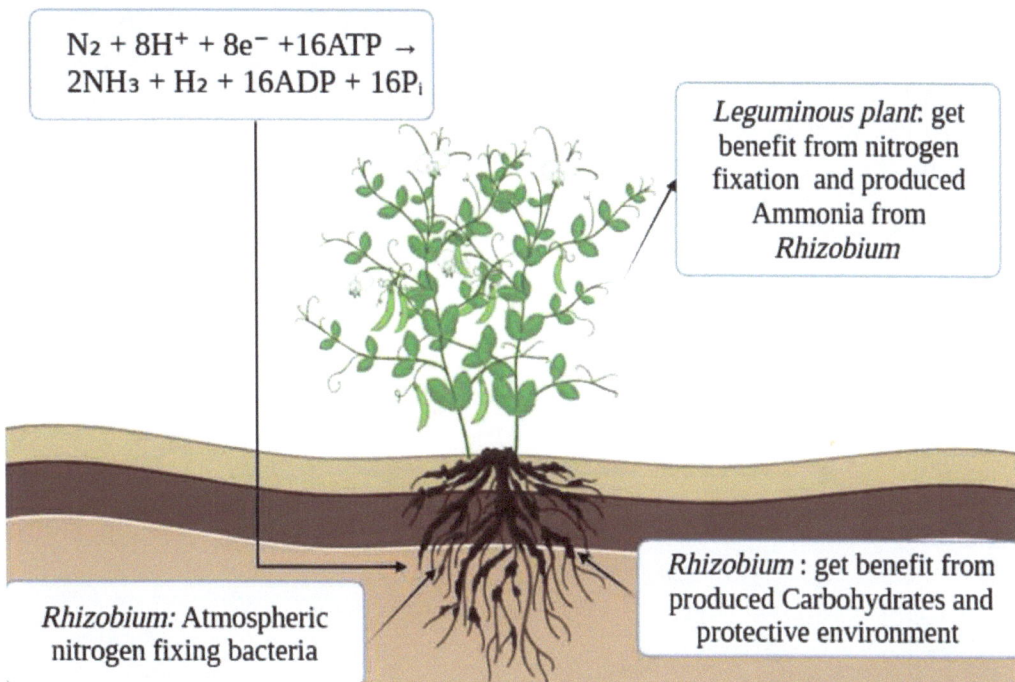

$$N_2 + 8H^+ + 8e^- + 16ATP \rightarrow 2NH_3 + H_2 + 16ADP + 16P_i$$

Leguminous plant: get benefit from nitrogen fixation and produced Ammonia from *Rhizobium*

Rhizobium: Atmospheric nitrogen fixing bacteria

Rhizobium : get benefit from produced Carbohydrates and protective environment

Fig. (2). The Mutualistic interactions between nitrogen-fixing bacteria, such as *Rhizobium* species, and the root nodules of leguminous plants, facilitating the conversion of atmospheric nitrogen (N_2) into ammonia (NH_3). This symbiotic process provides an essential nitrogen source to the plant while the bacteria receive carbon compounds and a favourable environment in return.

Well-known examples of mutualism include lichens, mycorrhizae, and symbiotic nitrogen fixation. Lichens represent a symbiotic association between a fungus and an alga or cyanobacterium. In this relationship, the fungal partner provides structural support, water retention, and protection from environmental stresses, whereas the algal or cyanobacterial partner performs photosynthesis, supplying organic nutrients to the fungus. If cyanobacteria are involved, the fungal partner also benefits from nitrogen fixation. This association improves the ecological resilience of both organisms, although the algal partner can often survive independently. Mycorrhizae represent another vital form of mutualism, where mycorrhizal fungi form associations with plant roots. Fungi increase the ability of plants to absorb water, phosphorus, nitrogen, and other minerals from the soil, while plants supply carbohydrates to fungi. This relationship also protects roots from pathogens and improves plant tolerance to abiotic stresses such as drought, salinity, and heavy metal toxicity.

Symbiotic nitrogen fixation is another essential mutualistic interaction that improves soil fertility and plant health. Nitrogen-fixing bacteria, such as *Rhizobium* species, colonize the root nodules of legumes, converting atmospheric nitrogen into ammonia, which is used by the plant for growth. Similarly, *Frankia* bacteria form nodules with the roots of nonleguminous trees such as *Alnus* and *Casuarina*, fixing nitrogen in exchange for carbohydrates provided by the plant. These mutualistic interactions demonstrate the cooperative nature of microbial ecosystems and their essential role in promoting nutrient cycling, enhancing resource utilization, and supporting biodiversity.

Commensalism

Commensalism is a form of microbial interaction in which one population benefits while the other remains unaffected, neither harmed nor benefited. This relationship is common in microbial ecosystems, where environmental conditions and available resources allow certain microbes to thrive without impacting others. It is usually unidirectional and not obligatory, meaning that the unaffected population adapts its habitat in a way that allows the benefiting population to survive. This subtle yet vital relationship helps maintain the ecological balance and supports microbial diversity across various ecosystems.

In soil environments, commensalism is often observed when one microorganism benefits from the metabolic byproducts of another. For example, fungi break down complex molecules such as cellulose and lignin through the production of extracellular enzymes, a process that enhances the nutritional properties of the soil. The resulting degraded products are utilized by other fungi and bacteria, which cannot directly degrade complex molecules. As a result, microbial activity is promoted, supporting plant growth and health. Similarly, facultative anaerobes and obligate anaerobes coexist in soil, with facultative anaerobes consuming oxygen and creating an environment where obligate anaerobes can thrive.

A well-known example of commensalism is the relationship between skin bacteria and the human host. Bacteria such as *Staphylococcus epidermidis* live on the skin and feed on oils and dead skin cells without harming the host. These microbes gain nutrients and a stable habitat, whereas the human host remains unaffected under normal conditions [14]. Similarly, in aquatic ecosystems, *Vibrio* species may attach to marine organisms such as algae, obtaining nutrients without impacting their host. In the gut microbiome of animals, certain microbes thrive on food residues or metabolites produced by others or the host again without directly affecting the host organism (Fig. **3**).

Although commensalism appears to be one-sided, it plays a crucial role in ecosystem dynamics. The microbes benefiting from this relationship may later engage in other interactions, such as mutualistic or synergistic interactions, increasing the complexity of microbial communities. This understanding is essential for applications in agriculture, medicine, and environmental management, including the development of probiotics, the promotion of soil health, and the management of microbial ecosystems.

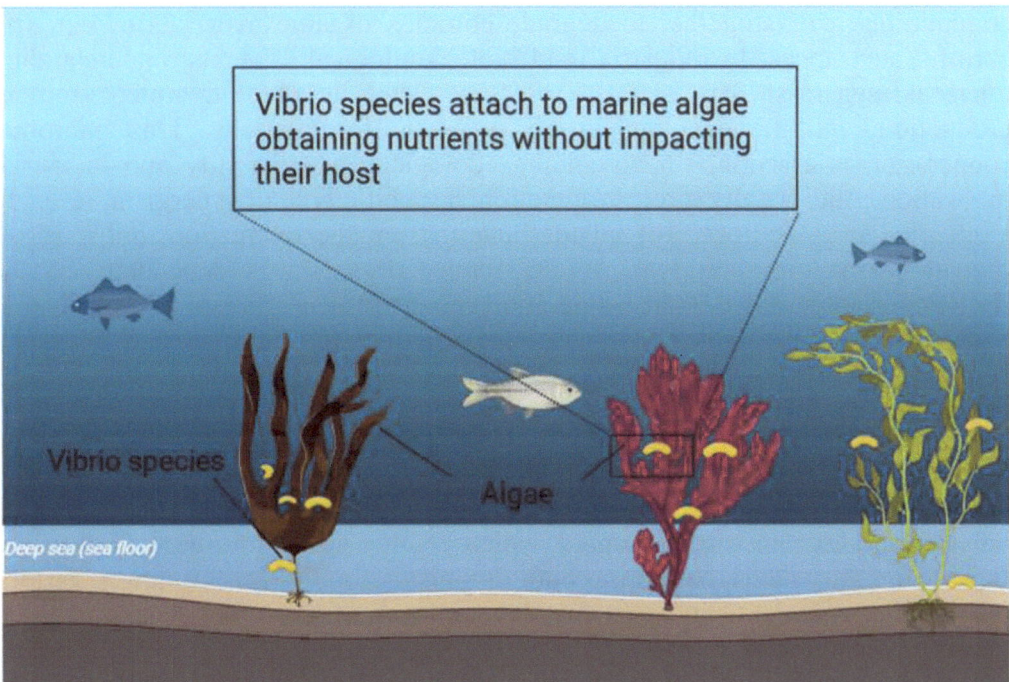

Vibrio species attach to marine algae obtaining nutrients without impacting their host

Vibrio species

Algae

Deep sea (sea floor)

Fig. (3). The Commensal interaction In aquatic ecosystems, *Vibrio* species can associate with marine organisms such as algae, deriving nutrients from their host without causing any harm to or affecting the host's normal functions.

Synergism (Protocooperation)

Synergism, also known as protocooperation, is a type of positive microbial interaction in which two or more organisms benefit from each other's presence, but the relationship is nonobligatory. Unlike mutualism, where the relationship is essential for survival, synergism allows organisms to coexist independently while achieving enhanced functionality when interacting. This interaction is particularly important in microbial ecosystems, where the combined activities of different species can drive complex biochemical processes that individual species cannot accomplish on their own. One of the best examples of synergism is anaerobic digestion, a process essential for biogas production. In this system, fermentative bacteria break down organic matter into simpler compounds such as volatile fatty acids, hydrogen, and carbon dioxide, which are then utilized by methanogenic archaea to produce methane gas [15]. Both groups of organisms benefit: fermentative bacteria maintain a favourable metabolic environment, while methanogens gain the necessary substrates for energy production, leading to efficient energy generation and organic waste recycling.

Synergism is also prevalent in natural environments such as soil, where microbial communities work together to degrade complex organic materials or detoxify harmful substances. In industrial and biotechnological contexts, understanding microbial synergism can optimize processes such as waste treatment, biofuel production, and the development of probiotic formulations. This microbial cooperation is essential for advancing ecological sustainability and industrial innovation. Additionally, protocooperation, a specific type of synergism, refers to a situation in which both organisms benefit from the interaction, but it is not obligatory for their survival. While neither organism is dependent on the relationship for survival, their cooperation enhances their efficiency or survival in a shared environment. Protocooperation is particularly significant in microbial communities, where diverse species coexist and complement each other's metabolic activities [16].

In the gut microbiome, for example, different bacterial species work together to break down complex dietary compounds. One species may degrade polysaccharides into simpler sugars, which another species ferments to produce short-chain fatty acids, benefiting both organisms. In biofilm formation, various microbial species contribute to the development of a protective extracellular matrix. Some microbes produce adhesive substances, whereas others secrete enzymes to trap nutrients or shield the community from environmental stressors. Although these species can survive individually, their cooperation within the biofilm enhances their resilience and access to nutrients. In soil ecosystems, protocooperation has also been observed between nitrogen-fixing bacteria and phosphate-solubilizing bacteria. The nitrogen fixers provide nitrogenous compounds that benefit the phosphate solubilizers, whereas the solubilizers release phosphorus, which aids the nitrogen fixers. Neither group is entirely dependent on the other, but their combined actions contribute to plant growth and improved soil fertility.

There are several types of protocooperation, particularly in terrestrial ecosystems, which have significant implications for agriculture. Nutritional protocooperation, or syntrophism, is the exchange of nutrients between microbial populations, which is highly beneficial for the organisms involved. Another type involves the metabolism of toxic end products, where one organism eliminates toxic substances, benefiting its partner, which in turn produces carbon products that are utilized by the other. A further example is the production of derivative enzymes, where organisms such as *Arthrobacter* and *Streptomyces* produce enzymes to collectively degrade toxic compounds, such as diazinon, an organophosphate pesticide (Fig. **4**). These types of microbial interactions highlight the flexible and dynamic nature of microbial ecosystems, where cooperation enhances resource efficiency and survival without obligating dependency. Understanding protocooperation offers valuable insights into microbial consortia for applications in agriculture, waste management, and bioremediation.

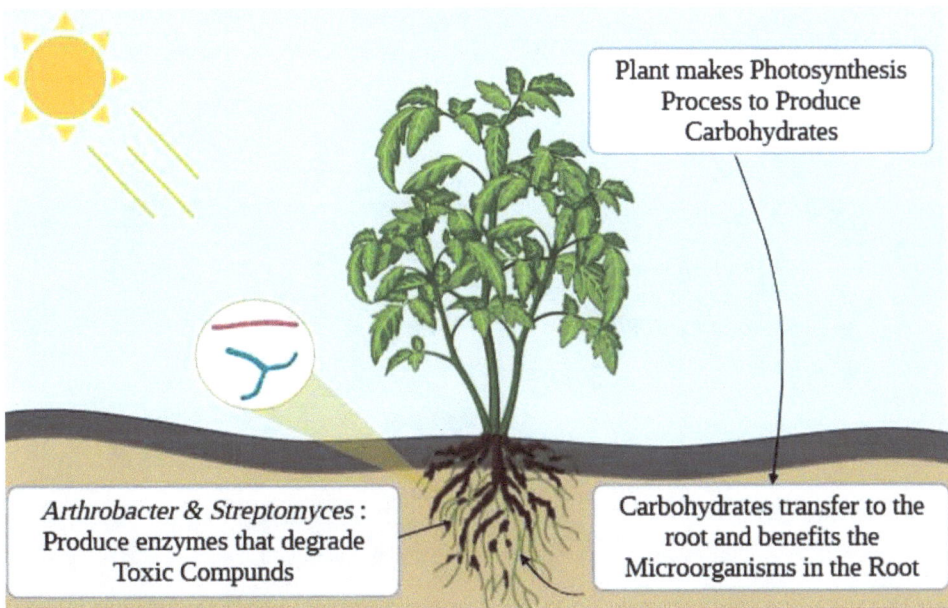

Fig. (4). Synergistic (protocooperative) interactions between *Arthrobacter* and *Streptomyces* work together by producing complementary enzymes that collaboratively break down toxic compounds, enhancing the degradation process more effectively than either could achieve alone.

Negative Microbial Interaction

Ammensalism (antagonism)

Amensalism and antagonism are types of negative microbial interactions where one population harms another without receiving a direct benefit. In amensalism, one organism suppresses or inhibits the growth and activities of another, often through the production of antimicrobial compounds or environmental modifications [17]. For example, certain microbes, such as *Streptomyces* species, produce antibiotics such as streptomycin, which inhibit soilborne pathogens such as *Phytophthora infestans*. Similarly, Penicillium species secrete penicillin, which affects bacterial growth in the rhizosphere, whereas *Lactobacillus* species lower the soil pH by producing organic acids, creating unfavourable conditions for competing microbes such as *Ralstonia solanacearum*. Although the producing organisms gain no direct benefit from these interactions, they can shape the microbial community and indirectly benefit plants by reducing harmful pathogens (Fig. **5**).

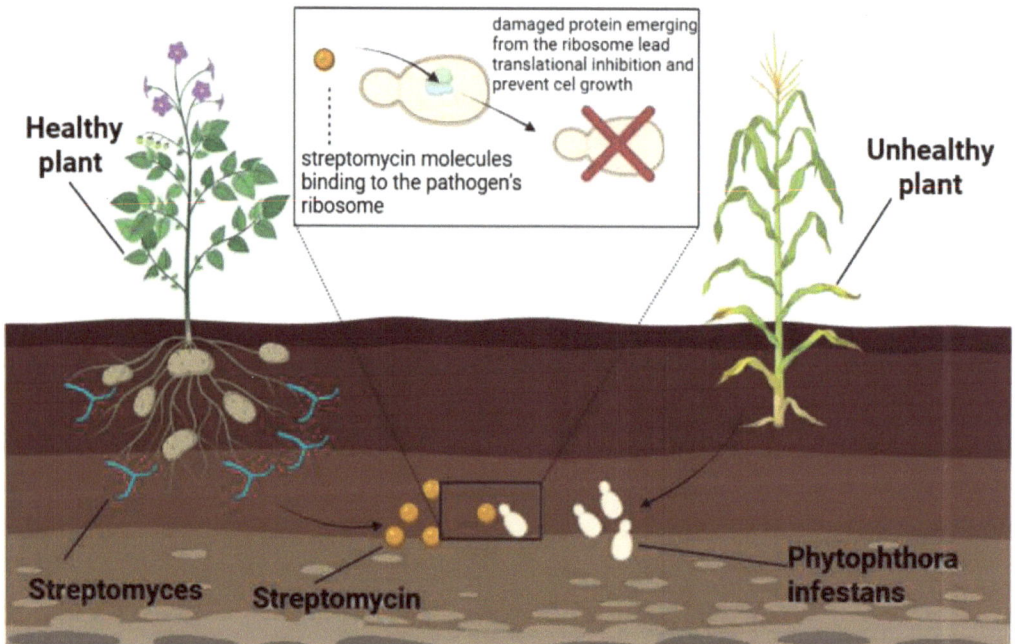

Fig. (5). Ammensalistic interactions, *Streptomyces* species produce antibiotics, such as streptomycin, which suppress the growth of soilborne pathogens such as *Phytophthora infestans*, creating an unfavourable environment for their survival without benefiting *Streptomyces* directly.

Antagonism, a more specific form of amensalism, involves the suppression or inhibition of one organism by another, typically benefiting plant health. This interaction is mediated through the production of inhibitory substances such as antibiotics, toxins, organic acids, or lytic enzymes. The organisms producing these substances are not harmed and gain a competitive advantage over their rivals. Common examples of antagonistic interactions include the production of antibiotics by *Bacillus subtilis*, *Pseudomonas fluorescens*, and Streptomyces species, which inhibit plant pathogens such as *Fusarium oxysporum* and

Rhizoctonia solani. Other antagonistic strategies involve the secretion of lytic enzymes such as glucanase, cellulase, and chitinase by soil microorganisms such as *Myxobacteria* and *Streptomyces*, which can degrade the cell walls of pathogens or inhibit their germination.

Antagonism can also manifest through compound antagonism, where an organism utilizes multiple mechanisms to suppress pathogens. For example, in lichens, algae produce oxygen, which inhibits the growth of anaerobic bacteria, whereas fungi release cyanide, which is toxic to other microorganisms. Similarly, Thiobacillus species can lower the soil pH to as low as 2 by oxidizing sulfide to sulfate, creating an environment inhospitable to pH-sensitive microbes. These antagonistic relationships play crucial roles in controlling plant diseases and maintaining ecological balance, particularly in the rhizosphere, and have important applications in sustainable agriculture and plant protection [18].

Parasitism

Parasitism is a type of negative microbial interaction in which one organism, the parasite, benefits at the expense of another organism, the host. In this relationship, the parasite derives nutrients, shelter, or other resources from the host, often causing harm. Parasitism is distinct from mutualism or commensalism, as it always has a detrimental effect on the host and can range from mild discomfort to severe disease or even death. This interaction is widespread and plays a significant role in shaping ecological and evolutionary dynamics [19].

In microbial ecosystems, parasitism involves a host-parasite relationship, where the parasite lives in close contact with the host, forming a metabolic association to feed on the host's cells, tissues, or fluids. The parasite benefits from this interaction, while the host is adversely affected. In some cases, this relationship may evolve into a pathogenic relationship, where the parasite causes significant disease or damage. For example, *Mycobacterium tuberculosis*, the bacterium responsible for tuberculosis, infects human lungs and uses host cells for replication while causing tissue damage and illness. Similarly, parasitic fungi such as *Candida albicans* can infect humans, leading to conditions such as oral thrush or systemic candidiasis.

In agriculture, parasitic relationships are observed in plant pathogens such as *Puccinia graminis*, a wheat rust fungus that extracts nutrients from plant tissues, weakening the host and reducing crop yield. This form of parasitism is harmful to plants, in contrast to beneficial parasitism, where parasites target harmful microbes or pathogens. For example, parasitism by bacteria, such as bacteriophages infecting harmful bacteria, can be useful for plant growth, as it may control plant disease-causing pathogens (Fig. **6**).

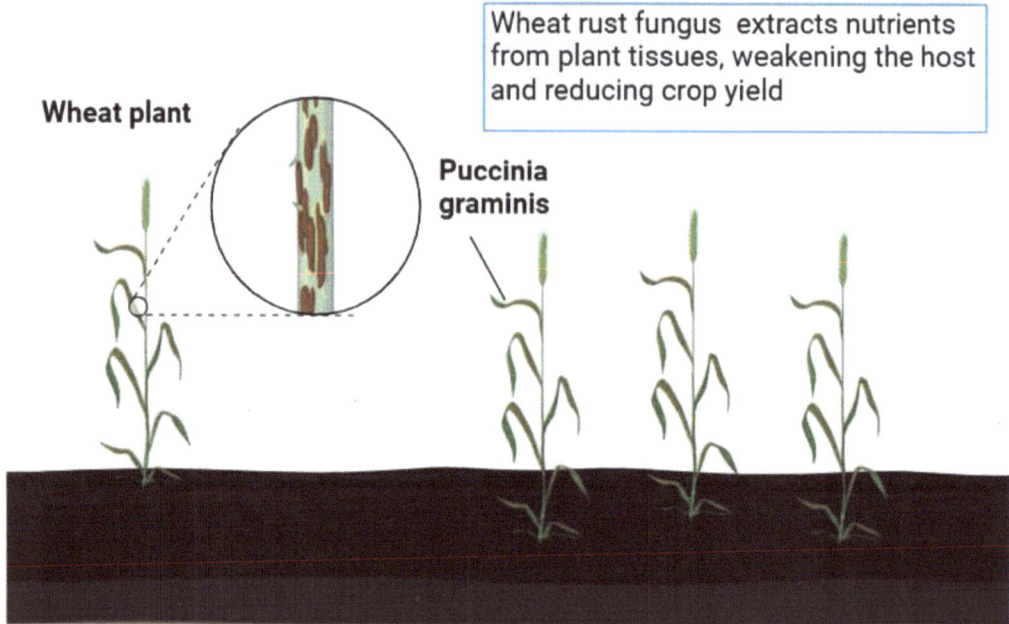

Fig. (6). Parasitism interaction. *Puccinia graminis*, commonly known as a wheat rust fungus, parasitizes plant tissues by extracting nutrients from its host. This interaction weakens the host plant and significantly reduces crop yields.

Parasitism can be classified on the basis of the nature of the interaction and the type of infection [20]. Obligate parasitism occurs when a parasite cannot survive without its host, whereas facultative parasitism allows the parasite to survive independently under certain conditions. Parasitism can also be categorized by infection form: ectoparasitism, where the parasite remains outside the host, and endoparasitism, where the parasite penetrates host cells.

Examples of parasitism include viruses such as bacteriophages, which infect bacteria; fungi that parasitize algae or other fungi; and chytrid fungi, which penetrate the host's tissues. These parasitic interactions are crucial in regulating microbial populations, influencing coevolutionary adaptations, and maintaining ecosystem balance. Despite their harmful nature, parasitism plays an important role in controlling populations, preventing overgrowth, and shaping ecological dynamics. Understanding parasitic relationships is vital for applications in medicine, agriculture, and biotechnology, where it aids in the development of treatments for parasitic infections, pest management, and even harnessing certain parasites for beneficial purposes.

Competition

Competition in plant-microbe interactions occurs when microorganisms exploit limited resources, such as nutrients, space, or plant-derived compounds, influencing the composition and dynamics of the microbial community in the rhizosphere and affecting plant health. Microorganisms compete for essential resources such as water, light, oxygen, and nutrients, which are crucial for their survival and reproduction [21]. In this competition, superior or better-adapted microorganisms typically dominate and can even eliminate others that rely on the same inadequate resources. Organisms with faster growth rates are often considered superior competitors, as they can outpace others in acquiring the necessary resources.

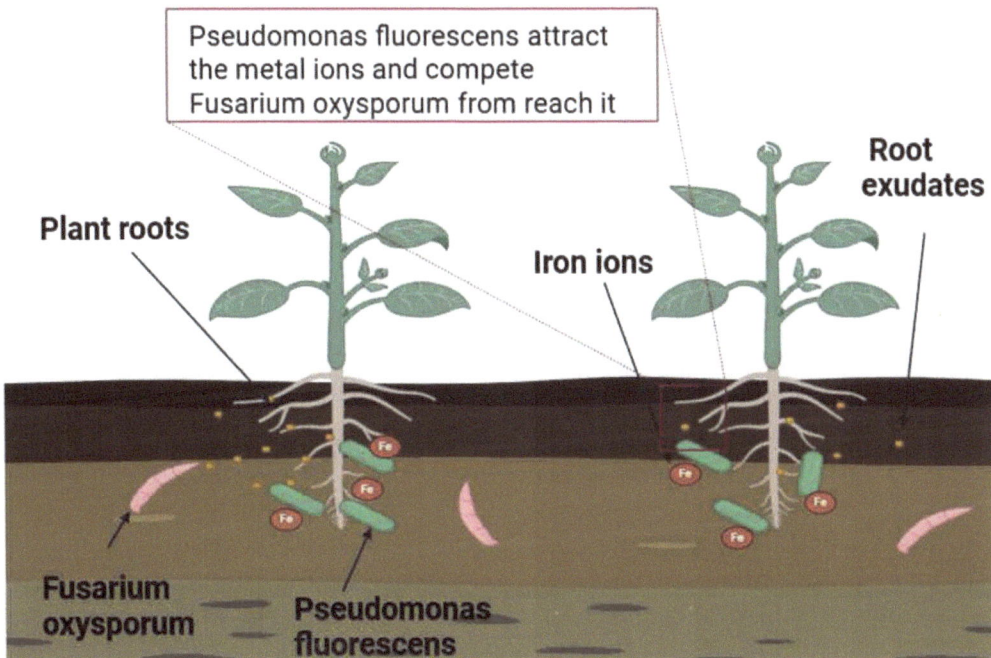

Fig. (7). Competition interaction *Pseudomonas fluorescens* inhibits the growth of pathogens such as *Fusarium oxysporum* by competing more effectively for essential nutrients and space in the root environment, thereby suppressing the ability of the pathogen to thrive.

Beneficial microbes can outcompete plant pathogens by rapidly utilizing root exudates, occupying available space on root surfaces, or producing siderophores that bind to and sequester essential nutrients such as iron, making them less available to pathogens. For example, *Pseudomonas fluorescens* suppresses pathogens such as *Fusarium oxysporum* by outcompeting them for nutrients and space (Fig. **7**), whereas *Trichoderma* spp. can dominate harmful fungi such as

Botrytis cinerea. Plants actively participate in this competitive process by releasing specific exudates that favour the growth of beneficial microbes, such as symbiotic rhizobia, over other microorganisms. This form of competition not only suppresses the growth of harmful pathogens but also promotes a microbiome that supports plant growth, making it an important tool for sustainable agricultural practices [22].

There are two primary types of competition in microbial communities: resource competition and interference competition. Resource competition, also known as indirect or exploitative competition, occurs when multiple populations compete for the same limiting resource and when one population reduces the availability of that resource for others. In contrast, interference competition, or direct competition, occurs when one population physically or chemically harms another, excluding it from the environment. For example, certain soil microorganisms can deplete essential nutrients required for the germination of plant pathogens such as *Fusarium* and *Aphanomyces*, thereby reducing the population of these pathogens in the soil and benefiting plant health.

Predation

Predation is one of the most intense relationships among microorganisms, in which a predator organism directly attacks and feeds on its prey. This interaction typically has a short duration, and while predators may or may not kill their prey before feeding, the result is generally the absorption of the prey's tissue, leading to its eventual death. The predator may be larger or smaller than the prey, depending on the species involved. Predation plays an important role in microbial ecosystems by facilitating nutrient cycling. Predators can mineralize organic compounds produced by autotrophs, preventing them from reaching higher consumers and thus enhancing nutrient cycling [23]. This process returns nutrients to primary producers, stimulating their activity and improving soil nutrition. Additionally, predators help control populations of potentially harmful organisms, such as plant pathogens, by consuming them.

Examples of predatory microorganisms in soil include *Bdellovibrio bacteriovorus*, a predatory bacterium that invades the cell wall of its prey and multiplies between the wall and plasma membrane, causing the prey to lyse and release its progeny. *Bdellovibrio* attacks various bacterial strains, including *Escherichia coli*, *Aquaspirillum serpens*, *Salmonella typhimurium*, and *Helicobacter pylori* (Fig. **8**). Another example is *Vampirococcus* spp., which attach to the surface of phototrophic bacteria such as *Chromatium* spp. (purple sulfur bacteria). Unlike *Bdellovibrio*, *Vampirococcus* does not penetrate prey cells but remains attached to the cell wall and consumes them externally. *Daptobacter* spp. also engages in predation by penetrating and degrading the cytoplasm of several *Chromatiaceae* genera, growing and propagating within the cytoplasm of their prey. These predatory microorganisms play significant roles in controlling microbial populations and influencing nutrient dynamics in the soil.

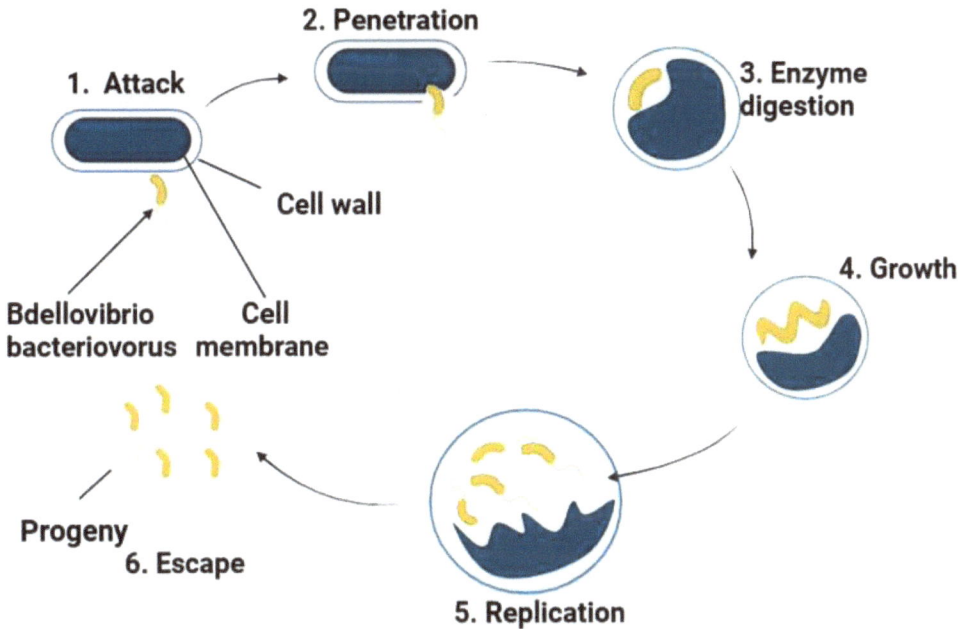

Fig. (8). Predatory interaction *Bdellovibrio bacteriovorus*, a predatory bacterium, invades the cell wall of its prey and multiplies in the periplasmic space between the wall and the plasma membrane. This process ultimately causes the prey cell to lyse, releasing the progeny of the predator.

Symbiotic Relationships

Symbiotic relationships are essential interactions between different species that live in close proximity, often resulting in mutual benefits for the organisms involved [24]. One of the most fascinating examples of such a relationship in nature is the association between plants and fungi, known as mycorrhizae. These symbiotic relationships have been vital to the evolution and success of terrestrial plants, helping them access critical nutrients from the soil. Mycorrhizal associations are so widespread that they are found in more than 90% of terrestrial plant species, illustrating their importance in plant survival and ecosystem functioning. Through these associations, plants and fungi engage in an exchange of resources that benefits both partners, ultimately increasing plant growth, improving soil structure, and increasing resilience to environmental stresses.

TYPES OF ENDOPHYTES

Endophytes can be broadly classified into two main categories: fungal endophytes and bacterial endophytes [25]. Both types interact with plants in distinct ways and contribute to plant health through different mechanisms, as shown in Fig. (**9**).

Fungal Endophytes

These are the most studied group of endophytes and include a wide range of fungi that colonize the interior of plant tissues. Fungal endophytes can be further divided into two categories:

1. **Systemic Endophytes**: These fungi spread throughout the entire plant, colonizing multiple tissues, including roots, stems, and leaves. Some systemic endophytes are vertically transmitted from parent to offspring *via* seeds, ensuring a continuous association across plant generations.
2. **Localized Endophytes:** These fungi colonize specific plant tissues, such as leaves or roots, and are typically horizontally transmitted, meaning that they are acquired from the environment during a plant's lifetime.

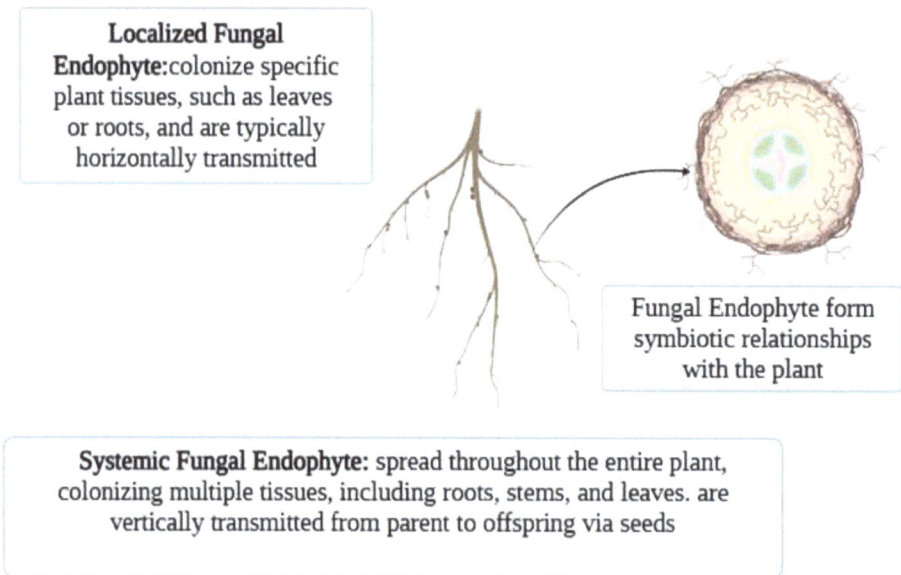

Localized Fungal Endophyte:colonize specific plant tissues, such as leaves or roots, and are typically horizontally transmitted

Fungal Endophyte form symbiotic relationships with the plant

Systemic Fungal Endophyte: spread throughout the entire plant, colonizing multiple tissues, including roots, stems, and leaves. are vertically transmitted from parent to offspring via seeds

Fig. (9). Types of endophytes involved in the colonization of plants by fungal and bacterial endophytes, which reside within plant tissues without causing harm. Fungal endophytes form symbiotic relationships with plants, often residing in intercellular spaces or within plant cells, contributing to plant health and stress resistance. Bacterial endophytes, on the other hand, inhabit plant tissues, aiding in nutrient cycling, disease resistance, and growth promotion.

Bacterial Endophytes

Bacterial endophytes are less well known than their fungal counterparts but are equally important in contributing to plant health. They inhabit the internal tissues of plants, particularly the roots, and can provide various benefits, such as increased nutrient uptake, the production of plant growth-promoting hormones, and protection against pathogens. These bacteria may enter plant tissues through natural openings such as stomata or root cracks or through root exudates that attract them to the rhizosphere [26].

Colonization Patterns

The colonization of plants by endophytes is a complex process that varies depending on the type of endophyte and the plant species [27]. In many cases, endophytes enter plant tissues through natural openings or injuries, but some endophytes have evolved specialized mechanisms to penetrate intact plant cells. Once inside the plant, endophytes establish themselves in various tissues, where they either remain localized or spread throughout the plant.

1. **Fungal Endophyte Colonization:** Fungal endophytes typically colonize plant tissues by growing hyphae that extend between plant cells or, in some cases, into plant cells. Systemic fungal endophytes often colonize the plant symplastically, meaning that they move within the plant's cells, whereas localized endophytes remain confined to certain tissues, such as leaves or roots. Some fungal endophytes can form structures such as *Appressoria* or *Haustoria* that allow them to penetrate plant cells and establish a symbiotic relationship.
2. **Bacterial Endophyte Colonization:** Bacterial endophytes typically colonize the apoplast, the space between plant cells, without penetrating the cells themselves. These bacteria form colonies inside the plant tissue and are often concentrated in the root cortex, vascular tissues, or other intercellular spaces. Some bacterial endophytes can also form biofilms inside plants, increasing their ability to persist and interact with the host over extended periods [28].

Colonization is not always uniform across the plant; different endophytes may specialize in certain tissues, leading to distinct patterns of distribution within the plant. For example, some endophytes may prefer root tissues, whereas others may dominate leaves or stems. These colonization patterns are influenced by factors such as plant species, environmental conditions, and the specific physiological needs of the endophyte (Fig. **10**).

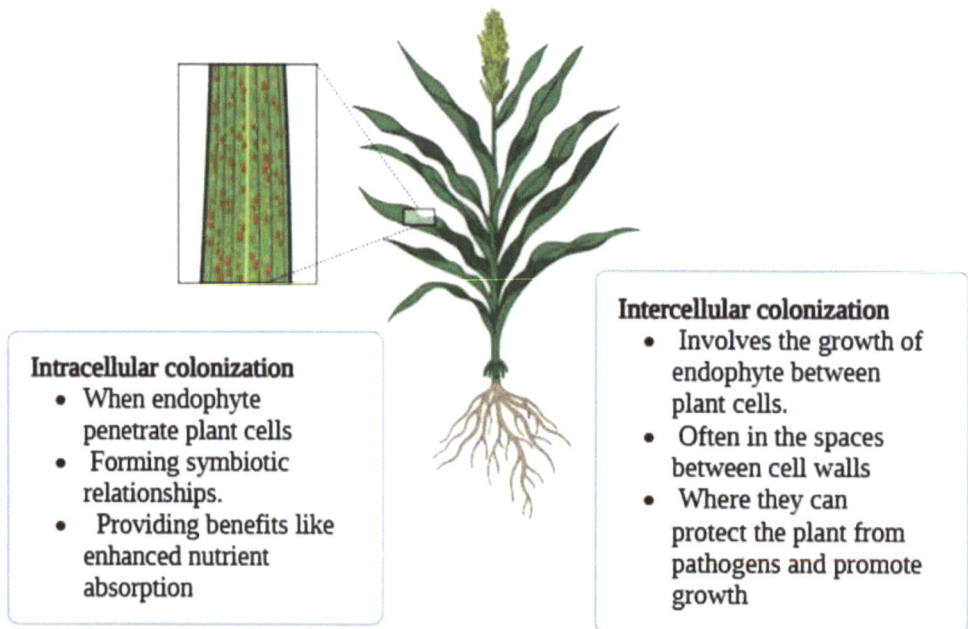

Intracellular colonization
- When endophyte penetrate plant cells
- Forming symbiotic relationships.
- Providing benefits like enhanced nutrient absorption

Intercellular colonization
- Involves the growth of endophyte between plant cells.
- Often in the spaces between cell walls
- Where they can protect the plant from pathogens and promote growth

Fig. (10). Intracellular and Intercellular Colonization Patterns of Endophytes in Plant Tissues This figure illustrates the different colonization patterns of endophytes within plant tissues. Intracellular colonization occurs when endophytes penetrate plant cells, forming symbiotic relationships and providing benefits such as enhanced nutrient absorption. Intercellular colonization involves the growth of endophytes between plant cells, often in the spaces between cell walls, where they can protect the plant from pathogens and promote growth.

Benefits to Host Plants

Endophytic associations offer a range of benefits to host plants, many of which contribute to the overall fitness, growth, and ability of plants to cope with environmental stressors. These benefits often depend on the type of endophyte involved and the nature of the symbiotic relationship between the endophyte and the plant.

1. **Enhanced Stress Tolerance:** One of the key benefits of endophytic associations is improved tolerance to environmental stresses such as drought, salinity, and extreme temperatures. Endophytes can help plants produce stress-responsive compounds such as osmoprotectants, which help the plant retain water during drought conditions or help manage the ion balance in saline environments (Fig. **11**).
2. **Pathogens and Pest Resistance:** Many endophytes produce antimicrobial compounds that protect the host plant from pathogenic bacteria, fungi, and viruses. They can also induce systemic resistance in plants, priming the plant's immune system to respond more effectively to future pathogen attacks.

Additionally, some endophytes produce metabolites that deter herbivores or attract natural enemies of pests, providing an extra layer of protection.

3. **Improved Nutrient Uptake:** Endophytes, particularly bacteria, can increase plant growth by facilitating nutrient acquisition. Some bacterial endophytes can fix atmospheric nitrogen, making it available to plants in a form that can be used for growth. Others produce enzymes that solubilize phosphate or release iron from soil minerals, increasing the availability of these critical nutrients to the plant.

4. **Growth Promotion:** Endophytic bacteria and fungi can produce plant hormones such as auxins, gibberellins, and cytokinins, which directly stimulate plant growth and development. These hormones can increase root growth, increase plant biomass, and improve the overall productivity of plants. Some endophytes can also alter a plant's hormonal balance to improve its tolerance to biotic and abiotic stresses.

Endophytes thus play a vital role in supporting plant health, growth, and resilience, making them valuable allies in natural ecosystems and agriculture. The potential for their use in sustainable agricultural practices, such as biocontrol agents and biofertilizers, is a growing area of research, with significant implications for increasing crop yields and reducing the need for chemical inputs.

Mycorrhizal associations

Mycorrhizal associations refer to the mutualistic relationships between fungi and plant roots. The fungi colonize the plant's root system, extending their hyphae into the soil, which significantly increases the surface area for water and nutrient absorption. In exchange for this expanded access to nutrients, plants provide fungi with carbohydrates, which the fungi cannot produce on their own. This interaction has developed over millions of years and is crucial for plant growth, especially in nutrient-poor soils. There are two main types of mycorrhizae, Arbuscular Mycorrhizae (AM) and ectomycorrhizae (EM), each of which plays a unique role in the nutrient cycling process [29].

Types of Mycorrhizae

1. **Arbuscular Mycorrhizae (AM):** These are the most common type of mycorrhizal association and are formed by fungi from the phylum Glomeromycota. AM fungi penetrate root cell walls and form arbuscules, which are small, treelike structures inside plant root cells. These arbuscules are the site of nutrient exchange, where fungi deliver nutrients such as phosphorus to the plant in return for sugars. AM fungi are particularly important in the uptake of phosphorus, a vital element that is often limited in many soils.

2. **Ectomycorrhizae (EMs):** These fungi do not penetrate the root cells but instead form a sheath or mantle around the root. The fungal hyphae extend into the soil, where they help with the uptake of nitrogen, phosphorus, and other essential nutrients. EM associations are commonly found in woody plants such as trees in temperate and boreal forests. Unlike AM fungi, EM fungi are involved in more complex nutrient exchanges and are particularly effective in decomposing organic matter, making them key players in nutrient cycling in forest ecosystems (Fig. **12**) [30].

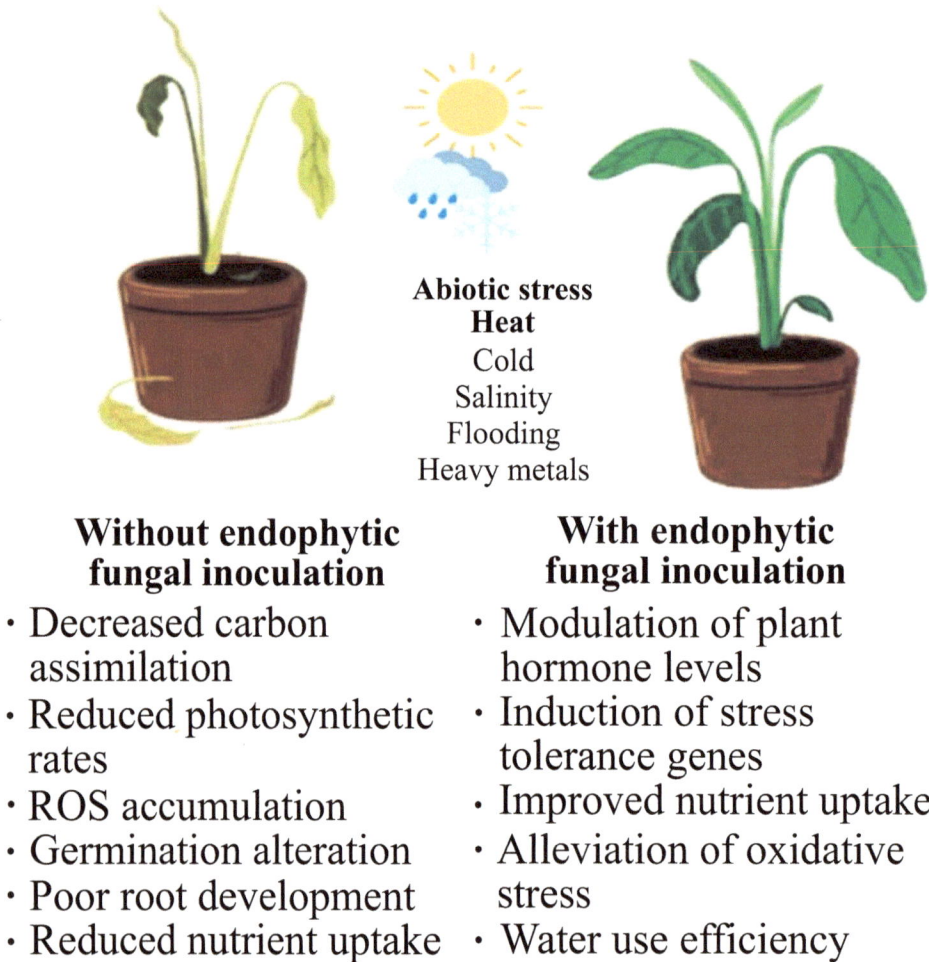

Abiotic stress
Heat
Cold
Salinity
Flooding
Heavy metals

Without endophytic fungal inoculation

· Decreased carbon assimilation
· Reduced photosynthetic rates
· ROS accumulation
· Germination alteration
· Poor root development
· Reduced nutrient uptake

With endophytic fungal inoculation

· Modulation of plant hormone levels
· Induction of stress tolerance genes
· Improved nutrient uptake
· Alleviation of oxidative stress
· Water use efficiency

Fig. (11). Stress tolerance mechanism. This figure depicts how endophytes help plants cope with various environmental stresses, such as drought, salinity, and extreme temperatures. Endophytes increase plant resilience by producing protective compounds, increasing water uptake, modulating stress-related genes, and increasing nutrient availability. Through these mechanisms, endophytes help maintain plant health and growth under challenging conditions.

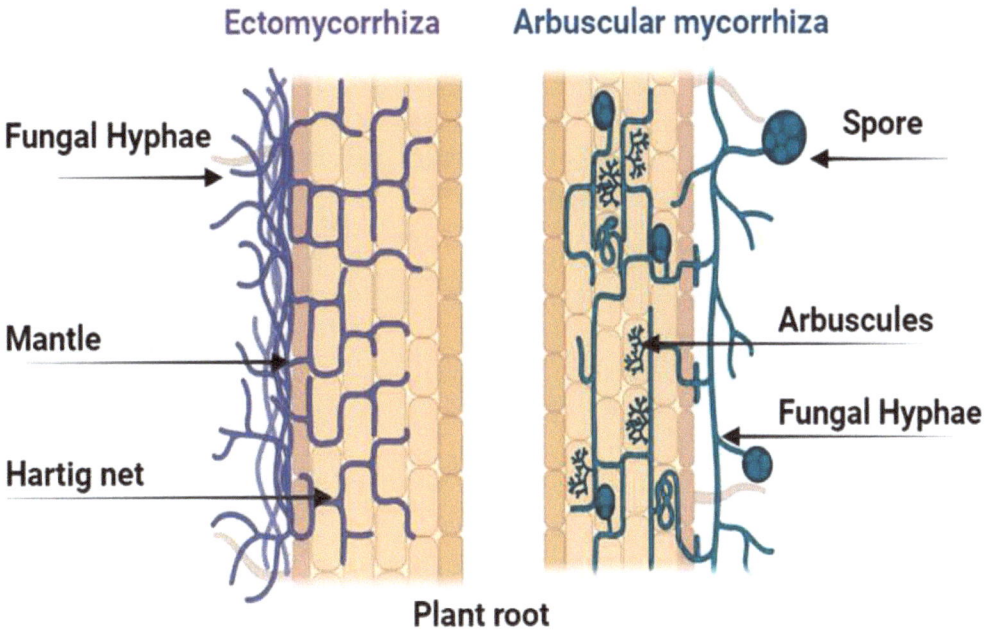

Fig. (12). Types of mycorrhizae: A comparative illustration showing the structural and functional differences between AM and Ectomycorrhizae (EM).

Arbuscular Mycorrhizae (AM): The AM fungi penetrate root cortical cells, forming arbuscules (branched structures) and vesicles within the cells. This enhances nutrient exchange between the fungi and the plant.

The EM fungi form a dense hyphal sheath around the roots and a Hartig net in the intercellular spaces without penetrating root cells, facilitating nutrient absorption and transfer mechanisms of nutrient exchange. The exchange of nutrients in mycorrhizal associations is a highly efficient process. The fungal hyphae extend far beyond the root zone of the plant, accessing areas of soil that would otherwise be unreachable [31]. In arbuscular mycorrhizae, the arbuscules formed inside the plant roots allow direct nutrient transfer between the plant and the fungus. Fungi absorb nutrients from the soil, particularly phosphorus, nitrogen, and micronutrients such as zinc, and transport them to plants. In return, the plant provides the fungus with carbon compounds, primarily in the form of sugars produced through photosynthesis. This mutualistic exchange not only benefits the plant but also ensures the survival and proliferation of the fungi.

Benefits to Host Plants

Mycorrhizal associations confer a wide range of benefits to host plants. First and foremost, they increase nutrient uptake, particularly of immobile nutrients such as phosphorus, which plants cannot easily access on their own. The extended reach of fungal hyphae into the soil allows for more efficient absorption of water and nutrients, improving plant growth and productivity. Mycorrhizae also increase plant resilience to environmental stresses, such as drought, salinity, and heavy metal toxicity. In some cases, mycorrhizal fungi can also protect plants from root pathogens by forming a physical barrier or by outcompeting harmful microbes in the rhizosphere [32]. Additionally, this association helps improve the soil structure by promoting the aggregation of soil particles, which enhances root penetration and water retention (Fig. **13**).

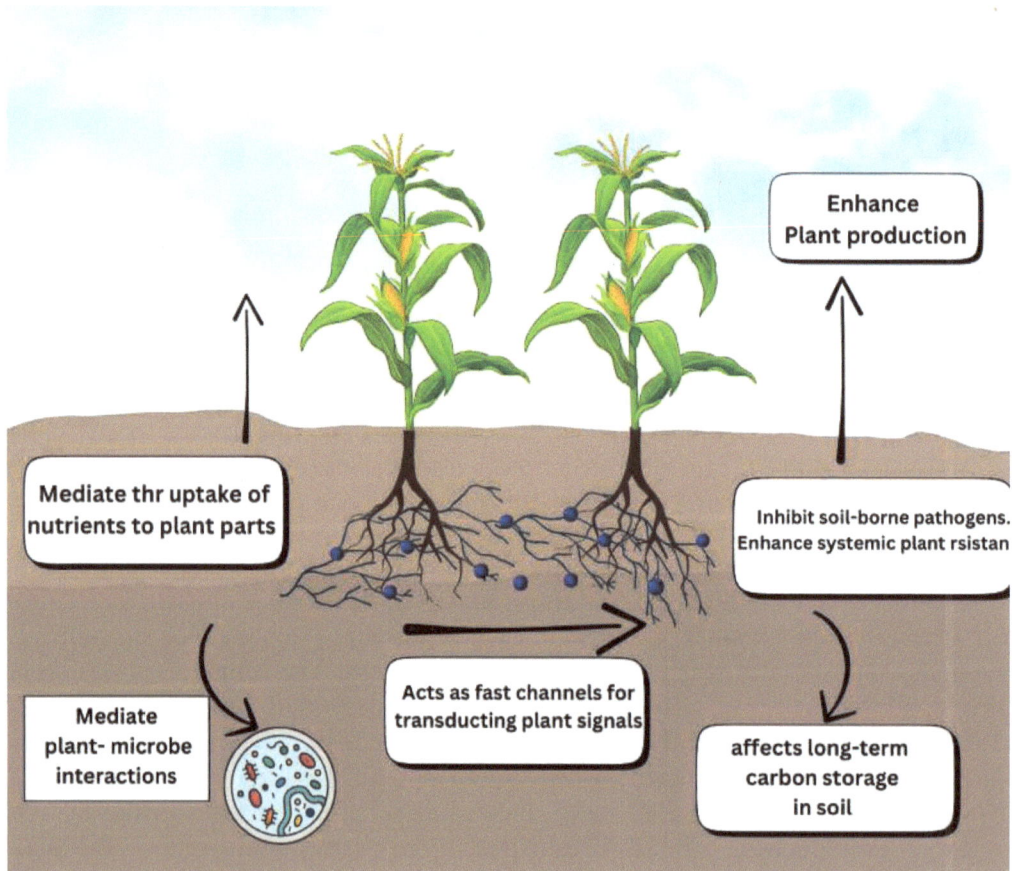

Fig. (13). Mycorrhizal associations provide benefits to host plant *mycorrhizal* fungi, establishing mutualistic relationships with host plants and increasing nutrient uptake, especially phosphorus and nitrogen. The fungi extend their hyphal networks into the soil, improving water and mineral absorption by the plant. In return, the plant supplies the fungi with organic carbon compounds. These associations contribute to plant health, growth, and stress resistance.

AGRICULTURAL APPLICATIONS

In agriculture, mycorrhizal fungi are increasingly recognized for their potential to improve crop yields and promote sustainable farming practices. By fostering these symbiotic relationships, farmers can reduce the need for chemical fertilizers, particularly phosphorus-based fertilizers, which are often overused and can lead to environmental problems such as eutrophication. Mycorrhizal inoculants, which contain beneficial fungi, are now commercially available and are used to treat crop seeds or soil to encourage these symbiotic partnerships. These practices increase nutrient uptake, improve soil health, and can lead to more resilient crops, reducing the need for synthetic inputs. Additionally, mycorrhizal fungi are being explored as tools for restoring degraded lands, as they help improve soil structure and fertility, creating more favourable conditions for plant growth in challenging environments.

Rhizobial Symbiosis

Nodule Formation Process

The symbiosis between rhizobia and legume roots begins with a complex and highly regulated process of nodule formation. Nodules are specialized structures on plant roots where nitrogen fixation occurs. The process starts with chemical signals exchanged between the legume plant and rhizobia [33]. The plant releases flavonoids into the soil, which act as chemical attractants for rhizobia. In response, rhizobia produce nodulation (Nod) factors, which trigger the plant to initiate root hair curling and the formation of a structure known as an infection thread. The infection thread allows the bacteria to invade the root cells and travel into the root cortex, where they stimulate the formation of nodules. Inside the nodules, Rhizobia species differentiate into Bacteroides, which actively fix nitrogen. These *bacteria* are housed within plant cells in structures called symbiosomes, where nitrogen fixation takes place. The nodule formation process is highly specific and tightly controlled to ensure the successful establishment of this symbiotic relationship (Fig. **14**).

Nitrogen Fixation Mechanisms

The central role of rhizobial symbiosis is nitrogen fixation, a process in which atmospheric nitrogen is converted into ammonia. The enzyme responsible for this conversion is nitrogenase, which is produced by Rhizobia inside the root nodules. Nitrogenase is highly sensitive to oxygen, so the nodule environment must be carefully regulated to allow nitrogen fixation to occur. To achieve this, legume nodules contain a protein called leghemoglobin, which binds oxygen and maintains the low-oxygen conditions necessary for nitrogenase to function while still supplying bacteroids with enough oxygen to carry out respiration. The ammonia produced by nitrogenase is then assimilated by the plant and used to synthesize amino acids and other nitrogen-containing compounds, fuelling plant growth. This process of biological nitrogen fixation is critical for reducing the need for synthetic nitrogen fertilizers, making legume crops environmentally sustainable and economically valuable in agriculture (Fig. **15**).

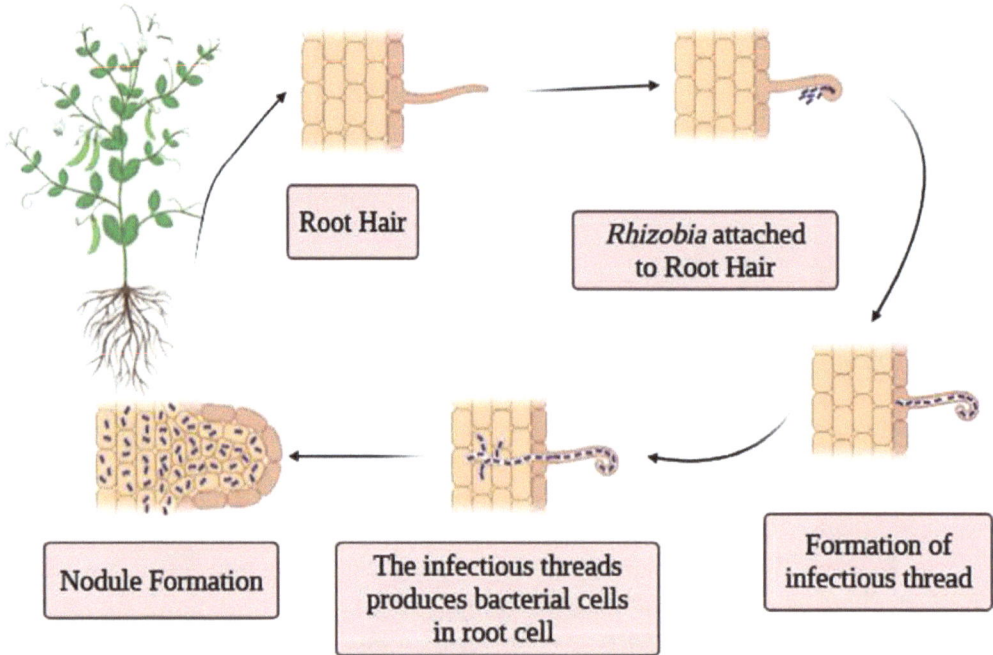

Fig. (14). Nodule formation process. Stages of the symbiotic relationship between legumes and Rhizobia, highlighting the key steps in the formation of root nodules for nitrogen fixation. This process begins with the recognition and attachment of Rhizobia to the legume roots, followed by the formation of infection threads, nodule development, and the establishment of a nitrogen-fixing environment inside the nodules.

Host Specificity

Rhizobial symbionts exhibit a high degree of host specificity, meaning that different species of rhizobia form symbiotic relationships with specific legume species. For example, *Rhizobium leguminosarum* is associated with peas and lentils, whereas *Bradyrhizobium japonicum* is associated with soybeans. This specificity is controlled by a complex interplay of chemical signals between the plant and Rhizobia. The flavonoids released by plant roots are specific to the plant species and determine which rhizobial species will respond. Similarly, the Nod factors produced by Rhizobia are recognized by specific receptors on plant roots, ensuring that only compatible bacteria initiate nodule formation. This specificity of Host rhizobia is crucial for the establishment of successful symbioses and for the effectiveness of nitrogen fixation in different legume species [34].

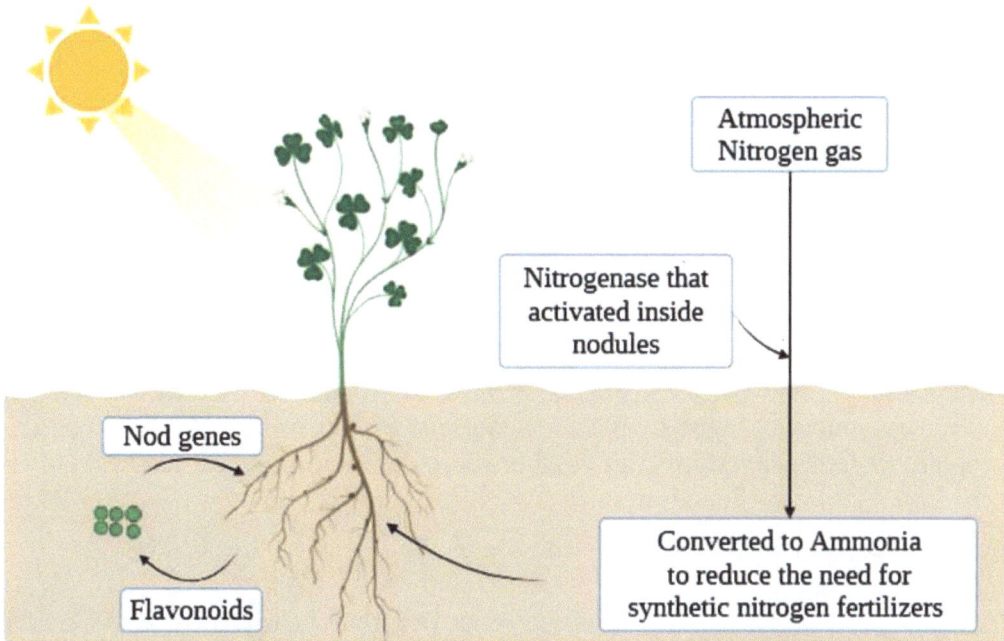

Fig. (15). The process of nitrogen fixation within a legume root nodule, where Rhizobia bacteria convert atmospheric nitrogen (N_2) into ammonia (NH_3). This process involves the enzyme nitrogenase, which is activated in a low-oxygen environment inside the nodule. The plant provides the necessary energy and reduces power for this conversion, allowing the plant to utilize fixed nitrogen for growth and development.

Agricultural Importance in Legume Crops

Rhizobial symbiosis has significant agricultural importance, especially in the cultivation of legume crops. Owing to their ability to fix atmospheric nitrogen, legumes require little to no nitrogen fertilizers, making them an important component of sustainable agriculture. The nitrogen fixed by rhizobia not only supports the growth of the legume plants themselves but also enriches the soil with nitrogen, benefiting subsequent crops grown in rotation. This process, known as green manuring, improves soil fertility and reduces the need for chemical fertilizers, which are costly and can contribute to environmental pollution. In addition to improving soil health, rhizobial symbiosis enhances the nutritional value of legume crops, making them a key source of protein in both human and animal diets. By reducing reliance on synthetic nitrogen fertilizers, rhizobial symbiosis supports more sustainable farming practices, lowers greenhouse gas emissions, and promotes biodiversity in agricultural systems.

PLANT GROWTH-PROMOTING RHIZOBACTERIA (PGPR)

PGPRs are a group of beneficial bacteria that colonize plant roots and increase plant growth and health. These bacteria live in the rhizosphere—the soil region immediately surrounding the plant roots—where they interact with plants in various ways, from improving nutrient availability to increasing resistance to diseases and stress [35]. PGPR is essential for sustainable agriculture, as it offers natural alternatives to chemical fertilizers and pesticides by promoting plant growth through biological means. The mechanisms by which PGPR benefit plants include nutrient solubilization, phytohormone production, nitrogen fixation, and protection against pathogens.

Major Groups of PGPR

Several key genera of PGPR are particularly well-studied and widely used in agricultural practices because of their beneficial effects on crops. These bacteria belong to different taxonomic groups but share common features that allow them to increase plant growth and productivity [36].

Pseudomonas

The genus *Pseudomonas* is one of the most prominent groups of PGPR and is known for its ability to promote plant growth and protect plants against diseases. These bacteria are characterized by their metabolic versatility and ability to colonize the rhizosphere efficiently. They can produce a wide range of secondary metabolites, including antibiotics, siderophores (iron-chelating compounds), and enzymes that degrade plant pathogenic microorganisms. Additionally, *Pseudomonas* species can produce phytohormones such as auxins, which promote root elongation and overall plant growth.

Disease suppression: *Pseudomonas* species are often used in biocontrol because of their ability to suppress plant pathogens by producing antimicrobial compounds that inhibit the growth of harmful fungi and bacteria [37].

Induced Systemic Resistance (ISR): Some strains of Pseudomonas can trigger a plant's immune system, activating its defense mechanisms against future pathogen attacks, even in parts of the plant that are not in direct contact with the bacteria.

Bacillus

The *Bacillus* genus includes several species of PGPR that are widely used in agriculture because of their ability to form endospores, which allows them to survive in harsh environmental conditions. *Bacillus* species are highly effective at promoting plant growth through multiple mechanisms, including nitrogen fixation, phosphate solubilization, and the production of growth-promoting hormones [38]. They also play a key role in protecting plants from pathogens by producing antibiotics and other antimicrobial compounds.

Phosphate solubilization: Many *Bacillus* species can solubilize phosphate, increasing its availability to plants. Phosphorus is often a limiting nutrient in soils, so this trait is particularly valuable in enhancing plant growth.

Growth Hormone Production: *Bacillus* species can produce phytohormones such as gibberellins and cytokinins, which promote root development and plant vigour.

Endospore Formation: The ability of *Bacillus* to form durable endospores allows them to remain viable under stressful conditions, ensuring their long-term presence and effectiveness in the soil [39].

Azospirillum

Azospirillum species are free nitrogen-fixing bacteria that associate with the roots of various plants, especially grasses and cereals. These bacteria do not form nodules like rhizobia do but instead increase nitrogen availability by fixing atmospheric nitrogen and releasing it into the soil, where the plant can take it up. *Azospirillum* species are also known to promote root growth by producing growth-promoting substances such as auxins, cytokinins, and gibberellins [40].

Nitrogen Fixation by *Azospirillum* can convert atmospheric nitrogen (N_2) into a form that plants can use, reducing the need for synthetic nitrogen fertilizers [41].

Root Growth Promotion could be stimulated by producing auxins and other growth hormones. *Azospirillum* encourages the development of a more extensive root system, improving the ability of plants to absorb water and nutrients.

Other important genera

a. ***Rhizobium and Bradyrhizobium***: These genera are well known for their symbiotic relationships with legumes, where they form root nodules and fix atmospheric nitrogen, directly supplying the host plant with nitrogen. This relationship significantly reduces the need for external nitrogen inputs in legume crops.

b. ***Enterobacter*** species are beneficial PGPR that promote plant growth by fixing nitrogen, solubilizing phosphate, and producing growth-promoting hormones.

c. ***Acinetobacter and Serratia:*** These genera include strains that can increase plant growth by solubilizing nutrients and producing bioactive compounds that improve root growth and plant vigour.

Each of these bacterial genera contributes to the overall health and productivity of plants through unique mechanisms, making PGPR essential components of integrated pest management and sustainable agriculture systems [42].

Direct Growth Promotion Mechanisms

PGPRs contribute to plant health and development through a variety of direct mechanisms that increase nutrient availability, improve stress tolerance, and stimulate growth. These mechanisms involve complex biochemical processes that enable PGPR to modify the soil environment and support the physiological needs of plants. By utilizing these direct mechanisms, PGPRs reduce the need for chemical fertilizers and other agricultural inputs, promoting more sustainable farming practices.

Nitrogen fixation

Nitrogen is a crucial macronutrient for plants, but most plants cannot directly utilize atmospheric nitrogen (N_2). PGPR, particularly nitrogen-fixing bacteria such as *Azospirillum, Rhizobium*, and *Bradyrhizobium*, convert atmospheric nitrogen into ammonia (NH_3), a form that plants can absorb and grow. Nitrogen fixation occurs *via* the enzyme nitrogenase, which catalyzes the reduction of atmospheric nitrogen into ammonia under low-oxygen conditions.

a. **Free-Living nitrogen fixers:** Bacteria such as *Azospirillum* and *Klebsiella* fix nitrogen independently in the rhizosphere, providing plants with a steady supply of usable nitrogen without forming symbiotic structures such as root nodules.

b. **Symbiotic nitrogen fixers:***Rhizobia* form nodules on leguminous plants, where they fix nitrogen in exchange for carbon compounds from the host plant. This symbiotic relationship significantly reduces the need for nitrogen fertilizers in legume crops and enhances soil fertility.

Through nitrogen fixation, PGPR not only improve plant growth but also contribute to the overall nitrogen economy of agricultural systems, making them essential for sustainable farming.

Phosphate Solubilization

Phosphorus is another vital nutrient for plant growth, but much of the phosphorus present in soil is present in insoluble forms that plants cannot access. Phosphate-solubilizing PGPR, such as *Bacillus, Pseudomonas,* and *Enterobacter*, produce organic acids such as gluconic and citric acids that convert insoluble phosphate into soluble forms (*e.g.*, orthophosphate), which can be easily absorbed by plant roots (Fig. **16**) [43].

a. **Mechanism:** Acids produced by PGPR decrease the soil pH, which helps dissolve phosphate compounds bound to soil particles, increasing the availability of phosphorus for plant uptake.

b. **Benefit:** By increasing the bioavailability of phosphorus, PGPR promotes root development, flowering, and fruiting, which are processes heavily reliant on an adequate phosphorus supply. This reduces the need for synthetic phosphate fertilizers and enhances nutrient efficiency in crops. Phosphate-solubilizing bacteria are particularly valuable in soils with high levels of bound phosphate, where their activity can significantly increase plant growth and productivity.

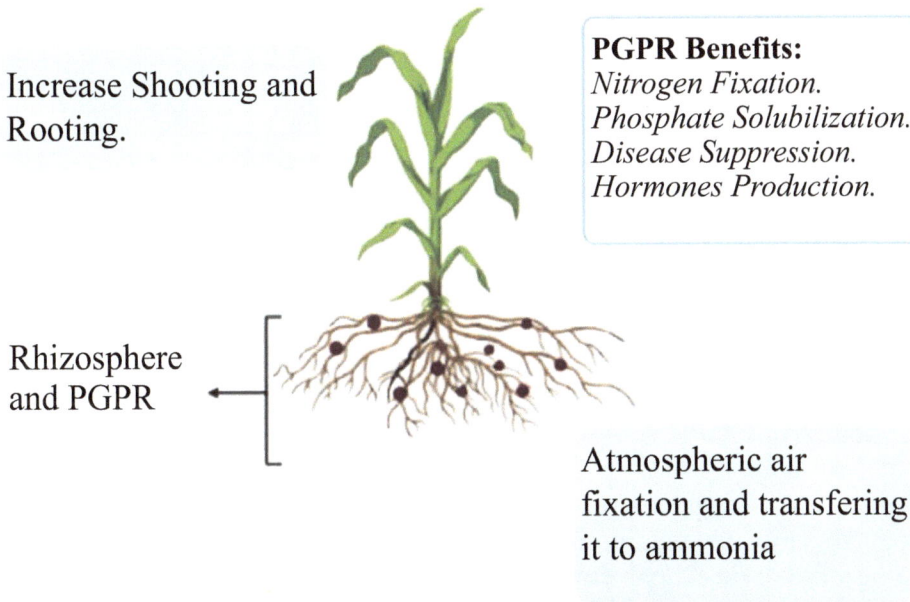

Increase Shooting and Rooting.

PGPR Benefits:
Nitrogen Fixation.
Phosphate Solubilization.
Disease Suppression.
Hormones Production.

Rhizosphere and PGPR

Atmospheric air fixation and transfering it to ammonia

Fig. (16). PGPR in the Rhizosphere. This figure illustrates the role of Plant Growth-Promoting Rhizobacteria (PGPR) in the rhizosphere. *Bacillus* species increase plant nutrition by solubilizing phosphate, making it available for plant uptake, whereas *Azospirillum* bacteria fix atmospheric nitrogen, converting it into ammonia, which plants can readily use. These interactions improve plant growth and nutrient availability in the rhizosphere.

Siderophore Production

Green Iron is an essential micronutrient for plants, but it often exists in insoluble forms in the soil, particularly in alkaline and calcareous soils. PGPR, such as *Pseudomonas*, produces siderophores—low-molecular-weight compounds with a high affinity for iron. Siderophores chelate (bind) iron and increase its availability to the plant by transporting it back to bacterial cells and the plant roots [44].

a. **Mechanism:** PGPR release siderophores into the soil, where they bind to ferric iron (Fe^{3+}). The siderophore-iron complex is then taken up by both the bacteria and the plant, allowing the plant to access iron that would otherwise be unavailable.

b. **Benefit:** Siderophore production by PGPR is especially important in iron-deficient soils, where plants struggle to access sufficient iron for vital processes such as chlorophyll synthesis, enzyme activation, and respiration. Improved iron availability enhances plant health and growth, reducing the risk of chlorosis (yellowing of leaves) caused by iron deficiency. Siderophore-producing PGPR indirectly protect plants from pathogens by competing with harmful microorganisms for iron resources while improving iron nutrition for them.

Phytohormone Synthesis

PGPR can synthesize and release plant hormones, known as phytohormones, which directly influence plant growth and development. These hormones regulate processes such as cell elongation, root formation, flowering, and fruiting [45]. The key phytohormones produced by PGPR include the following:

a. **Auxins:** Auxins, particularly Indole-3-Acetic Acid (IAA), are crucial for root development. PGPR that produce auxins stimulate root elongation and branching, leading to a more extensive and efficient root system. This enhances the ability of plants to absorb water and nutrients from the soil.

b. **Gibberellins:** These hormones promote stem elongation, seed germination, and flowering. PGPR that produce gibberellins can accelerate plant growth and increase crop yield.

c. **Cytokinins:** Cytokinins promote cell division and shoot growth. PGPR that produce cytokinins can increase shoot development and delay leaf senescence (aging), improving the overall health and longevity of the plant. By producing these phytohormones, PGPR modulate plant growth processes, leading to increased biomass, improved root architecture, and increased resilience to environmental stresses.

ACC Deaminase Activity

Ethylene is a plant hormone that regulates various aspects of growth and development, but excessive ethylene production, particularly under stress conditions (*e.g.*, drought, salinity, flooding), can inhibit plant growth and cause premature aging. PGPR, which produce the enzyme 1-amino-cyclo-propa-e-1-carboxylate (ACC) deaminase, help regulate ethylene levels in plants [46].

a. **Mechanism:** ACC deaminase breaks down ACC, the immediate precursor of ethylene in plants, reducing the synthesis of ethylene. By lowering ethylene levels, PGPR can prevent the negative effects of stress-induced ethylene production (Fig. **17**).

b. **Benefit:** Plants treated with ACC deaminase-producing PGPR exhibit improved stress tolerance, as lower ethylene levels allow them to continue growing and developing even under adverse conditions. This phenomenon is particularly important for crop plants exposed to drought, flooding, or high salinity. ACC deaminase-producing PGPR are valuable in helping plants cope with environmental stress, ensuring better growth and higher yields even under challenging conditions.

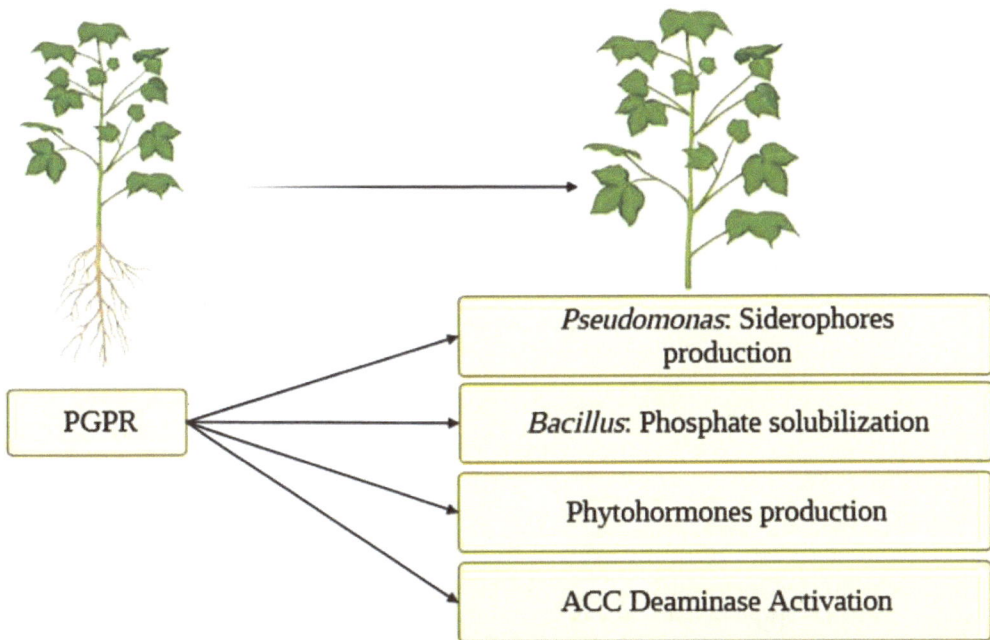

Fig. (17). Key PGPR mechanisms. This figure highlights the main mechanisms through which Plant Growth-Promoting Rhizobacteria (PGPR) increase plant growth and stress tolerance. ACC deaminase reduces ethylene levels, promoting stress tolerance; phytohormones produced by PGPR stimulate plant growth; *Pseudomonas* species produce siderophores that chelate iron, increasing its availability to plants; and Bacillus species increase the availability of phosphate, supporting overall plant nutrition and development.

Indirect Growth Promotion Mechanisms

In addition to the direct mechanisms through which PGPR increase plant growth, these bacteria also promote plant health and development through indirect mechanisms. These indirect mechanisms often involve protecting plants from pathogens, increasing their ability to resist stress, and optimizing root colonization. By reducing the impact of diseases and stress, PGPR indirectly improve plant growth, contributing to higher yields and overall plant resilience (Fig. **18**).

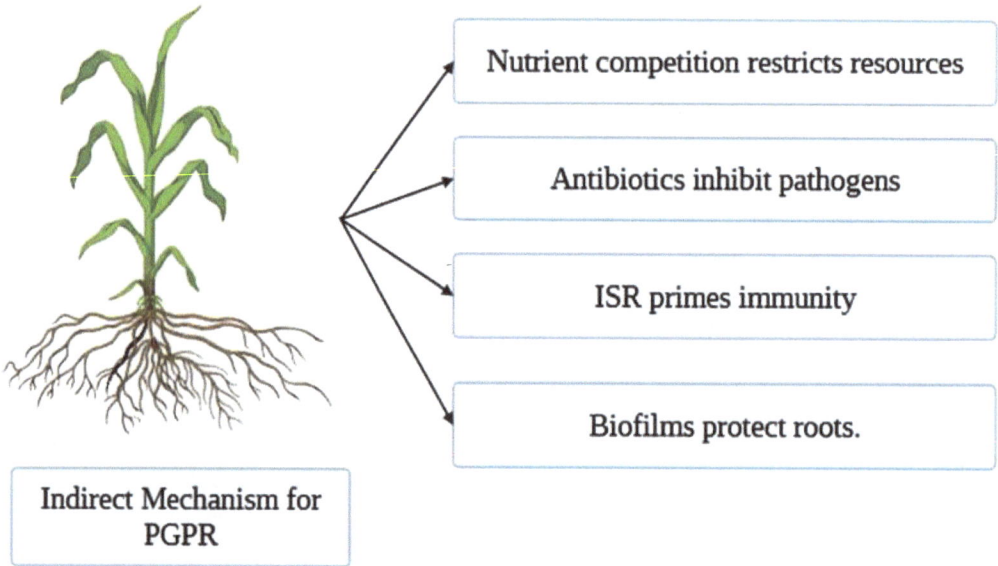

| Nutrient competition restricts resources |
| Antibiotics inhibit pathogens |
| ISR primes immunity |
| Biofilms protect roots. |

Indirect Mechanism for PGPR

Fig. (18). PGPR in plant defense. This figure illustrates how Plant Growth-Promoting Rhizobacteria (PGPR) contribute to plant defense mechanisms. Induced Systemic Resistance (ISR) primes a plant's immune system to respond more effectively to pathogen attacks. PGPR produce antibiotics that directly inhibit pathogens, whereas nutrient competition limits the resources available to harmful microbes. Additionally, the formation of biofilms by PGPR provides a protective layer around plant roots, reducing pathogen colonization.

INDUCED SYSTEMIC RESISTANCE (ISR)

Induced Systemic Resistance (ISR) is a defense mechanism activated by PGPR that primes the plant's immune system to respond more quickly and effectively to pathogen attacks. Unlike direct defense responses triggered by an infection, ISR does not involve the direct recognition of pathogens by a plant. Instead, PGPR produce signalling molecules (such as jasmonic acid and ethylene) that activate the plant's immune system, allowing the plant to respond more robustly when faced with future threats.

a. **Mechanism:** PGPR, particularly *Pseudomonas* and *Bacillus* species, colonize plant roots and release compounds that trigger ISR. This involves the production of phytohormones and signalling molecules that prepare the plant to defend itself against a wide range of pathogens, even in parts of the plant that are not in direct contact with the bacteria.

b. **Benefit:** ISR enables the plant to mount a faster and stronger response to pathogen invasion, reducing disease incidence and severity. This mechanism is effective against bacterial, fungal, and viral pathogens, making it a broad-spectrum form of plant protection. ISR does not involve the direct destruction of pathogens by bacteria but instead strengthens the plant's own defenses, increasing resistance to biotic stressors.

Antibiotic Production

Many PGPR, particularly *Pseudomonas* and *Bacillus* species, produce antibiotics and other antimicrobial compounds that inhibit the growth of plant pathogens in the rhizosphere. These antibiotics can directly kill or inhibit the growth of harmful fungi, bacteria, and even nematodes that threaten plant health. This form of biological control is crucial for PGPR-mediated disease suppression [47].

a. **Mechanism:** PGPR produce a variety of antimicrobial substances, including bacteriocins, lipopeptides, and volatile organic compounds. These antibiotics interfere with essential cellular processes in pathogens, such as protein synthesis, membrane integrity, or DNA replication, leading to pathogen death or inhibition.

b. **Benefit:** By producing antibiotics, PGPR can reduce pathogen populations in the soil, decreasing the likelihood of plant infection. This leads to healthier plants with reduced reliance on chemical pesticides, contributing to more sustainable agricultural practices.

The production of antibiotics by PGPR not only suppresses harmful organisms but also helps maintain a healthier microbial balance in the rhizosphere, promoting beneficial microbial communities [48].

Nutrient competition

In the rhizosphere, plants and microorganisms compete for limited resources such as nutrients, water, and space. PGPR, through rapid colonization and efficient nutrient uptake, can outcompete harmful pathogens for these resources, limiting the growth of pathogens and reducing their ability to infect plants.

a. **Mechanism:** PGPR, especially those that are highly competitive in the rhizosphere, efficiently utilize nutrients such as iron, nitrogen, and carbon, depriving pathogenic microorganisms of these essential resources. For example, siderophore-producing bacteria, such as *Pseudomonas*, sequester iron, making it unavailable to pathogens that require it for their survival.

b. **Benefit:** By monopolizing nutrients and space, PGPR prevent pathogens from establishing themselves in the rhizosphere or on plant roots. This competition is particularly important in disease suppression, as it reduces the pathogen load in the plant's immediate environment, thereby decreasing the risk of infection.

This form of competition, sometimes referred to as "niche exclusion," is a passive yet effective way for PGPR to protect plants by simply outcompeting harmful microbes for vital resources.

Biofilm Formation

Biofilms are structured communities of microorganisms that adhere to surfaces, such as plant roots, and are embedded in a protective extracellular matrix. Many PGPR form biofilms around plant roots, providing multiple benefits, including enhanced root colonization, protection against environmental stresses, and improved nutrient and water retention. Additionally, biofilms offer protection from pathogens by creating a physical barrier that prevents harmful microbes from reaching the plant root surface [49].

a. **Mechanism:** PGPR such as *Bacillus* and *Pseudomonas* secrete polysaccharides and proteins that help them adhere to plant roots and form biofilms. Once established, these biofilms create a microenvironment that supports bacterial growth and nutrient exchange with the plant. The extracellular matrix protects both bacteria and plant roots from desiccation, harmful chemicals, and pathogen invasion.
b. **Benefit:** Biofilm formation enhances the stability of PGPR populations in the rhizosphere, ensuring their continued presence and beneficial activity. Biofilms also create a protective barrier that prevents pathogens from reaching plant roots and establishes a more favourable environment for plant growth by retaining moisture and nutrients.

The formation of biofilms by PGPR is particularly advantageous in harsh environments, such as dry or nutrient-poor soils, where biofilms help bacteria and plants survive and thrive [50].

APPLICATIONS IN AGRICULTURE

PGPR have shown immense potential in revolutionizing agricultural practices by offering natural, sustainable alternatives to synthetic fertilizers and pesticides. Their ability to increase plant growth, improve soil health, and protect crops from pathogens makes them valuable tools in modern agriculture [51]. One of the primary ways in which PGPR are applied in agriculture is through the use of biofertilizers.

Biofertilizers (BF)

BF are formulations containing living microorganisms that, when applied to seeds, plant surfaces, or soil, colonize the rhizosphere or plant interior and promote growth by increasing the availability of essential nutrients to the host plant [52]. These microbial inoculants are used as an ecofriendly alternative to chemical fertilizers, helping to reduce the negative environmental impact associated with synthetic inputs while enhancing crop productivity.

1. Types and Formulations

BF can be classified on the basis of the type of microorganisms they contain and the benefits they provide to plants. There are several types of biofertilizers, each targeting specific nutrient needs or plant growth-promoting mechanisms.

1. **Nitrogen-fixing biofertilizers:** These biofertilizers include nitrogen-fixing bacteria such as *Rhizobium, Azospirillum,* and *Azotobacter*, which convert atmospheric nitrogen into forms that plants can use. They are especially valuable for legume crops that rely on symbiotic nitrogen-fixing bacteria, but free nitrogen fixers such as *Azospirillum* can also be used with nonlegumes.

2. **Phosphate-solubilizing biofertilizers:** These formulations contain bacteria or fungi (*e.g., Bacillus, Pseudomonas, Aspergillus*) that can solubilize insoluble phosphate in the soil, making it available to plants. Phosphorus is a critical nutrient for root development, flowering, and fruiting, and phosphate-solubilizing biofertilizers can significantly improve crop yield, especially in phosphorus-deficient soils.

3. **Potassium-solubilizing biofertilizers:** Certain bacteria, such as *Frateuria aurantia*, are capable of solubilizing potassium compounds in the soil, making this nutrient available to plants. Potassium is essential for processes such as water regulation, enzyme activation, and stress tolerance in plants.

4. **Biofertilizers for micronutrients:** Some biofertilizers are formulated to improve the availability of micronutrients such as iron, zinc, and copper. Siderophore-producing bacteria, such as *Pseudomonas*, can improve iron uptake, whereas other biofertilizers may be tailored for zinc or copper solubilization.

5. **Mycorrhizal biofertilizers:** These formulations contain fungi, such as Arbuscular Mycorrhizal (AM) fungi, which form symbiotic associations with plant roots. Mycorrhizae improve the absorption of water and nutrients, particularly phosphorus, and enhance plant resistance to drought and soil pathogens.

6. **Multifunctional biofertilizers:** These formulations combine multiple microorganisms, offering a broader range of benefits, including nitrogen fixation, phosphate solubilization, hormone production, and pathogen suppression. Multifunctional biofertilizers are increasingly popular in integrated nutrient management systems.

Application Methods

The success of biofertilizers in promoting plant growth and improving yield depends not only on the type of microorganisms used but also on how they are applied [53]. There are several methods for applying biofertilizers to crops:

1. **Seed treatment**: In this method, seeds are coated with biofertilizers before planting. The microorganisms adhere to the seed surface and begin to colonize the rhizosphere once the seed germinates. This method is commonly used with nitrogen-fixing and phosphate-solubilizing biofertilizers and is especially effective for leguminous crops.
2. **Soil application**: Biofertilizers can be applied directly to the soil, either by mixing them with compost or organic manure or applying them as a slurry or solution. This method allows microorganisms to colonize the root zone more quickly and is often used for crops that benefit from mycorrhizal fungi or potassium-solubilizing biofertilizers.
3. **Foliar Spray:** Some biofertilizers can be sprayed directly onto the leaves of plants. This method is used when the microorganisms produce phytohormones or other compounds that promote growth or protect the plant from pathogens. Foliar application is effective for delivering nutrients such as iron or zinc to plants with micronutrient deficiencies.
4. **Root Dipping:** For crops such as rice and vegetables, seedlings can be dipped in a biofertilizer suspension before transplanting. This method ensures that the roots are colonized by beneficial microorganisms from the beginning, promoting better root development and nutrient uptake.
5. **Drip Irrigation**: Biofertilizers can also be delivered through drip irrigation systems, allowing for even distribution of the microorganisms in the root zone. This method is efficient for large-scale farming and is particularly useful for mycorrhizal and multifunctional biofertilizers.

India's use of Rhizobial Biofertilizers in legume cultivation: In India, the widespread use of rhizobial biofertilizers in legume crops such as soybean, chickpea, and lentil has led to a significant reduction in the use of synthetic nitrogen fertilizers. These biofertilizers improve nitrogen fixation in the soil, resulting in higher yields and better soil fertility. Farmers have also reported improved crop resilience to diseases and environmental stresses. Brazil's Adoption of *Azospirillum* for Corn and Wheat: In Brazil, the application of *Azospirillum*-based biofertilizers in corn and wheat cultivation has been a major success. These nitrogen-fixing biofertilizers increase nitrogen availability, resulting in better root growth, increased yields, and reduced reliance on nitrogen fertilizers. Brazilian farmers have embraced biofertilizers as part of an integrated nutrient management system that promotes sustainable agriculture. Mycorrhizal

Biofertilizers in Wine Production in Italy: Italian vineyards have successfully used mycorrhizal biofertilizers to improve water uptake and nutrient availability in grapevines. In regions with poor soil fertility, the use of these biofertilizers has resulted in healthier vines, improved grape quality, and increased wine production. The introduction of mycorrhizal fungi has also helped vineyards cope with drought conditions, enhancing the long-term sustainability of wine production. Phosphate-solubilizing biofertilizers in Egypt have been applied to improve the productivity of crops growing in phosphorus-deficient soils. Farmers who grow crops such as wheat, maize, and potatoes have reported better root development, higher yields, and reduced fertilizer costs. The success of these biofertilizers has led to their increased adoption in both smallholder and commercial farming operations.

Integration with Farming Practices

The integration of biofertilizers with conventional farming practices is crucial for optimizing agricultural productivity and sustainability. While biofertilizers offer many benefits in terms of soil health, nutrient availability, and crop yield, their effectiveness is often enhanced when they are used in combination with chemical fertilizers and pesticides in a balanced and thoughtful way [54]. This section explores how biofertilizers can be incorporated into conventional farming systems, taking into account the combined use of chemical inputs and important soil management considerations.

Combined use with Chemical Fertilizers

Biofertilizers are often used alongside chemical fertilizers to achieve a more sustainable and efficient approach to crop nutrition. In addition to completely replacing chemical fertilizers, biofertilizers can increase nutrient availability, improve soil fertility, and reduce the overall need for synthetic inputs. This integrated approach—often referred to as Integrated Nutrient Management (INM)—helps optimize both plant productivity and soil health.

a. **Mechanism**: Biofertilizers, such as nitrogen-fixing bacteria *(Rhizobium, Azospirillum)* or phosphate-solubilizing microorganisms, improve the bioavailability of key nutrients such as nitrogen and phosphorus. This means that plants can make better use of the chemical fertilizers applied, requiring lower doses for the same effect. For example, when nitrogen-fixing biofertilizers are used with nitrogen-based fertilizers, microbial action reduces the amount of synthetic nitrogen needed by the plant.

b. **Benefits:** Combining biofertilizers with chemical fertilizers can lead to several advantages:
- **Improved nutrient use efficiency:** Biofertilizers increase the availability of nutrients, reduce the amount of synthetic fertilizer required and minimize nutrient loss through leaching or runoff.
- **Cost savings**: By reducing the need for expensive chemical fertilizers, biofertilizers help lower input costs for farmers.
- **Reduced environmental impact**: Less reliance on synthetic fertilizers translates to fewer environmental issues, such as soil degradation, water pollution from fertilizer run-off, and greenhouse gas emissions from fertilizer production. For example, integrating biofertilizers such as *Azotobacter* or *Azospirillum* with reduced doses of nitrogen fertilizers in cereal crops can maintain yields while improving soil organic matter and reducing chemical dependency.

Compatibility with Pesticides

In modern agriculture, pesticides play a vital role in protecting crops from pests and diseases. However, the combined use of biofertilizers and pesticides requires careful consideration to ensure compatibility and avoid negative interactions that could reduce the effectiveness of either input.

a. **Challenges of Compatibility:** Many pesticides are designed to kill harmful microorganisms, but they can also harm beneficial microbes, including those in biofertilizers. For example, fungicides can negatively impact phosphate-solubilizing bacteria or mycorrhizal fungi, whereas broad-spectrum insecticides might reduce the populations of beneficial rhizobacteria in the soil.

b. **Solution:** Selective use of pesticides: To integrate biofertilizers with pesticide use effectively, farmers need to choose pesticides that are less harmful to beneficial microorganisms. This can be achieved by:
- **Using biopesticides:** These are pest control agents derived from natural materials such as bacteria, fungi, and plant extracts. Biopesticides tend to be more compatible with biofertilizers, as they are often less toxic to beneficial microorganisms.
- **Targeted application:** Reducing the frequency and dosage of pesticide applications, or applying them in targeted areas, can minimize their impact on beneficial soil microorganisms [55].
- **Timing of application:** Applying pesticides when biofertilizer micro-organisms are less active or in a dormant state (*e.g.*, before seed germination) can help reduce negative interactions. By carefully managing pesticide use, farmers can preserve the beneficial effects of biofertilizers, ensuring a balanced approach to crop protection and soil fertility.

SOIL MANAGEMENT CONSIDERATIONS

Soil health is critical for the success of biofertilizer applications. The effectiveness of biofertilizers depends on the biological, chemical, and physical properties of the soil, and certain soil management practices can either enhance or diminish their impact [56].

a. **Soil pH**: The effectiveness of biofertilizers can be influenced by soil pH. Most beneficial microorganisms in biofertilizers prefer neutral to slightly acidic soils (pH 6–7). Extreme pH levels (too acidic or too alkaline) can inhibit the activity of biofertilizers, particularly those of phosphate-solubilizing bacteria. Soil pH should be managed through liming or other pH adjustment practices to create an environment conducive to microbial activity.

b. **Organic matter content**: Biofertilizers are more effective in soils with high organic matter contents, as organic matter provides a source of energy for soil microorganisms and enhances microbial activity. Practices such as crop rotation, cover cropping, and the addition of compost or manure can increase soil organic matter, improving the performance of biofertilizers.

c. **Tillage Practices**: Reduced tillage or no-till farming systems help preserve soil structure and microbial habitats, which can increase the survival and colonization of biofertilizer microorganisms. Excessive tillage can disturb the soil microbiota and reduce the effectiveness of biofertilizers by disrupting the rhizosphere environment.

d. **Moisture and Irrigation**: Adequate soil moisture is necessary for the survival and activity of biofertilizer microorganisms. Irrigation practices should be optimized to maintain consistent moisture levels in the soil, particularly during the early stages of plant growth when biofertilizer colonization is critical.

e. **Fertilizer application timing:** When both biofertilizers and chemical fertilizers are used, it is important to consider the timing of their application. BFs should be applied when plant roots are actively growing to ensure maximum colonization and nutrient uptake. Chemical fertilizers, on the other hand, can be timed to avoid any potential negative interactions with biofertilizer microbes [57].

CHALLENGES AND FUTURE PROSPECTS

While biofertilizers have great potential in enhancing sustainable agriculture, their widespread adoption still faces several challenges. Despite proven benefits in terms of nutrient availability, plant growth, and soil health, various limitations hamper their full utilization. This section explores current limitations and future prospects for overcoming these hurdles to make biofertilizers a more reliable and practical alternative in agriculture.

Current Limitations

The use of biofertilizers, despite their promising benefits, is often met with certain limitations that affect their consistency, storage, and field performance. These limitations need to be addressed for biofertilizers to reach their full potential.

Consistency Issues

One of the primary challenges with biofertilizers is the inconsistency in their performance across different environments and farming conditions. Unlike chemical fertilizers, which yield relatively uniform results, biofertilizers are living organisms whose activity can be influenced by several environmental factors.

1. **Soil Variability:** Soil composition and characteristics—such as pH, organic matter content, moisture level, and microbial population—can greatly influence the effectiveness of biofertilizers. The same biofertilizer may perform well in one region but poorly in another because of differences in soil health and composition.
2. **Environmental conditions:** Temperature and humidity also play significant roles in biofertilizer performance. Beneficial microbes in biofertilizers thrive within specific temperature and moisture ranges. Extremely hot or cold climates or soils that are too dry or too wet can hinder microbial survival and activity, leading to inconsistent results in terms of crop performance.
3. **Crop specificity:** Some biofertilizers work well only with specific crops or plant families. This limits their broad application and requires more targeted solutions, making it difficult for farmers with diverse crops to rely on biofertilizers as the sole solution for nutrient management.

Storage and Shelf Life

The storage and shelf-life of biofertilizers present significant challenges. Unlike chemical fertilizers, which have long shelf lives and can be stored under a wide range of conditions, biofertilizers contain living microorganisms that are sensitive to environmental changes.

Sensitivity to temperature and humidity: Biofertilizers need to be stored under specific conditions to maintain the viability of the microorganisms [58]. High temperatures, excessive moisture, or prolonged exposure to direct sunlight can kill or deactivate microbes. In some cases, this renders biofertilizers ineffective before they even reach the field.

Short Shelf Life: Many biofertilizers have relatively short shelf lives, ranging from a few months to a year. This can create logistical challenges in distribution and storage, especially in regions with limited access to cold storage facilities. Farmers may find it difficult to use biofertilizers within their effective time frame, leading to reduced efficacy.

Field effectiveness

While biofertilizers have shown excellent results in controlled environments such as laboratories or greenhouses, their performance in real-world agricultural settings can be inconsistent. Factors such as unpredictable weather patterns, pests, diseases, and varying soil conditions can impact their effectiveness in the field.

- **Competition with Native Microorganisms:** In soils with diverse microbial populations, the introduced biofertilizer organisms may struggle to establish themselves or outcompete native microbes. This is particularly true if the soil already contains microorganisms that perform similar functions, such as nitrogen fixation or phosphate solubilization. The biofertilizer microbes may fail to thrive or be outcompeted, reducing their overall effectiveness.
- **Compared** with chemical fertilizers, biofertilizers often act more slowly than chemical fertilizers do. While they enhance long-term soil health and sustainability, farmers may not see immediate benefits in terms of yield. This delay in visualization can make farmers hesitant to adopt biofertilizers, especially in regions where food security and crop productivity are critical concerns.

Emerging Technologies

As the demand for sustainable agriculture grows, new technologies are being developed to address the challenges associated with biofertilizers [59]. Innovations in microbial strain development, formulation improvements, and precision application methods are enhancing the effectiveness and reliability of biofertilizers, making them more practical and attractive for farmers. This section explores several emerging technologies that are promising for overcoming current limitations and paves the way for broader adoption of biofertilizers in modern agriculture.

New Strain Development

The development of new microbial strains is one of the most promising areas of innovation in biofertilizer technology [60]. By identifying and engineering strains that are more resilient, efficient, and effective, scientists are working to create biofertilizers that can perform better in a wider range of environmental conditions.

Strain Selection and Genetic Engineering: Advances in microbial genomics and biotechnology are enabling researchers to identify and select the most effective strains of bacteria and fungi for specific crops and environments. Genetic engineering techniques are also being employed to create microbes that are more efficient at nutrient fixation or solubilisation [61]. For example, nitrogen-fixing bacteria could be genetically modified to fix higher levels of nitrogen in the soil or to be more tolerant of extreme temperatures and pH levels.

Multifunctional Strains: Another innovation is the development of multifunctional strains that can perform multiple tasks simultaneously, such as nitrogen fixation, phosphate solubilization, and plant growth promotion. These "super strains" can offer a more comprehensive solution to nutrient management, reducing the need for multiple biofertilizers and improving overall soil health.

Microbial consortia: Instead of relying on a single strain, some biofertilizers now include consortia of different microorganisms that work together synergistically. These microbial communities are designed to mimic natural soil ecosystems, providing a broader range of benefits to plants and soil. By combining different strains that complement each other's functions, biofertilizers can achieve greater effectiveness and stability in the field.

Improved Formulations

The formulation of biofertilizers plays a critical role in their performance, particularly in terms of shelf-life, ease of use, and microbial viability. New advancements in formulation technologies are helping address issues related to storage, transportation, and field application.

Extended Shelf Life: One of the key goals in biofertilizer formulation is to extend the shelf-life of the product. New carrier materials, such as polymer-based carriers or biodegradable encapsulants, are being developed to protect microorganisms and keep them viable for longer periods. These carriers help maintain moisture and temperature stability, ensuring that the microbes remain active until they are applied in the field.

Encapsulation Technology: Encapsulation is an emerging technology that involves coating biofertilizer microorganisms in a protective layer, allowing them to survive harsh environmental conditions such as drought or extreme heat. This not only extends shelf-life but also improves the chances of successful colonization once applied to the soil [62].

Liquid biofertilizers: Traditionally, most biofertilizers have been in powder or granular form, which can be difficult to handle and apply. Liquid biofertilizers, on the other hand, offer easier application methods, longer shelf lives, and better microbial survival rates. These formulations can be more evenly distributed across fields and are easier to integrate with irrigation systems, making them more appealing for large-scale farming operations.

Precision Application Methods

Precision agriculture technologies are transforming the way farmers apply biofertilizers, ensuring that these valuable inputs are used efficiently and effectively [63]. Precision application methods can help optimize the placement, timing, and quantity of applied biofertilizer while reducing waste and maximizing plant growth and yield.

Drone and sensor technologies: Drones equipped with sensors and imaging technologies can be used to monitor soil health, plant growth, and nutrient deficiencies in real-time. These data allow farmers to apply biofertilizers precisely where they are needed rather than applying them uniformly across the field. Drones can also be used to deliver biofertilizers directly to targeted areas, minimizing waste and maximizing microbial colonization in the soil.

Variable Rate Application (VRA): VRA technologies use GPS and soil mapping to apply biofertilizers at variable rates across different parts of a field. This ensures that areas with poor soil health receive more biofertilizer, whereas areas with sufficient nutrients receive less biofertilizer. This precision reduces overapplication and improves the efficiency of biofertilizer use.

Seed inoculation technology: Another emerging precision application method is the use of seed inoculation technologies. With this method, biofertilizers are applied directly to the seeds before planting, ensuring that beneficial microbes are present in the root zone from the moment the plant starts growing. Automated seed coating machines can precisely inoculate each seed with the appropriate amount of biofertilizer, ensuring even distribution and efficient use of microbial inoculants.

Smart Irrigation Systems: Precision irrigation systems, such as drip irrigation combined with biofertilizer delivery, allow for precise control over the distribution of biofertilizers to plants. These systems ensure that the microorganisms are delivered directly to the root zone in controlled amounts, improving colonization and reducing losses due to runoff or evaporation.

Research Needs

To fully harness the potential of biofertilizers and integrate them into mainstream agricultural practices, more research is needed to address several key challenges. These include understanding the complex interactions between biofertilizers and the environment, improving their efficiency, and expanding their applicability to a broader range of crops. This section outlines the current research gaps and areas that require further investigation to improve the reliability and effectiveness of biofertilizers in diverse agricultural settings.

Understanding Complex Interactions

One of the most significant areas of research needs to gain a deeper understanding of the complex interactions among biofertilizers, soil microorganisms, plant roots, and the surrounding environment. These interactions are dynamic and influenced by a multitude of factors, including the soil type, climate, crop variety, and other microorganisms in the soil [64].

1. **Microbial Plant Interaction Dynamics:** While biofertilizers can increase plant growth through nutrient exchange, the exact mechanisms governing these processes are not fully understood. Further research is needed to explore how different microbial strains interact with specific plant species and the signals that plants and microbes exchange to initiate and regulate beneficial processes such as nitrogen fixation, phosphate solubilization, and root colonization.
2. **Rhizosphere Ecology:** The rhizosphere, the zone of soil directly influenced by plant roots, is a complex and dynamic environment. Researchers need to investigate how biofertilizers interact with native microbial communities in the rhizosphere and what factors influence the competition, cooperation, and survival of introduced microorganisms [65]. Understanding how to enhance biofertilizer colonization in the presence of native species is essential for improving field effectiveness.
3. **Environmental Influence:** Research must also address how different environmental conditions—such as pH, moisture, temperature, and nutrient availability—affect the performance of biofertilizers. In particular, climate change poses a challenge to the stability and predictability of biofertilizer performance [66]. More studies are needed to explore how biofertilizers can be optimized for use in diverse and changing environments.

Improving Efficiency

Another important area of research is improving the efficiency of biofertilizers to ensure consistent and predictable results. Current biofertilizers sometimes exhibit variable performance due to inconsistent colonization or low survival rates under field conditions. Research should focus on ways to increase their efficiency and ensure reliable outcomes.

1. **Microbial Survival and Colonization:** A major limitation of current biofertilizers is the difficulty in ensuring that the introduced microbes can successfully establish themselves in the soil and compete with native microorganisms. Research into improving the survival rates of biofertilizer microbes under harsh environmental conditions—such as drought, extreme temperatures, or poor soil fertility—is critical. This includes the exploration of advanced microbial formulations, protective carriers, and encapsulation technologies that increase microbial viability and survival.

2. **Optimization of Nutrient Use:** Further research is needed to optimize the nutrient use efficiency of biofertilizers. This includes investigating how biofertilizer strains can be enhanced to improve their ability to fix nitrogen, solubilize phosphate, or produce plant growth-promoting substances more effectively. This could involve genetic engineering, microbial consortia development, or breeding plants with improved compatibility with biofertilizers.

3. **Enhancing Microbial Consortia:** While single-strain biofertilizers are commonly used, microbial consortia—combinations of multiple beneficial microorganisms—have shown great promise for improving crop yield and soil health. Research into the design of more effective microbial consortia that work synergistically in the soil could lead to more robust and reliable biofertilizers that offer a broader range of benefits to crops.

Expanding the Host Range

Currently, many biofertilizers are designed for specific crops, such as legumes for nitrogen fixation or cereal crops for phosphate solubilization. Expanding the host range of biofertilizers to cover a wider variety of crops is crucial for increasing their utility in diverse agricultural systems [67].

1. **Broadening Crop Applicability:** Research is needed to develop biofertilizers that are effective for a wider range of crops, including vegetables, fruits, and nonleguminous cereals. Understanding how biofertilizers interact with different root systems, plant physiology, and nutrient needs is essential for expanding their application beyond their current niche.

2. **Host specificity research:** Certain biofertilizers, such as *Rhizobium* bacteria, have very specific host requirements, meaning that they only form symbiotic relationships with particular plant species. Research into reducing this host specificity or creating microbial strains that can work with multiple crop species would greatly expand the scope and impact of biofertilizers. This could

involve the breeding of microbial strains that are more generalist in nature or engineering plants to accept microbial colonization.

3. **Nontraditional Crops:** Additionally, there is a need to explore the use of biofertilizers in nontraditional or high-value crops, such as biofuel crops, medicinal plants, or plants grown in challenging environments such as saline or alkaline soils. Expanding the range of crops that can benefit from biofertilizers would make these technologies more applicable to a wider variety of farming systems and markets.

CONCLUSION

The integration of biofertilizers into agricultural systems represents a promising step toward more sustainable and environmentally friendly farming practices. As the global demand for food continues to increase with increasing environmental pressure, biofertilizers offer a valuable alternative to chemical fertilizers, improving soil health, reducing chemical inputs, and increasing plant growth. However, the successful implementation of biofertilizers requires overcoming several challenges, particularly in terms of consistency, effectiveness, and scalability.

Summary of Key Points: This chapter explores the potential of biofertilizers and their symbiotic relationships with plants. The key points include:

Symbiotic Relationships: Mycorrhizal associations and rhizobial symbioses are fundamental to plant nutrition, especially through nutrient exchange mechanisms that support plant growth. These relationships increase nutrient availability and uptake, particularly nitrogen and phosphorus, leading to better crop productivity.

Plant Growth-Promoting Rhizobacteria (PGPR): PGPR provide both direct and indirect growth-promoting mechanisms, including nitrogen fixation, phosphate solubilization, and induced systemic resistance, offering plants a holistic support system against pathogens and nutrient deficiencies.

Applications in agriculture: The use of biofertilizers in agricultural settings, particularly as biofertilizers integrated with other farming practices, shows great promise. Success stories and emerging technologies demonstrate that biofertilizers can play a crucial role in reducing reliance on chemical inputs while maintaining or improving crop yields.

Challenges and Future Prospects: Despite their benefits, biofertilizers face challenges related to storage, consistency, and field effectiveness. Emerging technologies such as new strain development, improved formulations, and precision application methods hold the potential to overcome these challenges

[68]. Further research is needed to understand microbial–plant interactions fully, improve the efficiency of biofertilizers, and expand their applicability to a broader range of crops.

FUTURE OUTLOOK

The future of biofertilizers is bright, with numerous advancements on the horizon. As researchers continue to develop more efficient strains and formulations, biofertilizers will become more reliable and adaptable to different environments and crops. Precision agriculture technologies, such as drone applications and soil sensors, make biofertilizers more practical and cost-effective for farmers, improving their uptake and overall impact [69].

Moreover, as the agricultural industry increasingly shifts towards sustainability, biofertilizers will likely play an integral role in reducing the environmental footprint of farming. They are well-positioned to complement organic farming practices, regenerative agriculture, and other eco-friendly farming methods. Government support, policy frameworks, and financial incentives will also be crucial in promoting the use of biofertilizers on a larger scale. Research Opportunities In the future, several areas of research remain critical for the advancement of biofertilizers.

Exploring Microbial Plant Interactions: Understanding the complex dynamics among plants, biofertilizers, and the surrounding soil environment is crucial. More research is needed to explore how different microbial species interact with plants, how they adapt to changing environmental conditions, and how to optimize these relationships for maximum benefit.

Improving biofertilizer efficiency: Research focused on increasing the efficiency of biofertilizers, particularly in terms of nutrient uptake, survival rates, and colonization, will be key to ensuring consistent results in the field [70]. This includes studying microbial consortia and developing more robust strains that can perform well in a variety of soil types and climates.

Expanding the Host Range: There is significant potential for expanding the use of biofertilizers to a wider variety of crops, including nonleguminous and high-value crops. Research that focuses on overcoming the barriers of host specificity and developing more generalist microbial strains will help make biofertilizers more versatile and applicable across different agricultural systems. In conclusion, while biofertilizers present exciting possibilities for sustainable agriculture, they require continued research and innovation to address existing limitations and unlock their full potential. By advancing our understanding of these living inputs and developing new technologies to improve their application, we can create a future where biofertilizers are a cornerstone of global agricultural practices.

REFERENCES

[1]   Hayat K, *et al.* Plant–microbe interaction in developing climate-resilient crop species. Environ Sci Eng. 2023; pp. 535-50.
        [http://dx.doi.org/10.1007/978-3-031-43729-8_20]

[2]   Shah M, Shah N. Environmental approach to remediate refractory pollutants from industrial wastewater treatment plant. 2024.

[3]   Tan C, Kalhoro M T, Faqir Y, Ma J, Osei M D, Khaliq G. Climate-resilient microbial biotechnology: a perspective on sustainable agriculture. Sustainability. 2022; 14: pp. (9)5574-4.
        [http://dx.doi.org/10.3390/su14095574]

[4]   Shah M. Application of sewage sludge in industrial wastewater treatment. 2024.
        [http://dx.doi.org/10.1002/9781119857396]

[5]   Griffiths BS, Philippot L. Insights into the resistance and resilience of the soil microbial community. FEMS Microbiol Rev 2013; 37(2): 112-29.
        [http://dx.doi.org/10.1111/j.1574-6976.2012.00343.x] [PMID: 22568555]

[6]   Jain H. From pollution to progress: Groundbreaking advances in clean technology unveiled. Innovation and Green Development 2024; 3(2): 100143.
        [http://dx.doi.org/10.1016/j.igd.2024.100143]

[7]   Bhardwaj LK, *et al.* COVID-19 and the interplay with antibacterial drug resistance Available from: https://services.igi-global.com/resolvedoi/resolve.aspx?doi=10.4018/979-8-3693-4139-1.ch010
        [http://dx.doi.org/10.4018/979-8-3693-4139-1.ch010]

[8]   Díaz-Muñoz SL, Koskella B. Bacteria-Phage interactions in natural environments. Advances in Applied Microbiology. Academic Press Inc. 2014; 89: pp. 135-83.
        [http://dx.doi.org/10.1016/B978-0-12-800259-9.00004-4]

[9]   Li L G, Zhang T. Plasmid-mediated antibiotic resistance gene transfer under environmental stresses: Insights from laboratory-based studies. Elsevier B.V. 2023.
        [http://dx.doi.org/10.1016/j.scitotenv.2023.163870]

[10]   Nizamani M M, *et al.* Microbial biodiversity and plant functional trait interactions in multifunctional ecosystems. Elsevier BV. 2024.
        [http://dx.doi.org/10.1016/j.apsoil.2024.105515]

[11]   Kaur M, Saxena R. Review on plant-microbe interactions, applications and future aspects. Microbial Technology for Agro-Ecosystems: Crop Productivity, Sustainability, and Biofortification. Elsevier 2024; pp. 133-51.
        [http://dx.doi.org/10.1016/B978-0-443-18446-8.00005-X]

[12]   Erin Chen Y, Fischbach M A, Belkaid Y. Skin microbiota-host interactions. Nature Publishing Group 2018.
        [http://dx.doi.org/10.1038/nature25177]

[13]   Kho Z Y, Lal S K. The human gut microbiome - A potential controller of wellness and disease. Frontiers Media SA. 2018.
        [http://dx.doi.org/10.3389/fmicb.2018.01835]

[14]   Egan S, Fernandes ND, Kumar V, Gardiner M, Thomas T. Bacterial pathogens, virulence mechanism and host defence in marine macroalgae. Blackwell Publishing Ltd 2014.
        [http://dx.doi.org/10.1111/1462-2920.12288]

[15]   Selim SM, Zayed MS. Microbial interactions and plant growth. Plant-microbe interactions in agro-ecological perspectives. Springer Singapore 2017; 1: pp. 17-34.

[http://dx.doi.org/10.1007/978-981-10-5813-4_1]

[16] Adamu KS, Bichi YH, Nasiru AY, *et al.* Review: Synthetic microbial consortia in bioremediation and biodegradation. International Journal of Research and Innovation in Applied Science 2023; VIII(VII): 232-41.
[http://dx.doi.org/10.51584/IJRIAS.2023.8727]

[17] Torres-Guardado R, Esteve-Zarzoso B, Reguant C, Bordons A. Microbial interactions in alcoholic beverages. Springer Science and Business Media Deutschland GmbH 2022.
[http://dx.doi.org/10.1007/s10123-021-00200-1]

[18] Kumar A, Alam A, Rani M, Ehtesham N Z, Hasnain S E. Biofilms: Survival and defense strategy for pathogens. Elsevier GmbH 2017.
[http://dx.doi.org/10.1016/j.ijmm.2017.09.016]

[19] Kemen A C, Agler M T, Kemen E. Host-microbe and microbe-microbe interactions in the evolution of obligate plant parasitism. Blackwell Publishing Ltd 2015.
[http://dx.doi.org/10.1111/nph.13284]

[20] Dheilly N M, Poulin R, Thomas F. Biological warfare: Microorganisms as drivers of host-parasite interactions. Elsevier. 2015.
[http://dx.doi.org/10.1016/j.meegid.2015.05.027]

[21] Ke PJ, Wan J. Effects of soil microbes on plant competition: a perspective from modern coexistence theory. Ecol Monogr 2020; 90(1): e01391.
[http://dx.doi.org/10.1002/ecm.1391]

[22] Machado D, Maistrenko OM, Andrejev S, *et al.* Polarization of microbial communities between competitive and cooperative metabolism. Nat Ecol Evol 2021; 5(2): 195-203.
[http://dx.doi.org/10.1038/s41559-020-01353-4] [PMID: 33398106]

[23] Mazorra-Alonso M, Tomás G, Soler J J. Microbially mediated chemical ecology of animals: A review of its role in conspecific communication, parasitism and predation. MDPI AG. 2021.
[http://dx.doi.org/10.3390/biology10040274]

[24] Hause B, Mrosk C, Isayenkov S, Strack D. Jasmonates in arbuscular mycorrhizal interactions. 2007.
[http://dx.doi.org/10.1016/j.phytochem.2006.09.025]

[25] Abu Taher M, Tong WY, Leong CR, Ab Rashid S, Tan WN. General characteristics of endophytes and bioprospecting potential of endophytic fungi. Adv Struct Mater. Springer Science and Business Media Deutschland GmbH 2023; 165: pp. 35-49.
[http://dx.doi.org/10.1007/978-3-031-21959-7_4]

[26] Dubey A, Saiyam D, Kumar A, Hashem A, Abd Allah EF, Khan ML. Bacterial root endophytes: Characterization of their competence and plant growth promotion in soybean (glycine max (L.) merr.) under drought stress. Int J Environ Res Public Health 2021; 18(3): 1-20.
[http://dx.doi.org/10.3390/ijerph18030931] [PMID: 33494513]

[27] ALKahtani MDF, Fouda A, Attia KA, *et al.* Isolation and characterization of plant growth promoting endophytic bacteria from desert plants and their application as bioinoculants for sustainable agriculture. Agronomy (Basel) 2020; 10(9): 1325.
[http://dx.doi.org/10.3390/agronomy10091325]

[28] Wu W, *et al.* Beneficial relationships between endophytic bacteria and medicinal plants. Frontiers Media S.A. 2021.
[http://dx.doi.org/10.3389/fpls.2021.646146]

[29] Shen Y, Xu L, You C, *et al.* Effects of ectomycorrhizae and hyphae on soil fungal community characteristics across forest gap positions. Forests 2024; 15(12): 2131.
[http://dx.doi.org/10.3390/f15122131]

[30] Teste F P, Jones M D, Dickie I A. Dual-mycorrhizal plants: their ecology and relevance. Blackwell Publishing Ltd. 2020.

[http://dx.doi.org/10.1111/nph.16190]

[31]   Hu W, Pan L. Applications of mycorrhizal fungi in agriculture and forestry. Microbial Bioprocesses: Applications and Perspectives. Elsevier 2023; pp. 1-20.
[http://dx.doi.org/10.1016/B978-0-323-95332-0.00012-0]

[32]   Porter SS, Dupin SE, Denison RF, Kiers ET, Sachs JL. Host-imposed control mechanisms in legume–rhizobia symbiosis. Nat Microbiol 2024; 9(8): 1929-39.
[http://dx.doi.org/10.1038/s41564-024-01762-2] [PMID: 39095495]

[33]   Yang J, Lan L, Jin Y, Yu N, Wang D, Wang E. Mechanisms underlying legume–rhizobium symbioses. John Wiley and Sons Inc. 2022.
[http://dx.doi.org/10.1111/jipb.13207]

[34]   Goyal R K, Mattoo A K, Schmidt M A. Rhizobial–host interactions and symbiotic nitrogen fixation in legume crops toward agriculture sustainability. Frontiers Media S.A. 2021.
[http://dx.doi.org/10.3389/fmicb.2021.669404]

[35]   de Andrade L A, Santos C H B, Frezarin E T, Sales L R, Rigobelo E C. Plant growth-promoting rhizobacteria for sustainable agricultural production. MDPI. 2023.
[http://dx.doi.org/10.3390/microorganisms11041088]

[36]   Wang Y, Pei Y, Wang X, Dai X, Zhu M. Antimicrobial metabolites produced by the Plant Growth-Promoting Rhizobacteria (PGPR): *Bacillus and Pseudomonas* KeAi Communications Co. 2024.
[http://dx.doi.org/10.1016/j.aac.2024.07.007]

[37]   Santoyo G, Urtis-Flores CA, Loeza-Lara PD, Orozco-Mosqueda MC, Glick BR. Rhizosphere colonization determinants by Plant Growth-Promoting Rhizobacteria (PGPR). Biology (Basel) 2021; 10(6): 475.
[http://dx.doi.org/10.3390/biology10060475] [PMID: 34072072]

[38]   Saxena A K, Kumar M, Chakdar H, Anuroopa N, Bagyaraj D J. *Bacillus* species in soil as a natural resource for plant health and nutrition John Wiley and Sons Inc. 2020.
[http://dx.doi.org/10.1111/jam.14506]

[39]   Bhat M A, *et al.* Plant Growth Promoting Rhizobacteria in plant health: A perspective study of the underground interaction. MDPI. 2023.
[http://dx.doi.org/10.3390/plants12030629]

[40]   Jehani MD, Singh S, Archana TS, Kumar D, Kumar G. *Azospirillum*—a free-living nitrogen-fixing bacterium. Rhizobiome: Ecology, Management and Application. Elsevier 2023; pp. 285-308.
[http://dx.doi.org/10.1016/B978-0-443-16030-1.00001-8]

[41]   Nag P, Shriti S, Das S. Microbiological strategies for enhancing biological nitrogen fixation in nonlegumes. John Wiley and Sons Inc. 2020.
[http://dx.doi.org/10.1111/jam.14557]

[42]   Kumawat KC, Singh I, Nagpal S, Sharma P, Gupta RK, Sirari A. Co-inoculation of indigenous *Pseudomonas oryzihabitans* and *Bradyrhizobium* sp. modulates the growth, symbiotic efficacy, nutrient acquisition, and grain yield of soybean. Pedosphere 2022; 32(3): 438-51.
[http://dx.doi.org/10.1016/S1002-0160(21)60085-1]

[43]   Zeng Q, Ding X, Wang J, Han X, Iqbal H M N, Bilal M. Insight into soil nitrogen and phosphorus availability and agricultural sustainability by plant growth-promoting rhizobacteria. Springer Science and Business Media Deutschland GmbH. 2022.
[http://dx.doi.org/10.1007/s11356-022-20399-4]

[44]   Sarwar S, Khaliq A, Yousra M, Sultan T, Ahmad N. Screening of siderophore-producing PGPRs isolated from groundnut (*Arachis hypogaea L.*) rhizosphere and their influence on iron release in soil. Commun Soil Sci Plant Anal. 2020; pp. 1680-92.
[http://dx.doi.org/10.1080/00103624.2020.1791159]

[45]   Khan N, Bano A, Ali S, Babar M A. Crosstalk amongst phytohormones from planta and PGPR under

biotic and abiotic stresses. Springer. 2020.
[http://dx.doi.org/10.1007/s10725-020-00571-x]

[46] Moon Y S, Ali S. Possible mechanisms for the equilibrium of ACC and role of ACC deaminase-producing bacteria. Springer Science and Business Media Deutschland GmbH. 2022.
[http://dx.doi.org/10.1007/s00253-022-11772-x]

[47] Etesami H, Adl SM. Plant Growth-Promoting Rhizobacteria (PGPR) and their action mechanisms in availability of nutrients to plants. 2020; pp. 147-203.
[http://dx.doi.org/10.1007/978-981-15-2576-6_9]

[48] Lahiri D, Nag M, Sayyed RZ, Gafur A, Ansari MJ, Ray RR. PGPR in biofilm formation and antibiotic production. 2022; pp. 65-82.
[http://dx.doi.org/10.1007/978-3-031-04805-0_4]

[49] Wang H, Liu R, You M P, Barbetti M J, Chen Y. Pathogen biocontrol using plant growth-promoting bacteria (PGPR): Role of bacterial diversity. MDPI. 2021.
[http://dx.doi.org/10.3390/microorganisms9091988]

[50] Ansari FA, Jabeen M, Ahmad I. *Pseudomonas azotoformans* FAP5, a novel biofilm-forming PGPR strain, alleviates drought stress in wheat plant. Int J Environ Sci Technol 2021; 18(12): 3855-70.
[http://dx.doi.org/10.1007/s13762-020-03045-9]

[51] Mohanty P, Singh P K, Chakraborty D, Mishra S, Pattnaik R. Insight into the role of PGPR in sustainable agriculture and environment. Frontiers Media S.A. 2021.
[http://dx.doi.org/10.3389/fsufs.2021.667150]

[52] Wahab A, *et al.* Plant growth-promoting rhizobacteria biochemical pathways and their environmental impact: a review of sustainable farming practices. Springer Science and Business Media B.V. 2024.
[http://dx.doi.org/10.1007/s10725-024-01218-x]

[53] Bharathula S, Sarvani B, Reddy RS, Prasad JS. Characterization of plant growth promoting rhizobacteria for compatibility with commonly used Agrochemicals Available from: https://www.researchgate.net/publication/360495116

[54] Patel JS, Kumar G, Bajpai R, Teli B, Rashid M, Sarma BK. PGPR formulations and application in the management of pulse crop health. Biofertilizers. Elsevier 2021; pp. 239-51.
[http://dx.doi.org/10.1016/B978-0-12-821667-5.00012-9]

[55] Abraham WR. Applications and impacts of stable isotope probing for analysis of microbial interactions. Springer Verlag 2014.
[http://dx.doi.org/10.1007/s00253-014-5705-8]

[56] Romeh AAA. Remedial Potential Of Plant Growth Promoting Rhizobacteria (PGPR) for pesticide residues: Recent trends and future challenges. Pesticides Bioremediation. Springer International Publishing 2022; pp. 381-97.
[http://dx.doi.org/10.1007/978-3-030-97000-0_14]

[57] Andreata M F L, *et al.* Microbial fertilizers: A study on the current scenario of brazilian inoculants and future perspectives. Multidisciplinary Digital Publishing Institute (MDPI). 2024.
[http://dx.doi.org/10.3390/plants13162246]

[58] Hossain M A, Hossain M S, Akter M. Challenges faced by plant growth-promoting bacteria in field-level applications and suggestions to overcome the barriers. Academic Press. 2023.
[http://dx.doi.org/10.1016/j.pmpp.2023.102029]

[59] Kumari B, Mallick MA, Solanki MK, Solanki AC, Hora A, Guo W. Plant Growth Promoting Rhizobacteria (PGPR): modern prospects for sustainable agriculture. Plant Health Under Biotic Stress. Springer Singapore 2019; pp. 109-27.
[http://dx.doi.org/10.1007/978-981-13-6040-4_6]

[60] Ahmad I, Pichtel J, Hayat S. Plant-bacteria interactions: strategies and techniques to promote plant growth. Wiley-VCH 2008.

[http://dx.doi.org/10.1002/9783527621989]

[61]   Gamalero E, Bona E, Glick B R. Current techniques to study beneficial plant-microbe interactions. MDPI. 2022.
[http://dx.doi.org/10.3390/microorganisms10071380]

[62]   Schoebitz M, López MD, Roldán A. Bioencapsulation of microbial inoculants for better soil–plant fertilization. A review. Agron Sustain Dev 2013; 33(4): 751-65.
[http://dx.doi.org/10.1007/s13593-013-0142-0]

[63]   Maggi F, Pistello M, Antonelli G. Future management of viral diseases: Role of new technologies and new approaches in microbial interactions. Elsevier B.V. 2019.
[http://dx.doi.org/10.1016/j.cmi.2018.11.015]

[64]   Faust K, Raes J. Microbial interactions: From networks to models. 2012.
[http://dx.doi.org/10.1038/nrmicro2832]

[65]   Agrawal R, Satlewal A, Varma A. Characterization of plant growth-promoting rhizobacteria (PGPR): A perspective of conventional versus recent techniques. 2015; pp. 471-85.
[http://dx.doi.org/10.1007/978-3-319-14526-6_23]

[66]   Weiland-Bräuer N. Friends or foes—microbial interactions in nature. MDPI AG. 2021.
[http://dx.doi.org/10.3390/biology10060496]

[67]   Verma P, Yadav AN, Kumar V, Singh DP, Saxena AK. Beneficial plant-microbes interactions: Biodiversity of microbes from diverse extreme environments and its impact for crop improvement. Plant-Microbe Interactions in Agro-Ecological Perspectives. Springer Singapore 2017; 2: pp. 543-80.
[http://dx.doi.org/10.1007/978-981-10-6593-4_22]

[68]   Tshikantwa T S, Ullah M W, He F, Yang G. Current trends and potential applications of microbial interactions for human welfare. Frontiers Media S.A. 2018.
[http://dx.doi.org/10.3389/fmicb.2018.01156]

[69]   Ibáñez A, Garrido-Chamorro S, Vasco-Cárdenas M F, Barreiro C. From lab to field: Biofertilizers in the 21st century. Multidisciplinary Digital Publishing Institute (MDPI). 2023.
[http://dx.doi.org/10.3390/horticulturae9121306]

[70]   Adesemoye A O, Kloepper J W. Plant-microbes interactions in enhanced fertilizer-use efficiency. 2009.
[http://dx.doi.org/10.1007/s00253-009-2196-0]

CHAPTER 8

Symbiotic Alliances in Nature: Microbial Roles in Plant Growth, Stress Tolerance, and Soil Health

Vrushali Desai[1], Anish Kumar Sharma[1] and Priyanka Chauhan[1,*]

[1] *School of Sciences, P P Savani University, Surat, Gujarat, India*

Abstract: Microbial interactions with plants are pivotal in enhancing environmental resilience and maintaining ecosystem health. The multifaceted relationships between plants and microbes such as mutualistic, commensal, and parasitic interactions, are necessary to plant productivity and stress management. Beneficial microorganisms such as rhizobia, mycorrhizal fungi, and plant growth-promoting rhizobacteria (PGPR) assist in nutrient acquisition, enhance plant growth, and fortify plants against biotic and abiotic stresses. This chapter examines the intricate links between plants and diverse microbial communities, including mutualistic, commensal, and parasitic interactions. Symbiotic partnerships, between plants and rhizobia or mycorrhizal fungi. This interaction facilitates critical processes like nitrogen fixation and nutrient uptake, which are essential for plant health and productivity. Furthermore, the chapter explores the molecular and biochemical mechanisms supporting these interactions, including signaling pathways, microbial metabolite production, and modulation of plant defense responses. It highlights how these interactions improve plant resilience to environmental challenges, including drought, salt, and disease threats, while also promoting soil health through nutrient cycling. Besides that, the applications of microbial inoculants and bio-stimulants in sustainable agriculture are also discussed. The chapter concludes with future perspectives, highlighting the potential of genetic engineering and advanced research to harness plant-microbe interactions for greater environmental resilience, biodiversity, and climate adaptation.

Keywords: Biosensing technology, Environmental monitoring, Microbial biosensors, Pollutant detection, Sustainable solutions.

INTRODUCTION

Microorganisms exhibit a diverse range of sizes, shapes, and complexities. All living things on Earth have developed over aeons by using sun energy and ambient materials to construct organised structures [1]. All microorganisms absorb matter and energy from their surroundings and use it to create structures and functions that enable life. Matter consists of unique chemical forms-elements,

---

* **Corresponding author Priyanka Chauhan:** School of Sciences, P P Savani University, Surat, Gujarat, India; E-mail: priyanka.chauhan@ppsu.ac.in

molecules, and compounds [2]. Certain elements are used by microorganisms in large quantities, whereas others are used in trace amounts or not at all. Some compounds are rapidly excluded from cells, while others remain hoarded within them. Carbon is very significant because it is arranged in chains and rings for creating the skeletons of organic compounds, the building blocks of biomolecules and living things [3]. Every cell has a thin, dynamic membrane made of proteins and lipids, which controls the passage of materials between the cell and its surroundings and receives information from the outside world. A unique class of proteins known as enzymes performs all of the chemical reactions required to generate these various structures and give them energy and materials to carry out their duties, dispose of wastes, and execute other functions of life at the cellular level [4]. The dynamic interaction between microbial communities and plant biodiversity serves as the foundation of the ecosystem. These microorganisms' main function is shaping critical processes including nutrient cycling, soil formation, and the maintenance of plant health and productivity.

Recent breakthroughs have highlighted the essential role of plant-microbe interactions in maintaining ecosystem resilience amidst rapid environmental shifts [5]. Innovative scientific methodologies, including remote sensing, molecular genetics, and advanced statistical modelling, are revolutionising our comprehension of these connections. These methodologies provide an exceptional understanding of the genetic processes underlying microbial adaptation and the consistent roles of core microbiota across diverse plant species, facilitating the development of novel techniques in sustainable agriculture and ecosystem management. Not withstanding considerable progress, a substantial gap persists in our understanding of the impact of certain unexamined microbial taxa on plant diversification. This disparity is further accentuated by the escalating difficulties of climate change and biodiversity loss, underscoring the urgent need for focused study into these complex interconnections. Plant functional diversity is the diversity of plant traits that influence ecosystem functions and processes, and it is important in understanding the variety of plant life forms and functional traits within ecosystems, which influence ecological processes and adaptability to environmental fluctuations [6].

Plants and microbes *i.e.* bacteria, fungi, and viruses, indulge in various complex interactions that influence nutrient solubilisation, disease resistance, and plant growth [7]. These interactions can be either beneficial, neutral, or harmful, and they often involve a variety of mechanisms [8]. Beneficial interactions include symbiotic relationships like nitrogen fixation, where some soil microorganisms transform atmospheric nitrogen into a form that plants can use, or mycorrhizal fungi that help plants to absorb water and nutrients in exchange for sugars [9]. However, pathogenic interactions occur when microorganisms cause diseases, negatively affecting plant health and reducing crop yields [10]. Additionally, plants may produce chemicals and phytohormones such as antimicrobial compounds, to defend against harmful microbes and influence plant growth [11]. Overall, the interaction between plants and microbes is an agile and essential component of the environment, with extensive implications for agriculture, biodiversity, and ecosystem sustainability.

ROLE OF MICROBIAL COMMUNITIES IN PLANT DEVELOPMENT

Microbial communities are vital in maintaining plant development and enhancing agricultural production [12]. These communities are composed of a diverse range of microorganisms, including bacteria, fungi, viruses, and protozoa (Fig. 1). These microbes can inhabit various parts of the plant, such as roots, stems, leaves, and flowers, as well as the surrounding soil and rhizosphere (the region of soil influenced by plant roots) [13]. These microbial relations can be positive, neutral, or harmful, and they influence plant growth, stress tolerance, nutrient uptake, disease resistance, and overall vitality [8]. Most of the microbes reside in the rhizospheric region soil and they are beneficial to plant growth [14]. Besides that, plant pathogenic microorganisms also colonize the rhizosphere region. To cause disease, these microbial pathogens are striving to defeat the innate plant defense mechanisms and the protective microbial shield. A third group of microorganisms that can be found in the rhizosphere are the true and opportunistic human pathogenic bacteria, which can be found on or in the plant tissue and may cause disease when introduced into weak humans. To enhance plant growth and health, it is essential to know which microorganisms are present in the rhizosphere microbes and what do they do [15].

Rhizospheric Microbes

A. **Symbiotic Nitrogen-Fixing Bacteria:** Nitrogen is an important macronutrient for plants, and their biological nitrogen fixation is a sustainable way to ensure its availability. Nitrogen is an essential nutrient for plant growth, and they rely on nitrogen from the soil in the form of nitrate or ammonium [16]. However, some plants especially legumes form symbiotic relationships with nitrogen-fixing bacteria, such as *Rhizobium* and *Bradyrhizobium*. These bacteria transform atmospheric nitrogen (N_2) into a bioavailable form, ammonia (NH_3). This ammonia can be used by plants for their growth through the nutrient cycle. In exchange, plants provide carbon and other nutrients to the bacteria [17].

B. **Plant Growth-Promoting Rhizobacteria (PGPR):** PGPR are valuable bacteria that inhabit plant roots and boost plant growth *via* various mechanisms. These mechanisms are basically involved in nutrients' solubilisation, phytohormones' production, and pathogen control [18].

○ *Nutrients' solubilisation*: PGPR can release biological catalyst that makes nutrients like phosphorus, nitrogen, and iron, which are more accessible to plants [19].

○ *Phytohormones production*: Some PGPRs produce plant growth regulators, which results in the promotion of root growth, seed germination, and stress resilience (*e.g.*, auxins, cytokinin) [20].

○ *Antagonistic against pathogen*: PGPR can produce antibiotics, siderophores (iron chelating compound), or other antimicrobial compounds, which prevent the growth of plant pathogens [21].

○ *Induced Systemic Resistance (ISR):* Some PGPRs can stimulate the plant's immune system, making it more resistant to diseases caused by pathogens [22].

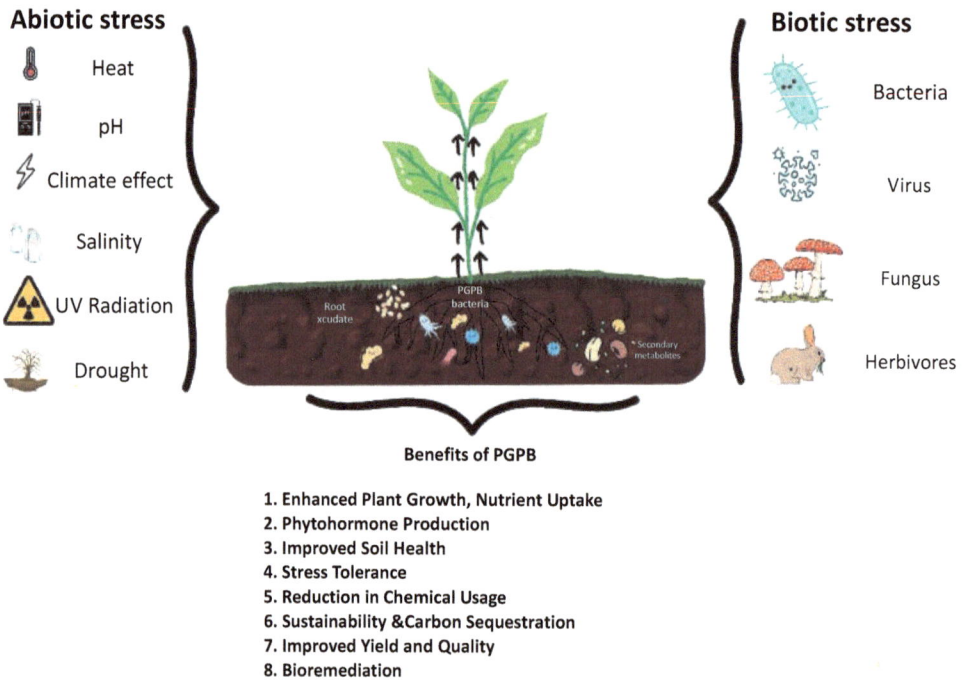

Abiotic stress

Heat

pH

Climate effect

Salinity

UV Radiation

Drought

Biotic stress

Bacteria

Virus

Fungus

Herbivores

Root xcudate   PGPB bacteria   Secondary metabolites

Benefits of PGPB

1. Enhanced Plant Growth, Nutrient Uptake
2. Phytohormone Production
3. Improved Soil Health
4. Stress Tolerance
5. Reduction in Chemical Usage
6. Sustainability &Carbon Sequestration
7. Improved Yield and Quality
8. Bioremediation

Fig. (1). Plant-Microbes interaction during environmental resilience.

Phyllosphere Microbes

Microbes that live in epiphytic or endophytic parts of the plant like leaves, stems and flowers are known as phyllosphere microbes. This microbial community plays an essential role in plants' health and growth, assuring different nutrients' cycling and protection against pathogens [23].

Pathogen Suppression

Many phyllosphere bacteria produce antibiotics, lytic enzymes, and siderophores that inhibit or kill pathogens. Similarly, *Pseudomonas fluorescens* produces phenazines and pyoluteroin to inhibit pathogens [24]. Phyllosphere microbes can trigger ISR, enhancing the plant's internal defense mechanisms. Activation of jasmonic acid and ethylene pathways increases resistance to necrotrophic pathogens and pests [25].

Plant growth promotion

a. Nutrient cycling: Phyllosphere microbes contribute to nutrient dynamics by recycling organic materials and facilitating nutrient uptake [26].

Decomposition of Organic Matter: Microbes on leaf surfaces degrade organic residues, releasing nutrients that plants can absorb [27].

Nitrogen Fixation: Some phyllosphere bacteria, such as *Methylobacterium,* fix atmospheric nitrogen, making it available to plants [28].

Carbon Cycling: Phyllosphere microbes metabolize carbon sources like methanol a byproduct of plant metabolism, promoting nutrient recycling [26]

a. Hormone Production: Synthesis of auxins, cytokinins, and gibberellins promotes growth. Auxins stimulate root and shoot growth. Cytokinins increase cell division and chlorophyll production. Gibberellins promotes stem elongations [29]. Some microbes such as *Bacillus subtilis*, *Pseudomonas*, *Bacillus tequilensis,* and *Trichoderma* have the ability to produce phytohormones, which improve plant growth and defense mechanism [30, 31].
b. Stress Alleviation: Microbes help mitigate abiotic stresses like drought, salinity, temperature, and UV radiation by producing protective metabolites [32].

Biofilm Formation

Many phyllosphere microbes form biofilms, creating a protective barrier against pathogen colonization [33]. Phyllosphere microbes can create biofilms on plant surfaces, which can act as a physical barrier against pathogens [34]. These biofilms consist of microbial communities enclosed with an extracellular polymeric substance, which acts as a protective matrix [26].

Plant Defence Mechanisms: Induced Systemic Resistance

Plants activate innate immune responses in response to external stimuli, including beneficial microorganisms during Induced Systematic Resistance (ISR) [35]. It is a form of plant immunity that provides protection against pathogens and stresses without the direct involvement of the immune system proteins such as pathogen-triggered immunity (*e.g.*, salicylic acid signaling, which is central to Systematic acquired resistance) [36]. ISR is generally linked with the activation of jasmonic acid (JA) and ethylene (ET) signaling pathways, which prime the plant's defense mechanisms [37].

Mechanisms of ISR

Priming: In ISR, the plant doesn't mount a full immune response immediately but "primes" its defence machinery, making it more responsive to subsequent pathogen attacks [39]. This is a metabolic shift that allows plants to respond more quickly and efficiently when challenged [38].

Increased Production of Defence Compounds: ISR leads to the synthesis of defence-related molecules like antimicrobial compounds, Pathogenesis-Related proteins (PR proteins), and defensive enzymes such as peroxidases, chitinases, and glucanases [35].

Activation of Hormonal Pathways: The activation of Jasmonic Acid (JA) and ethylene (ET) signaling pathways is important for ISR, and these pathways increase the production of defence-related compounds, help in structural changes in cell walls, and reduce pathogen entry [35].

Microbial Interactions in Induced Systemic Resistance

Beneficial microorganisms, particularly Plant Growth-Promoting Rhizobacteria (PGPR), fungi, and other soil microbes, play a pivotal role in activating ISR [40]. These microbes induce systemic resistance by interacting with plant roots and triggering various molecular and biochemical responses. Some well-known PGPRs that activate ISR include *Bacillus subtilis, Pseudomonas fluorescens, Azospirillum,* and *Rhizobium* species [41]. These bacteria interact with the plant's root system and produce metabolites that trigger the plant's immune system. Bacterial Signalling Molecules: PGPR produces signalling molecules like flavonoids, siderophores, and volatiles (*e.g.*, acetoin), which can activate plant immune responses, enhancing resistance to pathogens [42].

Microbial Volatile Organic Compounds (VOCs): Certain bacteria emit VOCs that induce ISR by activating the plant's defense mechanisms. These VOCs can trigger the plant's immune system through changes in hormonal balances, particularly JA and ET pathways [43].

Competition with Pathogens: PGPR can directly suppress pathogen growth by competing for nutrients and space on plant roots, thereby reducing the likelihood of pathogen infection. For example, *Bacillus* spp. produce lipopeptides that inhibit the growth of fungal pathogens [21].

Abiotic Stress Mitigation

Microbial communities influence plant resilience by alleviating abiotic stresses, such as drought, high salinity, and extreme temperatures [10]. Microbial communities in the phyllosphere (leaf surface) and the rhizosphere (root zone) play a pivotal role in enhancing plant resilience by mitigating various abiotic stresses like drought, salinity, and extreme temperatures [44]. These microbial communities include archaea, bacteria, yeast and fungi that interact with plants to alleviate stress through several mechanisms that directly or indirectly enhance plant health and survival under stressful conditions [45]. Some mechanisms of microbial-mediated stress tolerance and abiotic stress mitigation are given below.

Drought Stress

Drought stress is the worst severe abiotic stress on plants, leading to reduced water availability and compromised cellular functions [10]. Microbial communities can help alleviate drought stress through the following mechanisms such as the production of osmoprotectants (Compatible Solutes), Hydrogel and Exopolysaccharide (EPS), induction of plant water stress-responsive genes, and promotion of root growth and exploration [46]. Many beneficial microbes, such as bacteria from genera like *Pseudomonas*, *Bacillus*, and *Azospirillum*, produce osmoprotectants such as proline, trehalose, and glycerol. These solutes help microbes and plants maintain cellular integrity under osmotic stress (like drought) [47]. Osmoprotectants stabilize proteins and cellular membranes by maintaining cellular turgor and preventing cellular dehydration. These compounds can be transferred to plants, enhancing their ability to retain water, reduce osmotic stress, and protect cellular structures from drought-induced damage [48].

Certain soil and leaf-associated microbes can produce Exopolysaccharides (EPS) or hydrogels, which act as water-holding agents. EPS forms a protective layer around microbial cells or plant tissues, reducing water loss through evaporation and helping plants retain moisture during dry periods [49]. By creating a moisture reservoir around plant roots or leaf surfaces, microbes help plants avoid dehydration and maintain hydration during periods of water scarcity. Additionally, some microbes can induce the expression of drought-resistant genes in plants. These include genes that enhance water uptake, root elongation, or stomatal regulation to optimize water use efficiency under drought conditions [50]. *Azospirillum* and *Rhizobium* species have been shown to trigger phytohormones (like abscisic acid or cytokinin) that modulate plant water balance and stress response [51].

Rhizospheric and phyllospheric microbes boost root growth and architecture, allowing plants to access deeper water sources. Enhanced root development and a more extensive root system enable plants to access more water, making them more resilient to drought conditions [52].

Salinity Stress

Salinity is another major abiotic stress that affects plant growth and productivity, as high salt concentrations can lead to osmotic stress, ionic imbalance, and toxic effects. Microbial communities can assist plants in dealing with salt stress through several strategies such as the production of compatible solutes, excretion of plant growth regulators, salt stress tolerance genes, and sodium ion exclusion [53]. Like drought stress, microbes can produce osmolytes like proline, trehalose, and glycine betaine under saline conditions. These solutes help microbes and plants cope with the osmotic stress imposed by high salt concentrations. These osmoprotectants help maintain cellular water balance, stabilize proteins, and protect cell membranes under salinity stress [48]. Moreover, some microbes produce phytohormones, such as auxins and cytokinins, which help plants mitigate the adverse effects of salinity [54]. These hormones can improve root growth and water uptake, enhance salt exclusion mechanisms, and help plants maintain cellular processes even under saline conditions [55]. Beneficial microorganisms, particularly halotolerant or halophilic species, can introduce specific salt-tolerance genes into the plant through Horizontal Gene Transfer (HGT). These genes may enhance salt exclusion mechanisms or enable the plant to compartmentalize and detoxify excess salt within vacuoles, reducing the toxic effects of salinity [56]. Certain microbes help plants by promoting sodium ion exclusion from plant tissues. Microbes can stimulate the activity of plant Na+/K+ pumps in roots or leaves, thereby decreasing sodium uptake and increasing potassium uptake. By maintaining an optimal Na+/K+ ratio, plants are less susceptible to salt-induced toxicity [57].

Temperature Stress: Extreme temperatures (both high and low) are another source of abiotic stress for plants. Microbial communities contribute to temperature stress tolerance in the following ways: Cold-tolerant microbes, such as certain psychrophilic or psychrotolerant bacteria and fungi, produce Cold Shock Proteins (CSPs) and Antifreeze Proteins (AFPs), which prevent ice crystal formation inside plant cells and protect against cellular damage during freezing temperatures. These proteins inhibit ice crystal growth within plant tissues, and lower the possibilities of dehydration and physical injury to plant cells under cold conditions [58]. Heat-tolerant microbes, such as certain thermophilic bacteria, help plants tolerate high temperatures by producing Heat Shock Proteins (HSPs), which protect proteins and enzymes from thermal denaturation. These microbial HSPs can help plants better tolerate heat stress by stabilizing cellular structures and enzymes, promoting overall plant stability under high-temperature conditions. Besides that, microbial interactions can induce heat shock responses in plants [59]. For instance, microbes like *Bacillus subtilis* can produce compounds that trigger the plant's own Heat Shock Proteins (HSPs), which are crucial for coping with elevated temperatures [60]. By stimulating the plant's heat shock response, microbes enhance the plant's ability to protect cellular proteins and membranes under heat stress.

ROLE OF MICROBES IN PLANT RESILIENCE AND SOIL HEALTH

Soil health is fundamental to plant growth and resilience, as it directly influences water retention, nutrient availability, and disease resistance. Microbial communities in the soil comprising bacteria, fungi, archaea, protozoa, and viruses perform a pivotal function in retaining soil structure, richness, and biological activity. These microorganisms work synergistically with plants to improve soil quality and enhance plant resilience against abiotic and biotic stress [61]. This detailed note explores the mechanisms of microbial-mediated resilience and how microbial communities contribute to soil health, fostering an environment conducive to plant growth and survival.

Soil Structure Maintenance

Microbial activity is essential in maintaining soil structure, which affects water infiltration, root penetration, and overall soil aeration. Healthy soil structure improves the plant's ability to access water and nutrients, especially during stressful conditions like drought [62].

Formation of Soil Aggregates: Microorganisms, particularly bacteria and fungi, produce Extracellular Polymeric Substances (EPS), which help to bind soil particles together, forming soil aggregates. Soil aggregates improve soil porosity, creating microhabitats that facilitate water movement and root growth [63]. These aggregates also increase the stability of the soil, reducing erosion and improving the soil's capacity to store organic matter and nutrients. Aggregated soils provide better aeration and drainage, preventing root suffocation in waterlogged conditions and enhancing root penetration in dry soils [64].

Mycorrhizal Networks: Mycorrhizal fungi, such as Arbuscular Mycorrhizae (AMF), develop a symbiotic interaction with plant roots. They expand their hyphal nets through the soil, facilitating nutrient absorption, particularly phosphorus, and improving soil structure. Mycorrhizal networks increase soil porosity and help retain water in sandy soils, enhancing plant drought resilience. They also improve nutrient cycling, ensuring that plants receive a steady supply of essential minerals [65].

Reduction of Soil Compaction: Certain soil bacteria produce hydrogels or biofilms, which can reduce the effects of soil compaction. This is particularly beneficial in heavy clay soils or under intensive agricultural practices that often lead to compaction. Microbial activity loosens compacted soils, improving root growth and nutrient uptake while facilitating better water infiltration and root penetration [66].

Soil Fertility and Nutrient Cycling

Soil fertility is directly linked to the microbial communities that facilitate nutrient cycling. These microbial processes ensure that nutrients are continuously replenished, making them available for plant uptake. This is especially important in maintaining plant growth and resilience under stressful conditions like nutrient-poor or acidic soils [67].

Nitrogen Fixation: Diazotrophic bacteria (*e.g.*, *Rhizobium*, *Azospirillum*, *Frankia*) and certain cyanobacteria can convert atmospheric nitrogen (N_2) into ammonia (NH_3), a form that plants can absorb and use. These bacteria not only form symbiotic relationships with legumes but also associate with non-leguminous plants in the rhizosphere. Nitrogen fixation increases soil nitrogen availability, which is necessary for plant development, especially in nutrient-limited soils [68]. This is particularly valuable in sustainable farming systems that avoid synthetic nitrogen fertilizers.

Phosphorus Solubilization: Phosphate-solubilizing bacteria (*e.g.*, *Pseudomonas*, *Bacillus*) and mycorrhizal fungi can solubilize insoluble phosphorus compounds in the soil, making phosphorus available for plant uptake [69]. Phosphorus is an essential nutrient for root development, flowering, and stress resilience [70]. Improved phosphorus availability helps plants grow optimally, particularly in soils with low phosphorus content.

Decomposition of Organic Matter: Decomposer microorganisms, such as bacteria, fungi, and actinomycetes, convert organic matter into humus and other nutrient-rich compounds. This process releases carbon, nitrogen, phosphorus, and other essential nutrients into the soil. Decomposition replenishes soil nutrients, confirming a continuous allocation of organic matter for plant growth. Healthy microbial activity supports soil fertility by recycling nutrients that would otherwise remain locked in organic material.

Nutrient Mobilization: Soil microbes also mobilize other essential nutrients like iron, sulfur, and zinc, often less available to plants in certain soil conditions (*e.g.*, acidic or alkaline soils). Microbes enhance plant resilience by increasing the availability of critical micronutrients, particularly under nutrient-deficient conditions [71].

Soil Biological Activity and Disease Suppression

Soil-borne pathogens can significantly reduce plant health, especially under stressful environmental conditions. However, beneficial soil microbes help suppress these pathogens through direct and indirect mechanisms, thus contributing to plant resilience [73].

Antagonistic Microbial Interactions: Many beneficial bacteria and fungi produce antimicrobial compounds, such as antibiotics, volatile organic compounds (VOCs), and hydrolytic enzymes, that inhibit the growth of harmful soil pathogens (*e.g.*, *Fusarium*, *Pythium*, *Rhizoctonia*) [74]. By suppressing pathogens, beneficial microbes reduce the incidence of root rot, blight, and other soil-borne diseases, promoting healthy plant growth even in pathogen-rich environments [72].

Competition for Resources: Pathogens and beneficial microbes compete for nutrients and soil space. This reduces the ability of harmful microbes to establish themselves and thrive. This competition limits the impact of pathogenic microbes on plant health, thus enhancing resilience against disease [15].

Induced Systemic Resistance (ISR): Some soil microbes trigger Induced Systemic Resistance (ISR) in plants, which activates the plant's innate immune response. This strengthens the plant's defenses against any infections [75]. ISR enhances plant disease resistance, reducing the likelihood of infections under stressful conditions, such as periods of high humidity or after heavy rainfall that may promote pathogen growth [76].

Enhancing Soil Biodiversity

The presence of diverse microbial communities in soil supports soil biodiversity, which is crucial for maintaining a resilient ecosystem capable of withstanding environmental stress [77]. Diverse microbial communities can adapt to a range of conditions, ensuring that soil functions—such as nutrient cycling, disease suppression, and soil structure maintenance—remain intact, even when environmental conditions change [78]. High microbial diversity increases the likelihood that some microbes will thrive in a changing environment, ensuring that plant growth-promoting functions continue, even under stress [20].

MICROBES AND CLIMATE CHANGE CONDITION

Plant-microbe interaction may be disrupted by the climate change conditions at a local to global scale. The average worldwide surface temperature is anticipated to rise by 2.6-4.8 degrees Celsius by the year 2100 if no mitigation efforts are adopted [79]. This increased temperature will lower the water in the soil and upsurge the drought stress in several areas [80]. Therefore, higher temperatures and drought conditions affect the plant-microbe interaction and plant growth and development. To overcome this stress, plants have developed many changes in their physiology as well as their microbiome. Plant-microbe interactions may regulate resilience to environmental change to alter (capacity to retrieve from stress) and influence a plant or microbe's resilience (ability to remain unchanged) to climate upheaval [81]. Additionally, interactions between microbes and plants produce two significant climatic inputs. First, bacteria change primary production as mutualists or pathogens, which affects the size of the carbon (C) sink derived from plants [82]. Second, plant-associated microorganisms control respiration and the return of carbon to the atmosphere by influencing breakdown and soil carbon [83, 84].

Plants have an unbreakable connection to different, taxonomically structured groups of microbes. The plant microbiome consists of bacteria, fungi, protists, nematodes, and viruses that inhabit any available plant tissues [85]. The microbial community that lives in rhizospheric soil, roots, and other parts of plants has convoluted and active interaction with the host plant. These interactions are significantly impacted by the environment and can help plants withstand environmental challenges such as biotic and abiotic stresses [86 - 88] (Table **1**). Although the relevance of microbes to plant growth and development is becoming more widely acknowledged, it is still very difficult to use microbial interactions and features to increase plant resistance to climatic volatility.

In spite of this, several microbes providea defense mechanism to plants to copewith climate change conditions.

Microbial Synergies with Plants in a Changing Climate

Plants not only actively engage with the environmental microbiome to regulate their immune systems against diseases, they also attract beneficial bacteria from their surroundings to cope with various stressors, a strategy often referred to as the "cry for help" approach [112]. The interaction of plant microbes can influence how organisms respond to climate and linked stresses in different methods. Leaf endophytes can increase plant resistance to drought, heat, or salinity, enhance nutrient concentration in plant tissues, and repel herbivores or diseases [113 - 115]. Moreover, mycorrhizal fungi boost the intake of water from the soil and absorb up to 80% of phosphorus (P) for plants. Ascomycetes can absorb nitrogen from the soil and shield plants from heat stress or drought [113, 114].

Table 1. List of microbes and their effect on plant's growth and development.

S.No.	Microbes	Effect on plants	References
1.	*Flavimonas oryzihabitans*	Control root-knot nematode (*Meloidogyne* spp.) and *Fusarium wilt,* phosphate solubilisers, and plant growth-promoting bacteria.	[87]
2.	*Rhizobium radiobacter*	Potent Phosphate solubilizers, improved plant growth through secretion of PGP compounds.	[88]
3.	*Pseudomonas japonica*	Calcium ion binding.	[90]
4.	*Azospirillum humicireducens*	A free-living nitrogen fixer, functions in waterlogged heavy soils).	[89]
5.	*Azotobacter vinelandii*	A free-living nitrogen fixer secretes phytohormones and vitamins into the soil.	[91]
6.	*Bacillus cereus*	Prevents plant diseases; enhances plant growth.	[92]
7.	*Bacillus licheniformis*	Nutrient cycling in the soil.	[93]
8.	*Bacillus thuringiensis*	Acts as insecticidal.	[94]
9.	*Ensifer numidicus*	Symbiotic capacity to nodulate their host plant and fix atmospheric nitrogen.	[95]
10.	*Ensifer fredii*	A nitrogen fixer forms nodules on legumes.	[96]
11.	*Ensifer melilot*	Mitigates frought stress.	[97]
12.	*Trichoderma virens*	Increased biomass production and stimulates lateral roots.	[98]
13.	*Trichoderma aureoviridi*	Produce different extrolites and enzymes responsible for the bio-control activities against the harmful fungal phytopathogens.	[98]
14.	*Trichoderma harzianum*	Used for biological control of fungal diseases.	[98]
15.	*Trichoderma inhamatum*	Produces different extrolites and enzymes responsible for the bio-control activities against the harmful fungal phytopathogens, suitable against soil and seed-borne pathogens.	[98]
16.	*Sinorhizobium americanum*	Symbiont that nodulates and fixes nitrogen.	[99]

(Table 1) cont.....

S.No.	Microbes	Effect on plants	References
17.	*Rhizobia bacteria -Azospirillum, Enterobacter, Azorhizophillus, Sinorhizobium and Azotobacter sp*	Colonize the soil surface, the interior of roots, and occasionally even the stems and leaves, and can reduce nitrogen into ammonia.	[100]
18.	*Micrococcus luteus*	Improved seedling growth.	[101]
19.	*Escherichia coli*	Upregulate the expression of cell wall modification genes and downregulate heat shock proteins.	[102, 104]
20.	*Pseudomonas sp. Pantoea sp.*	Improved root gravitropism, root and shoot growth, and root hair formation.	[105]
21.	*Pseudomonas fluorescens*	Production of ethylene, and enhanced root cell growth.	[103]
22.	*Burkholderia sp.*	Nitrogen fixation.	[106]
23.	*Bacillus amyloliquefaciens*	Enhanced seedling growth and expression of nitrate transport genes and controlled fungal growth.	[107, 108]
24.	*Burkholderia australis*	Nitrogen fixation and increased seedling growth.	[109]
25.	*Gluconacetobacter diazotrophicus*	Nitrogen fixation, phytohormone, siderophore, and bacteriocin production.	[110]

It has been reported that certain fungi and PGPB produce hormone analogues that stimulate plant germination and growth [117]. *Streptomyces*, a rhizospheric bacteria enhances plant resilience to drought and disease [114, 118]. Nitrogen-fixing bacteria enrich nitrogen levels in plant leaves, roots, rhizospheres, and seeds [114, 119]. These microbial advantages can help plants withstand and retrieve from stress. Several microbes exhibit a neutral phase in their life cycle, such as certain endophytes, which are inactive in living leaves, act as primary decomposers in senesced tissue, obtaining early access to it [120]. Non-neutral effects often depend on specific stress conditions, leading to a spectrum of fitness outcomes even for the same plant-microbe pair.

Microbe-plant interactions typically range from mutually beneficial to parasitic, with varying degrees of specialization, from obligate to facultative. These consequences are influenced by the balance between benefits provided (such as resources) and their associated costs. Carbon (C) from photosynthesis is a key resource exchanged, but the extent of this exchange varies. For example, Arbuscular Mycorrhizal (AM) fungi depend entirely on plant-derived carbon, whereas mycorrhizal fungi attached to other host plants provide all carbon to achlorophyllous plants and orchid seedlings [121]. In parasitic relationships, some fungi and bacteria break down plant cell walls to access carbon (necrotrophy), while others extract carbon from living tissues (biotrophy) [122].

Based on resource utilization, a global database has categorized over 13,000 fungi into endophytes, pathogens, and mycorrhizal guilds in order to classify these interactions. About 50% were found to be pathogens, whereas about 16% were found to be helpful symbionts [123]. No similar database exists for plant-associated bacteria. Furthermore, these classifications depend on the stability of interaction outcomes, and many taxa may fit into multiple categories, such as both pathogens and endophytes.

Microbes in soil nutrient cycling

The activity of microbes in soil plays a pivotal role in the cycling of carbon and other greenhouse gases, profoundly influencing climate regulation. Soil microbes drive key activities such as decomposition, organic matter formation, and the transformation of carbon dioxide (CO_2), methane (CH_4), and nitrous oxide (N_2O). Through respiration and fermentation, these microorganisms release CO_2 and CH_4, while denitrifying bacteria produce N_2O, contributing to greenhouse gas emissions [124]. Microbes that are present in soil play a significant role in soil health and maintain ecosystems. Various microbes have played various functions such as nutrient cycling of N, P, K, and Zn, weathering of minerals, organic matter decomposition, and fossil fuel formation. Soil microorganisms, in addition to improving soil fertility, serve an important role in the natural cycles that allow life to exist on the planet. One gram of soil could contain hundreds of millions or billions of bacteria. Bacteria, fungi, actinomycetes, and algae are the most common microorganisms in the soil. A deeper understanding of soil microbiology is required if agricultural production is to meet the requirements of the expanding global population. Recycling is often referred to as the physiochemical recovery of resources for new applications. Plants absorb nutrients from the soil that are ingested by humans or animals and spit them out or release them back into the environment when living things die. Soil microbes play a crucial role in breaking down organic matter and converting it into mineral forms, allowing plants to absorb these nutrients. This nutrient cycling is essential for all living organisms, including plants and humans, as it ensures the continuous availability of vital elements. Key nutrients like nitrogen, phosphorus, and potassium are particularly important for water management and sanitation. During chemical transformations, living organisms impact their environment, with microbial processes like oxidation and reduction serving as significant drivers for nutrient conversion. Soil microbiomes contribute to the enhancement of essential nutrients by recycling a diverse range of elements, including carbon, phosphorus, potassium, nitrogen, zinc, silicon, calcium, and manganese [125 - 127]. This recycling supports the production of proteins, amino acids, and genetic materials such as DNA and RNA—key components of life. For instance, mineral weathering, facilitated by bacteria and fungi, transforms minerals into plant-usable nutrients. Processes like silicate weathering release calcium, potassium, and magnesium, while apatite weathering contributes phosphorus to the soil. These microbial activities are vital for nutrient cycling, ion exchange, and overall plant nutrition [128, 129].

MICROBIAL-ASSISTED BIOREMEDIATION AND PHYTOREME-DIATION

Heavy metals have an atomic number greater than 20 with metallic properties such as stability as cations, ligand specificity, conductivity, *etc.* The heavy metals which have a density greater than 5.0 g/cm^3 are clarified into three classes such as toxic metals, precious metals and radionuclides. However, the highly toxic heavy metals contaminants are lead, chromium, cadmium, nickel, zinc, mercury, and copper [130]. Moreover, heavy metals cannot be broken down either by chemical or physical processes. They only transform from one oxidation state to another state to make them water-soluble, less toxic, and precipitated. Therefore, the remediation of heavy metal-contaminated areas such as soil, sediment, and water can be a better option to reduce the toxic effect of heavy metals. This bioremediation approach is based on biologically encoded changes in the oxidation state. The use of microorganisms to remediate heavy-meta--contaminated sites is an eco-friendly and cost-effective approach to restoring polluted environments [131]. Bioremediation leverages living microorganisms to convert toxic pollutants into harmless or less toxic forms, requiring minimal energy, chemicals, and time. Microbes achieve this by utilizing heavy metals as terminal electron acceptors, altering their physical and chemical states, or reducing their presence in the environment. The success of this strategy depends on microorganisms' metabolic activity and their tolerance to heavy metals [132].

Microorganisms, which thrive in heavy-metal-contaminated environments, can detoxify metals through processes such as biosorption, biomineralization, bioaccumulation, bioleaching, and biotransformation [133]. They also mobilize metals from polluted sites using mechanisms such as chelation, leaching, redox transformation, and methylation. Microbial defense systems such as compartmentalization, complex formation, exclusion, and the production of binding proteins and peptides mitigate the stress caused by heavy metals [134, 135]. There are two primary types of bioremediations: *in situ* and *ex situ*, categorized based on factors like site geography, geology, pollution depth, pollutant type, and concentration.

In-situ Techniques

In-situ techniques eliminate the need to dig up polluted soil and transport it to off-site treatment facilities. This reduces soil disruption, lowers the risk of exposing workers and the community to toxic materials, and may save on treatment costs. It is important to consider key factors such as soil permeability, pollution depth, temperature, and potential chemical leaching [135]. *In-situ* techniques comprise bioaugmentation, bio attenuation, bio slurping, bioventing, and biosparging.

Ex-situ Techniques

The process involves excavating and transporting contaminants from polluted sites in a systematic manner for subsequent waste treatment. *Ex-situ* bioremediation technologies are typically assessed based on various factors, including the extent of pollution at the contaminated site, the level of contamination, the types of pollutants present, the specific region of interest, and the associated treatment costs. Performance criteria also determine the choice of *ex-situ* bioremediation. *Ex situ* techniques contain bio-pilling, landfarming, compositing, bioreactors, and windrows [137].

Microbes play a vital role in the bioremediation of heavy-metal-contaminated soils, often complementing techniques like phytoremediation. Phytoremediation employs methods such as phytostabilization, phytoextraction, and phyto-volatilization to manage soil contamination [136]. While effective at removing pollutants, phytoremediation is limited by its reliance on the bioavailable forms of toxic substances, making it ineffective for non-bioavailable contaminants [138]. Additionally, only a specific subset of plants can survive in highly metal-polluted soils, further constraining their application [139].

To overcome these challenges, researchers are exploring complementary approaches, such as the use of synthetic and organic amendments, the development of GMO plants, and the application of biofertilizers [140]. These strategies enhance phytoremediation by facilitating the metal-chelator complexes, increasing plant tolerance and resistance to heavy metals, and promoting the translocation and accumulation of metals within plant tissues. Such improvements significantly enhance the efficiency of phytoremediation [139]. Several bacterial genera, including *Acidovorax*, *Alcaligenes*, *Bacillus*, *Mycobacterium*, *Paenibacillus*, *Pseudomonas*, and *Rhodococcus*, have been widely studied for their contributions to phytoremediation [141]. According to research, Plant Growth-Promoting Bacteria (PGPB) have a variety of functions, including boosting nutrient intake, reducing plant diseases, improving soil fertility and fitness, and increasing resistance to harmful metals. Furthermore, PGPB aids plants in absorbing and excreting heavy metals [142].

Plants can be made more resistant to heavy metal toxicity, even at low concentrations of phytohormones such as gibberellic acid, salicylic acid, brassinosteroids, auxins, cytokinins, abscisic acid, 1-aminocyclopropane-1-carboxylate deaminase (ACCD), jasmonates, and others [143]. PGPB improves the plant's phytoremediation ability by increasing the plant's resistance to stresses [138]. Application of *Pseudomonas lurida* EOO26 protects the *H. annuus* against Cu toxicity and Cu accumulation in both roots and shoots and bacteria can produce IAA, phosphate solubilization, ACC deaminase, and siderophore [144]. Moreover, the inoculation of *Bacillus subtilis* improved the phytostabilization ability of ryegrass by decreasing the Cd bioavailability in soil and inhibiting the transmission from root to shoot [145]. Additionally, through a composite reaction, microbial siderophores can mobilize and chelate a variety of metals, such as Cd, Ni, Pb, Zn, Fe, Cu, U(IV), Pu (IV), and Th(IV), increasing the availability of metals in the rhizosphere. Furthermore, the endophyte *Streptomyces* sp. CoT10, which produces siderophores, improved the soil's microbial population and increased *Camellia oleifera's* productivity and absorption of nutrients [147].

Moreover, *Bacillus cereus* and *Brevundimonas* sp. are crucial for promoting plant growth. They produce plant growth-promoting compounds such as hydrogen cyanide, siderophores, nitrogen fixation, IAA, and phosphorus solubilization. Additionally, these bacteria effectively decontaminate metals from polluted soils [148, 149].

A CASE STUDY IN LEGUME CROPS

Legume crops are rich in protein, nutrition, and health benefits for humans. Legume crops are a global source of protein for humans as well as animals. These crops basically involve chickpeas, soybeans, green peas, cowpeas, common peas, faba beans, mungbeans, *etc.* Moreover, legumes are highly rich in several healthy compounds such as vitamins, minerals, chlorophyll, phenolic acid, and vitamin C [150]. These crops are being affected by several biotic and abiotic stresses. The legume crops are driving several defense strategies to mitigate these stresses. Among them, the unique capacity of legume crops to form a mutualistic relationship with rhizobacteria is the most important microbial defence system [151]. Therefore, many of them have plant growth-promoting characteristics, which could improve plant productivity by direct or indirect methods during stress conditions [152]. Numerous studies have shown that not only symbiotic microbes but also other microbes isolated from various environments enhance legume crops' growth and defense responses. Research has proved that the application of 26 nitrogen-fixing endophytic bacteria improves seed dormancy and germination in edible peas by providing a microbiome platform [153]. Similarly, a consortium of fungal endophytes (*Beauveria bassiana, Trichoderma asperellum, Metarhizium anisopliae, Purpureocillium lilacinum,* and *Pochonia chlamydosporia)* stimulate the accumulation of N, P, B, Cu, Fe, Mn and Zn in common bean plants. Moreover, they also increased the concentration of soil Mg, and fluorescein diacetate hydrolysis (FDA) [154]. Additionally, an endophytic bacteria *Bacillus subtilis* can colonize in soybean and improve plant health during Charcoal rot infection.

CONCLUSION

Microbial interactions with plants are fundamental to environmental resilience and sustainable agriculture. This chapter highlighted the diverse roles of microbes in nutrient solubilisation, stress tolerance, and bioremediation. Symbiotic associations, such as those with nitrogen-fixing bacteria and mycorrhizal fungi, enhance plant growth and stress adaptation. Microbial contributions to soil health and climate regulation further underscore their ecological significance. Despite progress, knowledge gaps remain in understanding unexplored microbial groups and their mechanisms. Advanced tools, including molecular genetics and bioinformatics, are essential to address these gaps. Harnessing plant-microbe interactions through microbial inoculants, bio-stimulants, and genetic engineering can improve agricultural productivity and ecosystem stability. Integrating these strategies is crucial for addressing climate change, biodiversity loss, and soil degradation.

REFERENCES

[1]   Odum HT. Environment, power, and society for the twenty-first century: the hierarchy of energy. Columbia University Press 2007.

[2]   Kirchman DL. Processes in microbial ecology. Oxford University Press 2018.
[http://dx.doi.org/10.1093/oso/9780198789406.001.0001]

[3]   Singh HP, Kalia S. Biomolecules in living organisms. Advances in Environment and Agriculture Biotechnology.

[4]   Katz E, Privman V. Enzyme-based logic systems for information processing. Chem Soc Rev 2010; 39(5): 1835-57.
[http://dx.doi.org/10.1039/b806038j] [PMID: 20419221]

[5]   Loiko N, Islam M. Plant–soil microbial interaction: Differential adaptations of beneficial *vs.* pathogenic bacterial and fungal communities to climate-induced drought. Agronomy (Basel) 2024; 14(9): 1949.
[http://dx.doi.org/10.3390/agronomy14091949]

[6]   Pii Y, Mimmo T, Tomasi N, Terzano R, Cesco S, Crecchio C. Microbial interactions in the rhizosphere: beneficial influences of plant growth-promoting rhizobacteria on nutrient acquisition process. A review. Biol Fertil Soils 2015; 51(4): 403-15.
[http://dx.doi.org/10.1007/s00374-015-0996-1]

[7]   Farrar K, Bryant D, Cope-Selby N. Understanding and engineering beneficial plant-microbe interactions: plant growth promotion in energy crops. Plant biotechnMartin, F.M., Uroz, S. and Barker, D.G., 2017. Ancestral alliances: plant mutualistic symbioses with fungi and bacteria.. Science 2014; 356(6340): eaad4501.

[8]   Martin FM, Uroz S, Barker DG. Ancestral alliances: Plant mutualistic symbioses with fungi and bacteria. Science 2017; 356(6340): eaad4501.
[http://dx.doi.org/10.1126/science.aad4501] [PMID: 28546156]

[9]   Kumar A, Verma JP. Does plant—Microbe interaction confer stress tolerance in plants: A review?.

Microbiol Res 2018; 207: 41-52.
[http://dx.doi.org/10.1016/j.micres.2017.11.004] [PMID: 29458867]

[10]　Ortíz-Castro R, Contreras-Cornejo HA, Macías-Rodríguez L, López-Bucio J. The role of microbial signals in plant growth and development. Plant Signal Behav 2009; 4(8): 701-12.
[http://dx.doi.org/10.4161/psb.4.8.9047] [PMID: 19820333]

[11]　Gupta A, Singh UB, Sahu PK, *et al.* Linking soil microbial diversity to modern agriculture practices: a review. Int J Environ Res Public Health 2022; 19(5): 3141.
[http://dx.doi.org/10.3390/ijerph19053141] [PMID: 35270832]

[12]　Mukerji KG, Manoharachary C, Singh J, Eds. Microbial activity in the rhizosphere. Springer Science & Business Media 2006; 7.
[http://dx.doi.org/10.1007/3-540-29420-1]

[13]　Prashar P, Kapoor N, Sachdeva S. Rhizosphere: its structure, bacterial diversity and significance. Rev Environ Sci Biotechnol 2014; 13(1): 63-77.
[http://dx.doi.org/10.1007/s11157-013-9317-z]

[14]　Mendes R, Garbeva P, Raaijmakers JM. The rhizosphere microbiome: significance of plant beneficial, plant pathogenic, and human pathogenic microorganisms. FEMS Microbiol Rev 2013; 37(5): 634-63.
[http://dx.doi.org/10.1111/1574-6976.12028] [PMID: 23790204]

[15]　Kebede E. Contribution, utilization, and improvement of legumes-driven biological nitrogen fixation in agricultural systems. Front Sustain Food Syst 2021; 5: 767998.
[http://dx.doi.org/10.3389/fsufs.2021.767998]

[16]　Monib AW, Niazi P, Barai SM, *et al.* Nitrogen cycling dynamics: Investigating volatilization and its interplay with N_2 fixation. Journal for Research in Applied Sciences and Biotechnology 2024; 3(1): 17-31.
[http://dx.doi.org/10.55544/jrasb.3.1.4]

[17]　Elshahat MR, Ahmed AA, Enas AH, Fekria MS. Plant growth promoting rhizobacteria and their potential for biocontrol of phytopathogens. Afr J Microbiol Res 2016; 10(15): 486-504.
[http://dx.doi.org/10.5897/AJMR2015.7714]

[18]　Hasan A, Tabassum B, Hashim M, Khan N. Role of plant growth promoting rhizobacteria (PGPR) as a plant growth enhancer for sustainable agriculture: A review. Bacteria 2024; 3(2): 59-75.
[http://dx.doi.org/10.3390/bacteria3020005]

[19]　Shah A, Nazari M, Antar M, *et al.* PGPR in agriculture: A sustainable approach to increasing climate change resilience. Front Sustain Food Syst 2021; 5: 667546.
[http://dx.doi.org/10.3389/fsufs.2021.667546]

[20]　Kenawy A, Dailin DJ, Abo-Zaid GA, *et al.* Biosynthesis of antibiotics by PGPR and their roles in biocontrol of plant diseases. Plant growth promoting rhizobacteria for sustainable stress management. 2019; pp. 1-35.

[21]　Santoyo G, Urtis-Flores CA, Loeza-Lara PD, Orozco-Mosqueda MC, Glick BR. Rhizosphere colonization determinants by plant growth-promoting rhizobacteria (PGPR). Biology (Basel) 2021; 10(6): 475.
[http://dx.doi.org/10.3390/biology10060475] [PMID: 34072072]

[22]　Chaudhary D, Kumar R, Sihag K, Kumari A. Phyllospheric microflora and its impact on plant growth: A review. Agric Rev (Karnal) 2017; 38(1): 51-9.
[http://dx.doi.org/10.18805/ag.v0iOF.7308]

[23]　Rana KL, Kour D, Kaur T, *et al.* Endophytic microbes: biodiversity, plant growth-promoting mechanisms and potential applications for agricultural sustainability. Antonie van Leeuwenhoek 2020; 113(8): 1075-107.
[http://dx.doi.org/10.1007/s10482-020-01429-y] [PMID: 32488494]

[24]　Veselova SV, Nuzhnaya TV, Maksimov IV. Role of jasmonic acid in interaction of plants with plant

growth promoting rhizobacteria during fungal pathogenesis. Jasmonic Acid: Biosynthesis, Functions and Role in Plant Development. 2015; pp. 33-66.

[25]   Sivakumar N, Sathishkumar R, Selvakumar G, Shyamkumar R, Arjunekumar K. Phyllospheric microbiomes: diversity, ecological significance, and biotechnological applications. Plant Microbiomes for Sustainable Agriculture. 2020; pp. 113-72.

[26]   Wei X, Lyu S, Yu Y, *et al.* Phylloremediation of air pollutants: exploiting the potential of plant leaves and leaf-associated microbes. Front Plant Sci 2017; 8: 1318.
[http://dx.doi.org/10.3389/fpls.2017.01318] [PMID: 28804491]

[27]   Kwak MJ, Jeong H, Madhaiyan M, *et al.* Genome information of Methylobacterium oryzae, a plant-probiotic methylotroph in the phyllosphere. PLoS One 2014; 9(9): e106704.
[http://dx.doi.org/10.1371/journal.pone.0106704] [PMID: 25211235]

[28]   Devi R, Kaur T, Kour D, Rana KL, Yadav A, Yadav AN. Beneficial fungal communities from different habitats and their roles in plant growth promotion and soil health. Microbial Biosystems 2020; 5(1): 21-47.
[http://dx.doi.org/10.21608/mb.2020.32802.1016]

[29]   Singh SP, Pandey S, Mishra N, *et al.* Supplementation of Trichoderma improves the alteration of nutrient allocation and transporter genes expression in rice under nutrient deficiencies. Plant Physiol Biochem 2019; 143: 351-63.
[http://dx.doi.org/10.1016/j.plaphy.2019.09.015] [PMID: 31541990]

[30]   Bhattacharya A, Giri VP, Singh SP, *et al.* Intervention of bio-protective endophyte *Bacillus tequilensis* enhance physiological strength of tomato during Fusarium wilt infection. Biol Control 2019; 139: 104074.
[http://dx.doi.org/10.1016/j.biocontrol.2019.104074]

[31]   Bashir I, War AF, Rafiq I, Reshi ZA, Rashid I, Shouche YS. Phyllosphere microbiome: Diversity and functions. Microbiol Res 2022; 254: 126888.
[http://dx.doi.org/10.1016/j.micres.2021.126888] [PMID: 34700185]

[32]   Xu N, Zhao Q, Zhang Z, *et al.* Phyllosphere microorganisms: sources, drivers, and their interactions with plant hosts. J Agric Food Chem 2022; 70(16): 4860-70.
[http://dx.doi.org/10.1021/acs.jafc.2c01113] [PMID: 35435673]

[33]   Yaron S, Römling U. Biofilm formation by enteric pathogens and its role in plant colonization and persistence. Microb Biotechnol 2014; 7(6): 496-516.
[http://dx.doi.org/10.1111/1751-7915.12186] [PMID: 25351039]

[34]   Abdul Malik NA, Kumar IS, Nadarajah K. Elicitor and receptor molecules: orchestrators of plant defense and immunity. Int J Mol Sci 2020; 21(3): 963.
[http://dx.doi.org/10.3390/ijms21030963] [PMID: 32024003]

[35]   Kaur S, Samota MK, Choudhary M, *et al.* How do plants defend themselves against pathogens-Biochemical mechanisms and genetic interventions. Physiol Mol Biol Plants 2022; 28(2): 485-504.
[http://dx.doi.org/10.1007/s12298-022-01146-y] [PMID: 35400890]

[36]   Nie P, Li X, Wang S, Guo J, Zhao H, Niu D. Induced systemic resistance against *Botrytis cinerea* by *Bacillus cereus* AR156 through a JA/ET-and NPR1-dependent signaling pathway and activates PAMP-triggered immunity in Arabidopsis. Front Plant Sci 2017; 8: 238.
[http://dx.doi.org/10.3389/fpls.2017.00238] [PMID: 28293243]

[37]   Pappas ML, Baptista P, Broufas GD, *et al.* Biological and molecular control tools in plant defense. Plant Defence: Biological Control. 2020; pp. 3-43.

[38]   Bagheri A, Fathipour Y. Induced resistance and defense primings. Molecular Approaches for Sustainable Insect Pest Management. 2021; pp. 73-139.

[39]   Rabari A, Ruparelia J, Jha CK, *et al.* Articulating beneficial rhizobacteria-mediated plant defenses through induced systemic resistance: A review. Pedosphere 2023; 33(4): 556-66.

[http://dx.doi.org/10.1016/j.pedsph.2022.10.003]

[40]   Surovy MZ, Islam T. Principle, diversity, mechanism, and potential of practical application of plant probiotic bacteria for the biocontrol of phytopathogens by induced systemic resistance. Food Security and Plant Disease Management 2021; pp. 75-94.
[http://dx.doi.org/10.1016/B978-0-12-821843-3.00004-0]

[41]   Seth K, Vyas P, Deora S, Gupta AK, Meena M, Swapnil P. Understanding plant-Plant Growth-Promoting Rhizobacteria (PGPR) interactions for inducing plant defense. Plant-Microbe Interaction-Recent Advances in Molecular and Biochemical Approaches. Academic Press 2023; pp. 201-26.

[42]   Bailly A. How plants might recognize rhizospheric bacterial volatiles. Bacterial Volatile Compounds as Mediators of Airborne Interactions. 2020; pp. 139-65.

[43]   Hussain SS, Mehnaz S, Siddique KH. Harnessing the plant microbiome for improved abiotic stress tolerance. Plant Microbiome: Stress Response. 2018; pp. 21-43.

[44]   Munir N, Hanif M, Abideen Z, *et al.* Mechanisms and strategies of plant microbiome interactions to mitigate abiotic stresses. Agronomy (Basel) 2022; 12(9): 2069.
[http://dx.doi.org/10.3390/agronomy12092069]

[45]   Bouremani N, Cherif-Silini H, Silini A, *et al.* Plant Growth-Promoting Rhizobacteria (PGPR): A rampart against the adverse effects of drought stress. Water 2023; 15(3): 418.
[http://dx.doi.org/10.3390/w15030418]

[46]   Buragohain K, Tamuly D, Sonowal S, Nath R. Impact of drought stress on plant growth and its management using plant growth promoting rhizobacteria. Indian J Microbiol 2024; 64(2): 287-303.
[http://dx.doi.org/10.1007/s12088-024-01201-0] [PMID: 39011023]

[47]   Singh M, Kumar J, Singh S, Singh VP, Prasad SM. Roles of osmoprotectants in improving salinity and drought tolerance in plants: a review. Rev Environ Sci Biotechnol 2015; 14(3): 407-26.
[http://dx.doi.org/10.1007/s11157-015-9372-8]

[48]   Gupta A, Gupta R, Singh RL. Microbes and environment. Principles and applications of environmental biotechnology for a sustainable future. 2017; pp. 43-84.

[49]   Aslam MM, Okal EJ, Idris AL, *et al.* Rhizosphere microbiomes can regulate plant drought tolerance. Pedosphere 2022; 32(1): 61-74.
[http://dx.doi.org/10.1016/S1002-0160(21)60061-9]

[50]   Ghosh D, Gupta A, Mohapatra S. Dynamics of endogenous hormone regulation in plants by phytohormone secreting rhizobacteria under water-stress. Symbiosis 2019; 77(3): 265-78.
[http://dx.doi.org/10.1007/s13199-018-00589-w]

[51]   Gupta S, Sinha S, Bhar A. Harnessing belowground interaction: Re-analyzing the role of rhizosphere microbiome in plant–pathogen interaction under water stress. J Plant Growth Regul 2024; 1-22.
[http://dx.doi.org/10.1007/s00344-024-11500-x]

[52]   Kumawat KC, Nagpal S, Sharma P. Potential of plant growth-promoting rhizobacteria-plant interactions in mitigating salt stress for sustainable agriculture: A review. Pedosphere 2022; 32(2): 223-45.
[http://dx.doi.org/10.1016/S1002-0160(21)60070-X]

[53]   Salwan R, Sharma A, Sharma V. Microbes mediated plant stress tolerance in saline agricultural ecosystem. Plant Soil 2019; 442(1-2): 1-22.
[http://dx.doi.org/10.1007/s11104-019-04202-x]

[54]   Ha-Tran DM, Nguyen TTM, Hung SH, Huang E, Huang CC. Roles of Plant Growth-Promoting Rhizobacteria (PGPR) in stimulating salinity stress defense in plants: A review. Int J Mol Sci 2021; 22(6): 3154.
[http://dx.doi.org/10.3390/ijms22063154] [PMID: 33808829]

[55]   Rayanoothala P, Hasibul Alam S, Mahapatra S, Gafur A, Antonius S. Rhizosphere microorganisms for

climate resilient and sustainable crop production. Gesunde Pflanzen 2023; 75(6): 2207-25.
[http://dx.doi.org/10.1007/s10343-023-00895-4]

[56]   Dong H, Wang Y, Di Y, *et al*. Plant growth-promoting rhizobacteria *Pseudomonas aeruginosa* HG28-5 improves salt tolerance by regulating Na⁺/K⁺ homeostasis and ABA signaling pathway in tomato. Microbiol Res 2024; 283: 127707.
[http://dx.doi.org/10.1016/j.micres.2024.127707] [PMID: 38582011]

[57]   Joshi D, Suyal DC, Singh J, Maithani D, Rajwar J. Plant growth-promoting psychrotrophic microbes: current research and future challenges. Microbial Inoculants: Applications for Sustainable Agriculture. 2024; pp. 247-80.

[58]   Mitra D, Díaz Rodríguez AM, Parra Cota FI, *et al*. Amelioration of thermal stress in crops by plant growth-promoting rhizobacteria. Physiol Mol Plant Pathol 2021; 115: 101679.
[http://dx.doi.org/10.1016/j.pmpp.2021.101679]

[59]   Hanaka A, Ozimek E, Reszczyńska E, Jaroszuk-Ściseł J, Stolarz M. Plant tolerance to drought stress in the presence of supporting bacteria and fungi: An efficient strategy in horticulture. Horticulturae 2021; 7(10): 390.
[http://dx.doi.org/10.3390/horticulturae7100390]

[60]   Kissoudis C, van de Wiel C, Visser RGF, van der Linden G. Enhancing crop resilience to combined abiotic and biotic stress through the dissection of physiological and molecular crosstalk. Front Plant Sci 2014; 5: 207.
[http://dx.doi.org/10.3389/fpls.2014.00207] [PMID: 24904607]

[61]   Farooqi ZUR, Qadir AA, Alserae H, Raza A, Mohy-Ud-Din W. Organic amendment–mediated reclamation and build-up of soil microbial diversity in salt-affected soils: fostering soil biota for shaping rhizosphere to enhance soil health and crop productivity. Environ Sci Pollut Res Int 2023; 30(51): 109889-920.
[http://dx.doi.org/10.1007/s11356-023-30143-1] [PMID: 37792186]

[62]   Guhra T, Stolze K, Totsche KU. Pathways of biogenically excreted organic matter into soil aggregates. Soil Biol Biochem 2022; 164: 108483.
[http://dx.doi.org/10.1016/j.soilbio.2021.108483]

[63]   Crowley D. Managing soils for avocado production and root health. California Avocado Society Yearbook 2007; 90: 107-30.

[64]   Wahab A, Muhammad M, Munir A, *et al*. Role of arbuscular mycorrhizal fungi in regulating growth, enhancing productivity, and potentially influencing ecosystems under abiotic and biotic stresses. Plants 2023; 12(17): 3102.
[http://dx.doi.org/10.3390/plants12173102] [PMID: 37687353]

[65]   Gamalero E, Bona E, Glick BR. Current techniques to study beneficial plant-microbe interactions. Microorganisms 2022; 10(7): 1380.
[http://dx.doi.org/10.3390/microorganisms10071380] [PMID: 35889099]

[66]   Rashid MI, Mujawar LH, Shahzad T, Almeelbi T, Ismail IMI, Oves M. Bacteria and fungi can contribute to nutrients bioavailability and aggregate formation in degraded soils. Microbiol Res 2016; 183: 26-41.
[http://dx.doi.org/10.1016/j.micres.2015.11.007] [PMID: 26805616]

[67]   Dobbelaere S, Vanderleyden J, Okon Y. Plant growth-promoting effects of diazotrophs in the rhizosphere. Crit Rev Plant Sci 2003; 22(2): 107-49.
[http://dx.doi.org/10.1080/713610853]

[68]   Aslani borj M, Etesami H, Alikhani HA. Silicon improves the effect of phosphate-solubilizing bacterium and arbuscular mycorrhizal fungus on phosphorus concentration of salinity-stressed alfalfa (Medicago sativa L.). Rhizosphere 2022; 24: 100619.
[http://dx.doi.org/10.1016/j.rhisph.2022.100619]

[69]   Chaffai R, Ganesan M, Cherif A. Plant–microbe interactions for enhanced plant growth and stress resilience. Plant adaptation to abiotic stress: From signaling pathways and microbiomes to molecular mechanisms. Singapore: Springer Nature Singapore 2024; pp. 497-514.
       [http://dx.doi.org/10.1007/978-981-97-0672-3_18]

[70]   Osman KT, Osman KT. Plant nutrients and soil fertility management. 2013; pp. 129-59.
       [http://dx.doi.org/10.1007/978-94-007-5663-2_10]

[71]   Liggitt J. Studies on the chemical control of Fusarium ear blight of winter wheat (*Triticum aestivum L.*). United Kingdom: Open University 1997.

[72]   Dixon GR, Tilston EL. Soil-borne pathogens and their interactions with the soil environment. Soil Microbiology and Sustainable Crop Production. 2010; pp. 197-271.

[73]   Naz R, Khushhal S, Asif T, Mubeen S, Saranraj P, Sayyed RZ. Inhibition of bacterial and fungal phytopathogens through volatile organic compounds produced by *Pseudomonas* sp. Secondary metabolites and volatiles of PGPR in plant-growth promotion. Cham: Springer International Publishing 2022; pp. 95-118.
       [http://dx.doi.org/10.1007/978-3-031-07559-9_6]

[74]   Zhu L, Huang J, Lu X, Zhou C. Development of plant systemic resistance by beneficial rhizobacteria: Recognition, initiation, elicitation and regulation. Front Plant Sci 2022; 13: 952397.
       [http://dx.doi.org/10.3389/fpls.2022.952397] [PMID: 36017257]

[75]   Elad Y, Pertot I. Climate change impacts on plant pathogens and plant diseases. J Crop Improv 2014; 28(1): 99-139.
       [http://dx.doi.org/10.1080/15427528.2014.865412]

[76]   Nizamani MM, Hughes AC, Qureshi S, *et al.* Microbial biodiversity and plant functional trait interactions in multifunctional ecosystems. Appl Soil Ecol 2024; 201: 105515.
       [http://dx.doi.org/10.1016/j.apsoil.2024.105515]

[77]   Fierer N. Embracing the unknown: disentangling the complexities of the soil microbiome. Nat Rev Microbiol 2017; 15(10): 579-90.
       [http://dx.doi.org/10.1038/nrmicro.2017.87] [PMID: 28824177]

[78]   IPCC. Climate change 2014: synthesis report Contribution of working groups I, II and III to the fifth assessment report of the Intergovernmental Panel on Climate Change. Pachauri RK, Meyer LA, Eds. Geneva: IPCC 2014.

[79]   Samaniego L, Thober S, Kumar R, *et al.* Anthropogenic warming exacerbates European soil moisture droughts. Nat Clim Chang 2018; 8(5): 421-6.
       [http://dx.doi.org/10.1038/s41558-018-0138-5]

[80]   Sulman BN, Shevliakova E, Brzostek ER, *et al.* Diverse mycorrhizal associations enhance terrestrial C storage in a global model. Global Biogeochem Cycles 2019; 33(4): 501-23.
       [http://dx.doi.org/10.1029/2018GB005973]

[81]   Rudgers JA, Afkhami ME, Bell-Dereske L, *et al.* Climate disruption of plant-microbe interactions. Annu Rev Ecol Evol Syst 2020; 51(1): 561-86.
       [http://dx.doi.org/10.1146/annurev-ecolsys-011720-090819]

[82]   Song J, Wan S, Piao S, *et al.* A meta-analysis of 1,119 manipulative experiments on terrestrial carbon-cycling responses to global change. Nat Ecol Evol 2019; 3(9): 1309-20.
       [http://dx.doi.org/10.1038/s41559-019-0958-3] [PMID: 31427733]

[83]   Classen AT, Sundqvist MK, Henning JA, *et al.* Direct and indirect effects of climate change on soil microbial and soil microbial-plant interactions: What lies ahead?. Ecosphere 2015; 6(8): 1-21.
       [http://dx.doi.org/10.1890/ES15-00217.1]

[84]   Trivedi P, Leach JE, Tringe SG, Sa T, Singh BK. Plant–microbiome interactions: from community assembly to plant health. Nat Rev Microbiol 2020; 18(11): 607-21.

[http://dx.doi.org/10.1038/s41579-020-0412-1] [PMID: 32788714]

[85]   Trivedi P, Mattupalli C, Eversole K, Leach JE. Enabling sustainable agriculture through understanding and enhancement of microbiomes. New Phytol 2021; 230(6): 2129-47.
[http://dx.doi.org/10.1111/nph.17319] [PMID: 33657660]

[86]   Naylor D, Coleman-Derr D. Drought stress and root-associated bacterial communities. Front Plant Sci 2018; 8: 2223.
[http://dx.doi.org/10.3389/fpls.2017.02223] [PMID: 29375600]

[87]   Vagelas IK, Pembroke B, Gowen SR, Davies KG. The control of root-knot nematodes (*Meloidogyne spp.*) by Pseudomonas oryzihabitans and its immunological detection on tomato roots. Nematology 2007; 9(3): 363-70.
[http://dx.doi.org/10.1163/156854107781352061]

[88]   Johnston AWB, Todd JD, Curson AR, *et al.* Living without Fur: the subtlety and complexity of iron-responsive gene regulation in the symbiotic bacterium Rhizobium and other α-proteobacteria. Biometals 2007; 20(3-4): 501-11.
[http://dx.doi.org/10.1007/s10534-007-9085-8] [PMID: 17310401]

[89]   Doroshenko EV, Bulygina ES, Spiridonova EM, Turova TP, Kravchenko IK. [Isolation and characterization of nitrogen-fixing bacteria of the genus *Azospirillum* from the soil of a Sphagnum peat bog]. Mikrobiologiia 2007; 76(1): 107-15.
[PMID: 17410881]

[90]   Iqbal A, Mushtaq MU, Khan AHA, *et al.* Influence of *Pseudomonas japonica* and organic amendments on the growth and metal tolerance of *Celosia argentea L.* Environ Sci Pollut Res Int 2020; 27(20): 24671-85.
[http://dx.doi.org/10.1007/s11356-019-06181-z] [PMID: 31428967]

[91]   Requena N, Baca TM, Azcdn R. Evolution of humic substances from unripe compost during incubation with lignolytic or cellulolytic microorganisms and effects on the lettuce growth promotion mediated by *Azotobacter chroococcum.* Biol Fertil Soils 1997; 24: 59-65.
[http://dx.doi.org/10.1007/BF01420221]

[92]   Zhou H, Ren Z, Zu X, *et al.* Efficacy of plant growth-promoting bacteria *Bacillus cereus* YN917 for biocontrol of rice blast. Front Microbiol 2021; 12: 684888.
[http://dx.doi.org/10.3389/fmicb.2021.684888] [PMID: 34354684]

[93]   Sukkasem P, Kurniawan A, Kao TC, Chuang H. A multifaceted rhizobacterium *Bacillus licheniformis* functions as a fungal antagonist and a promoter of plant growth and abiotic stress tolerance. Environ Exp Bot 2018; 155: 541-51.
[http://dx.doi.org/10.1016/j.envexpbot.2018.08.005]

[94]   Broderick NA, Raffa KF, Handelsman J. Midgut bacteria required for *Bacillus thuringiensis* insecticidal activity. Proc Natl Acad Sci USA 2006; 103(41): 15196-9.
[http://dx.doi.org/10.1073/pnas.0604865103] [PMID: 17005725]

[95]   Sami D, Mokhtar R, Peter M, Mohamed M. *Rhizobium leguminosarum* symbiovar *trifolii, Ensifer numidicus* and *Mesorhizobium amorphae* symbiovar *ciceri* (or *Mesorhizobium loti*) are new endosymbiotic bacteria of *Lens culinaris* Medik. FEMS Microbiol Ecol 2016; 92(8): fiw118.
[http://dx.doi.org/10.1093/femsec/fiw118] [PMID: 27267929]

[96]   Le Quéré A, Tak N, Gehlot HS, *et al.* Genomic characterization of *Ensifer aridi*, a proposed new species of nitrogen-fixing rhizobium recovered from Asian, African and American deserts. BMC Genomics 2017; 18(1): 85.
[http://dx.doi.org/10.1186/s12864-016-3447-y] [PMID: 28088165]

[97]   Miao Y, Shi S, Zhang J, Mohamad OA. Migration, colonization and seedling growth of rhizobia with matrine treatment in alfalfa (*Medicago sativa* L.). Acta Agric Scand B Soil Plant Sci 2018; 68(1): 26-38.
[http://dx.doi.org/10.1080/09064710.2017.1353131]

[98] Kubicek CP, Bissett J, Druzhinina I, Kullnig-Gradinger C, Szakacs G. Genetic and metabolic diversity of Trichoderma: a case study on South-East Asian isolates. Fungal Genet Biol 2003; 38(3): 310-9.
[http://dx.doi.org/10.1016/S1087-1845(02)00583-2] [PMID: 12684020]

[99] Mnasri B, Saïdi S, Chihaoui SA, Mhamdi R. *Sinorhizobium americanum symbiovar mediterranense* is a predominant symbiont that nodulates and fixes nitrogen with common bean (*Phaseolus vulgaris L.*) in a Northern Tunisian field. Syst Appl Microbiol 2012; 35(4): 263-9.
[http://dx.doi.org/10.1016/j.syapm.2012.04.003] [PMID: 22633818]

[100] Zahran H. Rhizobia from wild legumes: diversity, taxonomy, ecology, nitrogen fixation and biotechnology. J Biotechnol 2001; 91(2-3): 143-53.
[http://dx.doi.org/10.1016/S0168-1656(01)00342-X] [PMID: 11566386]

[101] White JF, Kingsley KL, Verma SK, Kowalski KP. Rhizophagy cycle: an oxidative process in plants for nutrient extraction from symbiotic microbes. Microorganisms 2018; 6(3): 95.
[http://dx.doi.org/10.3390/microorganisms6030095] [PMID: 30227634]

[102] Paungfoo-Lonhienne C, Rentsch D, Robatzek S, *et al.* Turning the table: plants consume microbes as a source of nutrients. PLoS One 2010; 5(7): e11915. a
[http://dx.doi.org/10.1371/journal.pone.0011915] [PMID: 20689833]

[103] Chang X, Kingsley K, White J. Chemical interactions at the interface of plant root hair cells and intracellular bacteria. Microorganisms 2021; 9(5): 1041.
[http://dx.doi.org/10.3390/microorganisms9051041] [PMID: 34066008]

[104] Paungfoo-Lonhienne C, Schmidt S, Lonhienne TGA. Uptake of non-pathogenic *E. coli* by Arabidopsis induces down-regulation of heat shock proteins. Plant Signal Behav 2010; 5(12): 1626-8. b
[http://dx.doi.org/10.4161/psb.5.12.13760] [PMID: 21139429]

[105] Verma SK, Kingsley K, Bergen M, *et al.* Bacterial endophytes from rice cut grass (*Leersia oryzoides L.*) increase growth, promote root gravitropic response, stimulate root hair formation, and protect rice seedlings from disease. Plant Soil 2018; 422(1-2): 223-38.
[http://dx.doi.org/10.1007/s11104-017-3339-1]

[106] Roley SS, Duncan DS, Liang D, *et al.* Associative Nitrogen Fixation (ANF) in switchgrass (*Panicum virgatum*) across a nitrogen input gradient. PLoS One 2018; 13(6): e0197320.
[http://dx.doi.org/10.1371/journal.pone.0197320] [PMID: 29856843]

[107] Irizarry I, White JF. Application of bacteria from non-cultivated plants to promote growth, alter root architecture and alleviate salt stress of cotton. J Appl Microbiol 2017; 122(4): 1110-20.
[http://dx.doi.org/10.1111/jam.13414] [PMID: 28176460]

[108] White JF Jr, Torres MS, Sullivan RF, *et al.* Occurrence of *B acillus amyloliquefaciens* as a systemic endophyte of vanilla orchids. Microsc Res Tech 2014; 77(11): 874-85.
[http://dx.doi.org/10.1002/jemt.22410] [PMID: 25060609]

[109] Paungfoo-Lonhienne C, Lonhienne TGA, Yeoh YK, *et al.* A new species of *Burkholderia* isolated from sugarcane roots promotes plant growth. Microb Biotechnol 2014; 7(2): 142-54.
[http://dx.doi.org/10.1111/1751-7915.12105] [PMID: 24350979]

[110] Cocking E, Dent D. The prospect of N_2-fixing crops galore!. Biochemist (Lond) 2019; 41(4): 14-7.
[http://dx.doi.org/10.1042/BIO04104014]

[111] Liu H, Li J, Carvalhais LC, *et al.* Evidence for the plant recruitment of beneficial microbes to suppress soil-borne pathogens. New Phytol 2021; 229(5): 2873-85.
[http://dx.doi.org/10.1111/nph.17057] [PMID: 33131088]

[112] Kivlin SN, Winston GC, Goulden ML, Treseder KK. Environmental filtering affects soil fungal community composition more than dispersal limitation at regional scales. Fungal Ecol 2014; 12: 14-25.
[http://dx.doi.org/10.1016/j.funeco.2014.04.004]

[113] Hardoim PR, van Overbeek LS, Berg G, *et al.* The hidden world within plants: ecological and evolutionary considerations for defining functioning of microbial endophytes. Microbiol Mol Biol Rev 2015; 79(3): 293-320.
[http://dx.doi.org/10.1128/MMBR.00050-14] [PMID: 26136581]

[114] Busby PE, Ridout M, Newcombe G. Fungal endophytes: modifiers of plant disease. Plant Mol Biol 2016; 90(6): 645-55.
[http://dx.doi.org/10.1007/s11103-015-0412-0] [PMID: 26646287]

[115] Porras-Alfaro A, Bayman P. Hidden fungi, emergent properties: endophytes and microbiomes. Annu Rev Phytopathol 2011; 49(1): 291-315.
[http://dx.doi.org/10.1146/annurev-phyto-080508-081831] [PMID: 19400639]

[116] Rana KL, Kour D, Sheikh I, *et al.* Endophytic fungi: biodiversity, ecological significance, and potential industrial applications. Recent Advancement in White Biotechnology Through Fungi. 2019; pp. 1-62.

[117] Fitzpatrick CR, Copeland J, Wang PW, Guttman DS, Kotanen PM, Johnson MTJ. Assembly and ecological function of the root microbiome across angiosperm plant species. Proc Natl Acad Sci USA 2018; 115(6): E1157-65.
[http://dx.doi.org/10.1073/pnas.1717617115] [PMID: 29358405]

[118] Frank A, Saldierna Guzmán J, Shay J. Transmission of bacterial endophytes. Microorganisms 2017; 5(4): 70.
[http://dx.doi.org/10.3390/microorganisms5040070] [PMID: 29125552]

[119] Osono T. Role of phyllosphere fungi of forest trees in the development of decomposer fungal communities and decomposition processes of leaf litter. Can J Microbiol 2006; 52(8): 701-16.
[http://dx.doi.org/10.1139/w06-023] [PMID: 16917528]

[120] Merckx V, Bidartondo MI, Hynson NA. Myco-heterotrophy: when fungi host plants. Ann Bot (Lond) 2009; 104(7): 1255-61.
[http://dx.doi.org/10.1093/aob/mcp235] [PMID: 19767309]

[121] Berger S, Sinha AK, Roitsch T. Plant physiology meets phytopathology: plant primary metabolism and plant pathogen interactions. J Exp Bot 2007; 58(15-16): 4019-26.
[http://dx.doi.org/10.1093/jxb/erm298] [PMID: 18182420]

[122] Nguyen NH, Song Z, Bates ST, *et al.* FUNGuild: An open annotation tool for parsing fungal community datasets by ecological guild. Fungal Ecol 2016; 20: 241-8.
[http://dx.doi.org/10.1016/j.funeco.2015.06.006]

[123] Gupta K, Kumar R, Baruah KK, Hazarika S, Karmakar S, Bordoloi N. Greenhouse gas emission from rice fields: a review from Indian context. Environ Sci Pollut Res Int 2021; 28(24): 30551-72.
[http://dx.doi.org/10.1007/s11356-021-13935-1] [PMID: 33905059]

[124] Mondal S, Halder SK, Yadav AN, Mondal KC. Microbial consortium with multifunctional plant growth-promoting attributes: future perspective in agriculture. Advances in Plant Microbiome and Sustainable Agriculture: Functional Annotation and Future Challenges. 2020; pp. 219-58.

[125] Singh B, Boukhris I, Pragya , *et al.* Contribution of microbial phytases to the improvement of plant growth and nutrition: A review. Pedosphere 2020; 30(3): 295-313.
[http://dx.doi.org/10.1016/S1002-0160(20)60010-8]

[126] Yadav AN, Kumar V, Dhaliwal HS, Prasad R, Saxena AK. Microbiome in crops: diversity, distribution, and potential role in crop improvement. Crop improvement through microbial biotechnology. Elsevier 2018; pp. 305-32.
[http://dx.doi.org/10.1016/B978-0-444-63987-5.00015-3]

[127] Aislabie J, Deslippe JR, Dymond J. Soil microbes and their contribution to soil services. Ecosystem Services in New Zealand–Conditions and Trends. Lincoln, New Zealand: Manaaki Whenua Press 2013; 1: pp. (12)143-61.

[128] Rajawat MVS, Singh R, Singh D, *et al.* Spatial distribution and identification of bacteria in stressed environments capable to weather potassium aluminosilicate mineral. Braz J Microbiol 2020; 51(2): 751-64.
[http://dx.doi.org/10.1007/s42770-019-00210-2] [PMID: 31898251]

[129] Dawoud MMA, Hegazy MM, Helew WK, Saleh HM. Overview of environmental pollution and clean management of heavy metals and radionuclides by using microcrystalline cellulose. J Nucl Energy Sci Power Gener Technol 2021; 3(2)

[130] Lavanya MB, Viswanath DS, Sivapullaiah PV. Phytoremediation: An eco-friendly approach for remediation of heavy metal-contaminated soils-A comprehensive review. Environ Nanotechnol Monit Manag 2024; 22: 100975.
[http://dx.doi.org/10.1016/j.enmm.2024.100975]

[131] Kuppan N, Padman M, Mahadeva M, Srinivasan S, Devarajan R. A comprehensive review of sustainable bioremediation techniques: eco friendly solutions for waste and pollution management. Waste Management Bulletin 2024.

[132] Zhou B, Zhang T, Wang F. Microbial-based heavy metal bioremediation: Toxicity and eco-friendly approaches to heavy metal decontamination. Appl Sci (Basel) 2023; 13(14): 8439.
[http://dx.doi.org/10.3390/app13148439]

[133] Kapahi M, Sachdeva S. Bioremediation options for heavy metal pollution. J Health Pollut 2019; 9(24): 191203.
[http://dx.doi.org/10.5696/2156-9614-9.24.191203] [PMID: 31893164]

[134] Fasani E, Manara A, Martini F, Furini A, DalCorso G. The potential of genetic engineering of plants for the remediation of soils contaminated with heavy metals. Plant Cell Environ 2018; 41(5): 1201-32.
[http://dx.doi.org/10.1111/pce.12963] [PMID: 28386947]

[135] Awa SH, Hadibarata T. Removal of heavy metals in contaminated soil by phytoremediation mechanism: a review. Water Air Soil Pollut 2020; 231(2): 47.
[http://dx.doi.org/10.1007/s11270-020-4426-0]

[136] Hussain A, Rehman F, Rafeeq H, *et al. In-situ, Ex-situ,* and nano-remediation strategies to treat polluted soil, water, and air – A review. Chemosphere 2022; 289: 133252.
[http://dx.doi.org/10.1016/j.chemosphere.2021.133252] [PMID: 34902385]

[137] Bhanse P, Kumar M, Singh L, Awasthi MK, Qureshi A. Role of plant growth-promoting rhizobacteria in boosting the phytoremediation of stressed soils: Opportunities, challenges, and prospects. Chemosphere 2022; 303(Pt 1): 134954.
[http://dx.doi.org/10.1016/j.chemosphere.2022.134954] [PMID: 35595111]

[138] Alves ARA, Yin Q, Oliveira RS, Silva EF, Novo LAB. Plant growth-promoting bacteria in phytoremediation of metal-polluted soils: Current knowledge and future directions. Sci Total Environ 2022; 838(Pt 4): 156435.
[http://dx.doi.org/10.1016/j.scitotenv.2022.156435] [PMID: 35660615]

[139] Kafle A, Timilsina A, Gautam A, Adhikari K, Bhattarai A, Aryal N. Phytoremediation: Mechanisms, plant selection and enhancement by natural and synthetic agents. Environ Adv 2022; 8: 100203.
[http://dx.doi.org/10.1016/j.envadv.2022.100203]

[140] Sharma P, Tripathi S, Chaturvedi P, Chaurasia D, Chandra R. Newly isolated *Bacillus* sp. PS-6 assisted phytoremediation of heavy metals using *Phragmites communis*: Potential application in wastewater treatment. Bioresour Technol 2021; 320(Pt B): 124353.
[http://dx.doi.org/10.1016/j.biortech.2020.124353] [PMID: 33202343]

[141] Manoj SR, Karthik C, Kadirvelu K, *et al.* Understanding the molecular mechanisms for the enhanced phytoremediation of heavy metals through plant growth promoting rhizobacteria: A review. J Environ Manage 2020; 254: 109779.
[http://dx.doi.org/10.1016/j.jenvman.2019.109779] [PMID: 31726280]

[142] Ur Rahman S, Li Y, Hussain S, *et al.* Role of phytohormones in heavy metal tolerance in plants: A review. Ecol Indic 2023; 146: 109844.
[http://dx.doi.org/10.1016/j.ecolind.2022.109844]

[143] Kumar M, You S, Beiyuan J, *et al.* Lignin valorization by bacterial genus Pseudomonas: State-of-t-e-art review and prospects. Bioresour Technol 2021; 320(Pt B): 124412.
[http://dx.doi.org/10.1016/j.biortech.2020.124412] [PMID: 33249259]

[144] Li Q, Xing Y, Huang B, *et al.* Rhizospheric mechanisms of *Bacillus subtilis* bioaugmentation-assisted phytostabilization of cadmium-contaminated soil. Sci Total Environ 2022; 825: 154136.
[http://dx.doi.org/10.1016/j.scitotenv.2022.154136] [PMID: 35218830]

[145] Khan A, Gupta A, Singh P, Mishra AK, Ranjan RK, Srivastava A. Siderophore-assisted cadmium hyperaccumulation in *Bacillus subtilis*. Int Microbiol 2020; 23(2): 277-86.
[http://dx.doi.org/10.1007/s10123-019-00101-4] [PMID: 31655918]

[146] Shah M. Application of sewage sludge in industrial wastewater treatment. 2024.

[147] Bhardwaj LK, *et al.* COVID-19 and the interplay with antibacterial drug resistance Available from: https://services.igi-global.com/resolvedoi/resolve.aspx?doi=10.4018/979-8-3693-4139-1.ch010
[http://dx.doi.org/10.4018/979-8-3693-4139-1.ch010]

[148] Sharma P, Chaturvedi P, Chandra R, Kumar S. Identification of heavy metals tolerant *Brevundimonas* sp. from rhizospheric zone of *Saccharum munja L.* and their efficacy in *in-situ* phytoremediation. Chemosphere 2022; 295: 133823.
[http://dx.doi.org/10.1016/j.chemosphere.2022.133823] [PMID: 35114263]

[149] Dave K, Kumar A, Dave N, *et al.* Climate change impacts on legume physiology and ecosystem dynamics: A multifaceted perspective. Sustainability (Basel) 2024; 16(14): 6026.
[http://dx.doi.org/10.3390/su16146026]

[150] Oldroyd GED. Speak, friend, and enter: signalling systems that promote beneficial symbiotic associations in plants. Nat Rev Microbiol 2013; 11(4): 252-63.
[http://dx.doi.org/10.1038/nrmicro2990] [PMID: 23493145]

[151] Brígido C, Glick BR, Oliveira S. Survey of plant growth-promoting mechanisms in native Portuguese chickpea Mesorhizobium isolates. Microb Ecol 2017; 73(4): 900-15.
[http://dx.doi.org/10.1007/s00248-016-0891-9] [PMID: 27904921]

[152] Jain H. From pollution to progress: Groundbreaking advances in clean technology unveiled. Innovation and Green Development 2024; 3(2): 100143.
[http://dx.doi.org/10.1016/j.igd.2024.100143]

[153] Alves GS, Bertini SCB, Barbosa BB, *et al.* Fungal endophytes inoculation improves soil nutrient availability, arbuscular mycorrhizal colonization and common bean growth. Rhizosphere 2021; 18: 100330.
[http://dx.doi.org/10.1016/j.rhisph.2021.100330]

[154] Chauhan P, Bhattacharya A, Giri VP, *et al. Bacillus subtilis* suppresses the charcoal rot disease by inducing defence responses and physiological attributes in soybean. Arch Microbiol 2022; 204(5): 266.
[http://dx.doi.org/10.1007/s00203-022-02876-z] [PMID: 35437612]

CHAPTER 9

Advanced Molecular Techniques in Microbial Research

Divya Bharti¹, Harshita Jain[1,*], Niketa Bhati¹ and **Renu Dhupper¹**

Amity Institute of Environmental Sciences, Amity University, Noida, Gautam Budh Nagar, Uttar Pradesh-201313, India

Abstract: Recent advances in advanced molecular techniques have revolutionized microbial research by providing a detailed understanding of the genetic and functional properties of microorganisms. In this chapter, three important methods, metagenomics, proteomics, and DNA microarrays, applications in environmental microbiology, namely, bioremediation, are studied. The last ten years have witnessed the revolution of metagenomics and proteomics in understanding microbial diversity particularly in polluted environments allowing us to get first a diversity of microbial communities and the set of their functional genes. They are used for the investigation of metabolic pathways, stress response, and degradation of pollutants, giving a basis for designing efficient bioremediation strategies. However, proteomics complements metagenomics with protein expression analysis to go below the species level and reveal microbial function at the molecular level. These methods, together, provide a global perspective on microbial ecosystems necessary for addressing environmental challenges. The chapter also presents DNA microarray technology as a robust method for observing microbial activity and monitoring gene expression in complex environments. Microarrays can be used to monitor the temporal dynamics of microbial communities and how they interact with contaminants, providing real-time environmental monitoring. The integration of these molecular techniques with bioinformatics to manage and interpret large datasets is demonstrated in this chapter and provides a path for systems biology approaches. Technological limitations, data complexity, and scaling-up challenges are explored, along with emerging trends and innovations in the area. As a bridge between molecular research and practical applications, this chapter presents useful ideas to general researchers and environmental practitioners interested in using microbial capabilities for sustainable environmental management.

Keywords: Bioremediation, DNA microarrays, Metagenomics, Proteomics.

*** Corresponding author Harshita Jain:** Amity Institute of Environmental Sciences, Amity University, Noida, Gautam Budh Nagar, Uttar Pradesh-201313, India; E-mail: hjain@amity.edu

Harshita Jain & Maulin P. Shah (Eds.)

INTRODUCTION

Microbes are everywhere, and a healthy system depends on them to keep it in balance. Among the most important drivers of essential ecosystem processes including nutrient cycling, organic matter decomposition and pollutant degradation, they play a major role in determining ecosystem resilience and stability [1]. These microbial communities possess important ecological roles yet the complexity of microbial communities and their interactions in the environment remains a difficult problem to understand because these systems are diverse, variable, and dynamic. In polluted environments, microbial communities must learn to adapt to and interact with a myriad of different contaminants, much of which is complex; this makes the complexity even greater. Innovative tools and techniques are required to harness the potential of these microbes for environmental restoration, especially in bioremediation. In the last two decades, microbe biotechnology has advanced and has made great progress in discovering microbial systems. As traditional culture-dependent methods fail to capture the diversity and functionality of unculturable microbes, techniques like metagenomics, proteomics and DNA microarrays have allowed researchers to proceed beyond [2]. These molecular approaches provide high-resolution data that can be used to explore microbial diversity, community dynamics, and functional potential in new detail.

The study of collective genetic material extracted directly from environmental samples has been a pillar of microbial research: metagenomics. We have found that much of this genetic diversity lies among microorganisms that were previously not known or were thought to be unculturable. In polluted environments, metagenomics helps scientists to identify microbial taxa that are potentially involved in pollutant degradation and assessing the functional genes and community-level metabolic pathways involved in it [3]. This technique reveals microbial community responses to environmental stressors and thus becomes a powerful bioremediation design tool based on targeted strategies. Since proteomics analyses the protein expression profiles of microbial communities complementing metagenomics, the latter enables the study of microbial metabolisms. The functional molecules that mediate cellular processes are proteins, and examining their expression is a direct readout to the metabolic activities of microbes in that environment. Proteomics can show which proteins respond to the presence of contaminants in bioremediation and highlight microbe proteins that help adapt to the contaminant, as well as enzymes that degrade contaminants. Therefore, this information is very valuable to allow optimization of bioremediation approaches, providing a deeper insight into the biochemical pathways at play [4].

DNA microarray technology has become another pivotal tool in molecular microbial research. Gene expression is detected and measured using microarrays and represents a high-throughput method of monitoring microbial activity in realtime. Microarrays are used in environmental microbiology to identify which specific genes or pathways are activated in response to pollutants, to trace changes in microbial community composition, and to monitor progress in bioremediation [5]. Because microarrays allow the simultaneous analysis of thousands of genes they are a powerful method to study microbial dynamics in complex environments. These molecular techniques have integrated with each other to offer a multi-dimensional understanding of microbial systems that bridged the gap between microbial ecology and environmental applications. For example, combining metagenomics and proteomics enables the linkage of genetic potential to functional activity and a view of the full 'ecosystem' through a single lens. In the same way, combining microarray data to metagenomic ones can validate gene expression patterns, and correlate them with environmental factors. Therefore, the integration of such multiscale approaches is critical to tackling the challenges of environmental pollution within the multiple dimensions [6 - 9].

These molecular techniques, however, also have challenges. Vast amounts of data are produced by metagenomics and robust bioinformatics tools for analysis and interpretation are needed. However, proteomics, while increasing function insight, can be hindered by the complexity of protein extraction and identification in environmental samples. Like DNA microarrays, prior knowledge of the target genes also limits the application of DNA chips to less characterized environments. Given these technical and analytical challenges, the methodology must adapt and platforms need to be designed for integrating the analysis of data. Therefore, it is also necessary in addition to solving these challenges that researchers address the issues regarding the scalability and utility of molecular techniques in real-world applications [10]. Lab studies do help but it is still a big challenge to translate discoveries in the laboratory to the field. Implementation of bioremediation strategies can be complicated by the effects of environmental variability and microbial community dynamics and by the presence of multiple pollutants. Thus, it is necessary to integrate molecular techniques with ecological and engineering approaches in order to make these practical and sustainable. But looking ahead, there are trends and innovations that will serve to bolster the field of microbial research. The resolution and accuracy of metagenomic analyses are improved by advances in sequencing technologies, including long-read sequencing and single-cell genomics [11 - 14]. The evolution of more sensitive mass spectrometry techniques and better protein databases will increase the use of proteomics. As with other areas of microbiology, the advent of next-generation microarrays and CRISPR-based diagnostics has extended the range of microbial monitoring and functional analysis [15]. Driven by the next wave of innovation in environmental microbiology, these innovations, combined with bioinformatics and systems biology, are making waves in the research field.

METAGENOMICS AND PROTEOMICS APPROACHES FOR BIORE-MEDIATION

Exploring Microbial Diversity in Polluted Environments

Microbes are a foundation of ecosystem function because they are abundant and diverse, and they are found in environments that are polluted. Many industrial wastewater, oil spillage, and heavy metal contaminated sites have specialized microbial communities that reside in polluted environments to make their living. Breaking down pollution, sealing heavy metals, and restoring ecosystem health depend on these microbes. But increasingly, the work of studying these microbial ecosystems has been propelled by advanced molecular techniques like metagenomics [16]. Traditionally, through culture-based methods, genetic material from environment samples is analyzed; metagenomics allows for bypassing such traditional methods and allows the direct analysis of genetic material from environment samples. By using this approach, the genetic potential of a vast, unculturable array of microorganisms can be identified. Knowing microbial diversity through metagenomics not only provides insight into pollutant degradation and stress tolerance gene, but these genes are also responsible for bioremediation metabolic pathways. These are key insights for designing targeted interventions to augment nature's natural remediation capacity of polluted environments (Table 1).

Table 1. Metagenomics and proteomics approaches for bioremediation from previous studies.

Microbial Species/Consortium	Pollutant	Metagenomics Approach	Proteomics Approach	Key Findings
Pseudomonas putida [17]	Polycyclic Aromatic Hydrocarbons (PAHs)	Identification of catabolic genes (*e.g.*, *nah* and *phn*) involved in PAH degradation using metagenomic sequencing.	Quantification of proteins involved in PAH metabolism, including dioxygenase enzymes.	Elucidated the genetic and enzymatic basis for efficient PAH degradation under aerobic conditions.

(Table 1) cont.....

Microbial Species/Consortium	Pollutant	Metagenomics Approach	Proteomics Approach	Key Findings
Geobacter sulfurreducens [18]	Heavy metals (*e.g.*, uranium)	Analysis of functional genes involved in metal reduction pathways using shotgun metagenomics.	Characterization of c-type cytochromes and other metal-reducing proteins through mass spectrometry.	Identified key proteins and genes critical for uranium reduction, enhancing bioremediation strategies.
Dehalococcoides mccartyi [19]	Chlorinated solvents (*e.g.*, TCE, PCE)	Discovery of reductive dehalogenase genes responsible for chlorinated solvent degradation.	Profiling of protein expression linked to dechlorination processes, focusing on energy metabolism proteins.	Demonstrated the organism's specialized metabolic pathways for complete dechlorination of solvents.
Sphingomonas sp [20].	Aromatic compounds (*e.g.*, benzene, toluene)	Metagenomic exploration of gene clusters encoding monooxygenase and dioxygenase enzymes.	Identification of adaptive stress-response proteins *via* proteomic studies.	Revealed enzymatic pathways and stress mechanisms enabling aromatic compound degradation.
Mixed microbial consortia (anaerobic sludge) [21]	Industrial dyes	Metagenomics analysis to uncover dye-degrading microbial communities and their functional genes.	Proteomics to identify enzymes like azoreductase and laccase involved in dye decolorization.	Highlighted synergistic microbial interactions and enzyme systems for effective dye degradation.
Bacillus subtilis [22]	Pesticides (*e.g.*, organophosphates)	Identification of organophosphate hydrolase (OPH) genes *via* metagenomic studies.	Quantitative analysis of OPH and related proteins using proteomic profiling.	Demonstrated robust pesticide degradation pathways with implications for agricultural bioremediation.

(Table 1) cont.....

Microbial Species/Consortium	Pollutant	Metagenomics Approach	Proteomics Approach	Key Findings
Nitrosomonas europaea [23]	Ammonia and nitrogen compounds	Analysis of nitrification genes (*e.g.*, *amoA*) using targeted metagenomics.	Proteomic studies identifying ammonia monooxygenase and associated proteins.	Advanced understanding of nitrification and its application in nitrogen pollutant mitigation.
Alcanivorax borkumensis [24]	Hydrocarbons (*e.g.*, crude oil)	Metagenomic sequencing to reveal hydrocarbon degradation gene clusters.	Proteomic analysis of alkane monooxygenases and other hydrocarbon-degrading proteins.	Identified key genes and enzymes enabling hydrocarbon degradation in marine environments.

Functional Genomics in Environmental Microbiology

Whereas metagenomics seeks to understand microbial communities without visualization at the molecular level, functional genomics emphasizes the roles of particular genes and gene expression in microbial communities. This approach has been used in environmental microbiology to understand how microbes respond to pollution at the molecular level. Functional genomics is a study of gene expression patterns under stress conditions, and the regulation of patterns, which is performed by techniques such as transcriptomics and proteomics. For example, genes that encode enzymes that degrade hydrocarbons or resist heavy metals can be identified and their mechanisms of activation studied [25]. In addition to improving our understanding of microbial metabolism, functional genomics helps to inform the development of bioengineered microbes or consortia for increased bioremediation applications.

Role of Proteomics in Decoding Microbial Functionality

The large-scale study of proteins is a powerful tool to understand microbial functionality, that is proteomics. Metagenomics offers insights into what genetic potential microbial communities have, while proteomics reveals what biological activities these microbial communities actually perform. Proteins are the functional molecules of cells, mediating, directly, the process of pollutant degradation, stress adaptation, and energy metabolism. In the context of polluted environments, proteomics will detect stress-responsive proteins, pollution transformation related enzymes and signaling molecules responsible for controlling microbial response [26]. Because mass spectrometry and bioinformatics have advanced to the point where complex protein mixtures from environmental samples can be analyzed, we have a window into the biochemical workings involved in microbial activity. This information is crucial for bioremediation for optimizing conditions for enhancing the expression of important enzymes and metabolic pathways.

Case Studies: Successful Bioremediation Applications

Furthermore, metagenomics, functional genomics, and proteomics are used together in several successful bioremediation applications. In these case studics, the practical benefits of advanced molecular techniques to solving real-world environmental problems are illustrated. Studies of oil-contaminated marine environments through metagenomics have demonstrated the presence of hydrocarbon-degrading microbial communities, including Alcanivorax and Cycloclasticus. Enzyme analyses including alkane monooxygenases and dioxygenases, essential in the breakdown of hydrocarbons, were performed. Further functional genomics studies further optimised biostimulation techniques in these settings and increased the efficiency of bioremediation [27].

Metagenomics in heavy metal-contaminated soils extracted microbial taxa with resistance to and sequestration of metal such as *Pseudomonas* and *Acidithiobacillus*. This proteomic data revealed metalloproteins that bind and detoxify metals and enabled insight into the adaptive mechanisms of microbial communities. The identification of these findings led to the construction of microbial consortia for the bioremediation of mining sites and to the application of functional genomics and proteomics to study microbial communities in wastewater treatment plants. For example, chlorinated compound degrading microbes were identified and the corresponding enzyme systems were characterized [28]. The optimization of treatment processes was enabled by these insights and ultimately aided by them in achieving greater removal efficiency of hazardous contaminants. The examples illustrate how molecular techniques are not only advancing our understanding of microbial ecosystems but are being applied practically to the restoration of the environment using microbes. With these approaches, bioremediation practice can be steered more effectively, targeted, and sustainable. In Fig. (1), the basic workflow of extracting metagenomes from contaminated soil is explained. This was initiated with contaminated soil collected from the environment. Two processing routes of the collected dirt soil sample are possible: one is by direct cell lysis and DNA purification, and the other, the separation of cells from the contaminated soil and cell lysis followed by DNA purification. Specific cloning vectors are used to then clone the isolated DNA. After that, the cloned contaminated soil DNA is delivered into host cells through different gene delivery systems. Metagenome library formed by multiplied contaminated host cells with contaminated soil DNA were screened with the contaminated soil metagenomes. Recently, biosurfactant producers were screened from petroleum hydrocarbon-contaminated sources in cold marine environments [29].

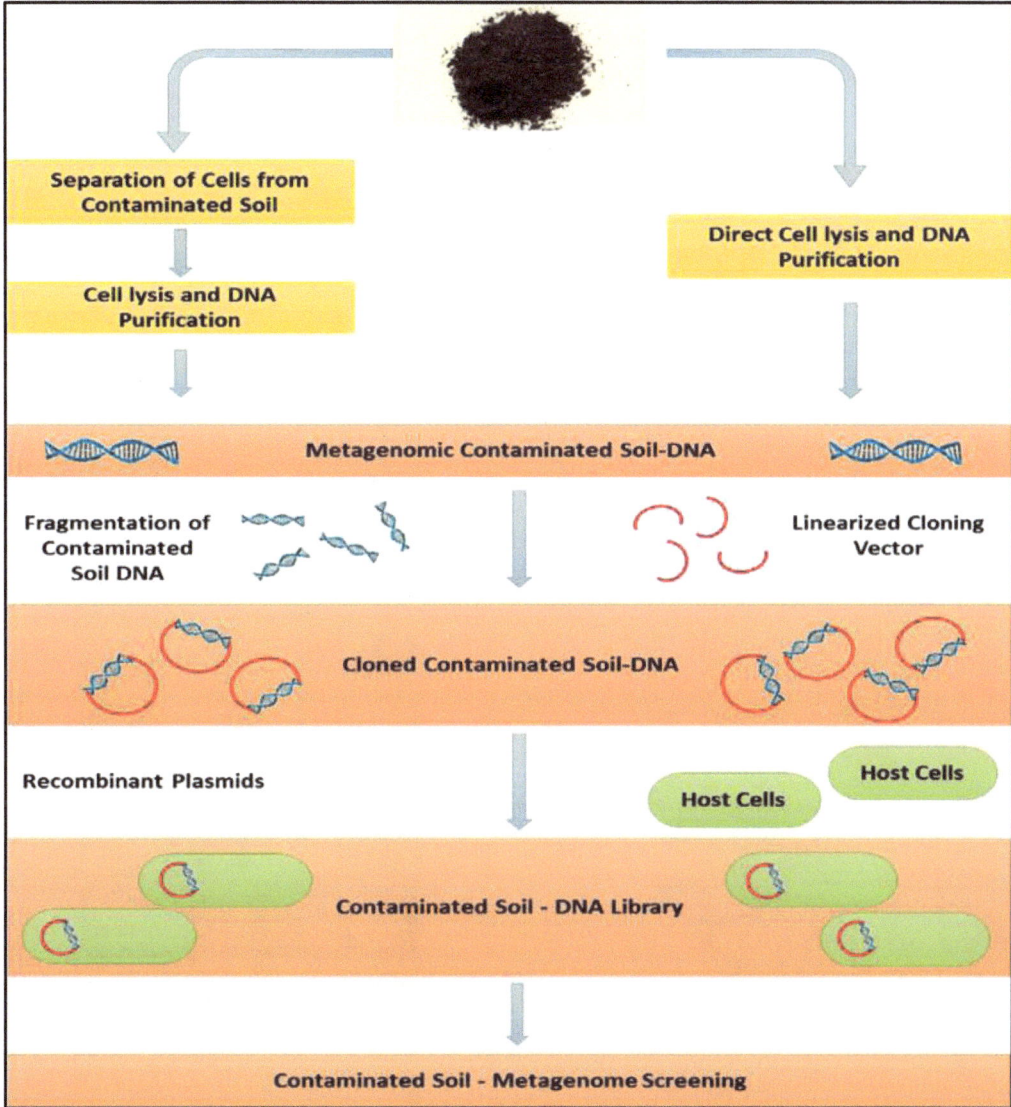

Fig. (1). Towards an ideal systematic workflow of steps involved in contaminated-soil metagenomics [29].

DNA MICROARRAY APPLICATIONS IN ENVIRONMENTAL MICRO-BIOLOGY

Principles and Methodology of DNA Microarrays

DNA microarrays, also termed gene chips, are high throughput tools that allow the study of changes in expression for thousands of genes at one time. The basic principle behind DNA microarrays is hybridization: A biological system process where a fluorescent-labeled cDNA or RNA is combined with complementary DNA probes that can be fixed on a solid surface such as a glass slide or silicon chips [30]. The array contains each DNA probe representing a single gene or a part of a gene and each spot contains a specific DNA probe, permitting an expression pattern detection of a huge number of genes by a single experiment (Fig. **2**).

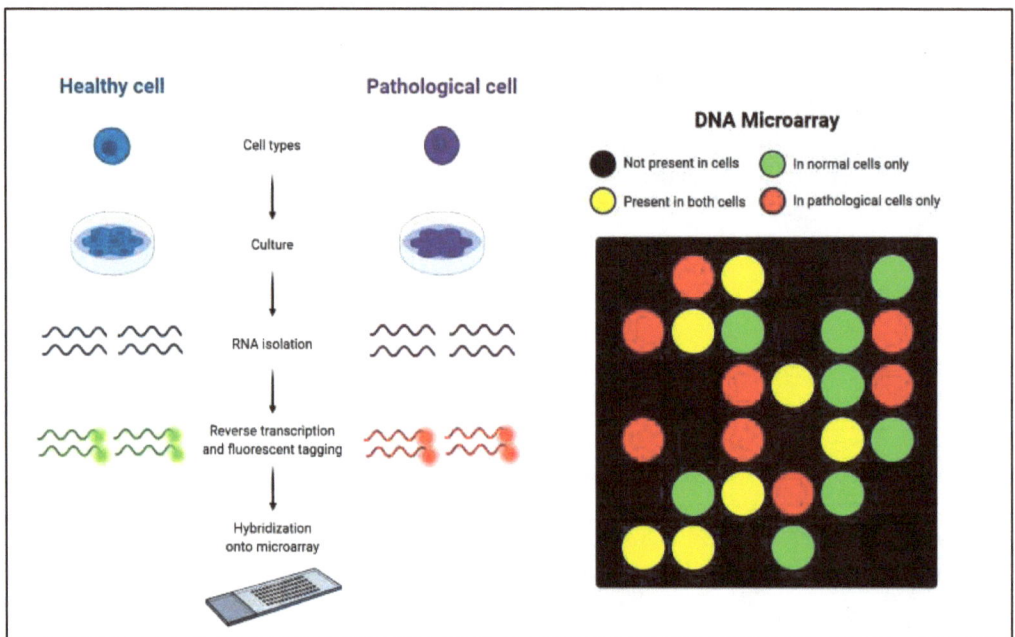

Fig. (2). Principles and Methodology of DNA Microarrays.

The methodology [31] typically involves several steps:

1. **Sample Collection:** A microbial community or specific microorganism is sampled from environmental or biological samples.
2. **RNA Extraction:** The RNA of the sample is extracted, so that it is of the highest quality and is without contaminating matter.

3. **cDNA Synthesis:** Reverse transcription of the RNA often with fluorescent dyes yields complementary DNA (cDNA).
4. **Hybridization:** The cDNA is labelled, the labelled cDNA is applied to the microarray, and it hybridises with the complementary probes fixed on the array.
5. **Scanning and Data Analysis:** The array is hybridized after which fluorescent signals are then scanned with a laser. The signals correspond to the relative abundance of the gene expression in the sample. The data is analyzed using bioinformatics tools that identify genes that are upregulated or downregulated under some given conditions.

High throughput and comprehensive study of gene expression can be done by microarrays, and thus, monitoring of complex microbial activities in various environments.

Applications in Monitoring Microbial Activity

DNA microarrays have great value in environmental microbiology because of their ability to describe changes in gene expression patterns brought about by environmental changes or pollutants in microbial communities. Microbial responses to environmental pollutants, including heavy metals, organic compounds, and xenobiotics are being studied widely *via* microarrays. For example, specific genes associated with detoxification, stress response, and degradation pathways can be upregulated when microbes are exposed to pollutants. These changes in gene expression can be tracked in real-time using microarrays, giving researchers a picture of how microbial activities occur and what pollutant degradation may be going on [32]. Microarrays can be used to monitor the metabolic pathways involved in the breakdown of contaminants in bioremediation applications. For example, researchers can analyze the expression of genes that are involved in hydrocarbon degradation, nitrogen fixation, or metal sequestration on how well microbial bioremediation is performing so that they can change the conditions to improve their performance.

We can use microarrays to determine shifts in the gene expression of microbial communities over time. An example is the observation of the change in microbial activity in a bioreactor or natural remediation system due to changing environmental conditions or a pollutant [33]. Most importantly, this type of information tells us about the functional roles of particular microbes and how they impact ecosystem function. Microarrays can also detect specific pathogenic microorganisms in environmental samples by identifying pathogen-enriched DNA containing genes specific for the pathogen. Monitoring microbial water quality or soil contamination with microbial pathogens that put human and environmental health at risk is particularly useful for this measure.

Advancements in Microarray Technology for Environmental Studies

Advancements in microarray technology over recent years have greatly improved microarray's applications in environmental studies. Thousands of individual probes, which are the probes themselves, form modern microarrays that are able to analyze a greater number of genes simultaneously, an especially good trait for studying complex microbial communities. Therefore, high-density arrays can now cover entire genomes or gene set targets of interest, allowing for more comprehensive studies of microbial responses [34]. Now there are developed environmental microarrays to target genes in environmental strains of bacteria, fungi, and archaea. A set of these arrays designed to detect environmental microorganisms is used to detect environmental microbial diversity and functionality without a prior culture of the sample.

As real-time microarray platforms have been approached by advances in microarray technology. These platforms are always observing microbial gene expression and changes in the environment temperature, pollutant levels, or nutrient availability over time. Microarrays offer their users the real-time capability necessary to use them as a tool for dynamic environmental monitoring and for the success of bioremediation. With the advances in microarray technology, miniaturization has allowed the development of portable microarray systems [35]. The compact systems permit on-site analysis of the environmental samples, *e.g.* water or soil, thereby reducing the need for extended lab-based processing. The increased feasibility of using microarrays in field-based environmental monitoring was the result of this development.

Limitations and Challenges in Microarray Applications

While DNA microarrays offer significant advantages in environmental microbiology, there are several limitations and challenges associated with their use. The high quality of probes used in microarray technology, in particular, is one of the major challenges of the technology. Microarrays often make use of pre-designed probes for known sequences. That makes them unable to detect novel genes or new organisms that have not been characterized before. New techniques such as environmental DNA microarrays will help to address this, however design of effective probes for uncharacterized environments persists. Because microarray data can be highly complex and difficult to interpret, complex microbial communities represented by many species or nonmodel organisms can be especially hard to study using microarrays. For example, amino acid variation can be quite large and is hard to map through visualization, big datasets need sophisticated bioinformatics tools and expertise to analyse [36].

Although gene expression profiling using microarrays can be effective, these microarrays may be insensitive in the detection of low abundance genes or microorganisms in low concentration. Background noise can also complicate the interpretation of gene expression changes resolved on microarrays, and can affect the resolution of microarrays. Samples from the environment are frequently presented with complex matrices at varying levels of contaminants, organic matter, and other interfering substances. The complexity can affect the accuracy and reproducibility of the microarray results. For overcoming these challenges, appropriate controls, and sample preparation techniques are needed. Microarray analysis is expensive, meaning it is expensive, and more severely limits access to poor resource settings [37]. The costs over time have decreased but an initial investment in equipment and reagents still may be prohibitive for many laboratories.

INTEGRATION OF MOLECULAR TECHNIQUES FOR ENVIRO-NMENTAL SOLUTIONS

An intriguing approach to address big environmental challenges is the integration of molecular techniques such as metagenomics, proteomics, DNA microarrays, *etc*. In combination, these techniques offer a complete picture of microbial communities and their functions that could inform the development of more effective and targeted environmental solutions. With the concern over environmental issues such as pollution, climate change, and loss of habitat needs urgent action, these molecular techniques can contribute towards the information required to move forward with bioremediation strategies, sustainability initiatives, and conservation. When combined, researchers can monitor changes in microbial diversity, metabolic functions, and gene expression profiles: a breadth of information about microbial activity in polluted or stressed environments [38]. Through this integration, we gain better insights into how to make better decisions and how to carry out interventions to regulate environmental restoration. These techniques together improve bioremediation processes, waste management strategies, and conservation practices (Table **2**).

Combining Metagenomics, Proteomics, and Microarrays

Metagenomics, proteomics, and microarrays, when combined, provide a more global approach to studying environmental microbiomes. However, each technique provides a distinct perspective on the patterns of microbial diversity and activity that, when combined, generate a comprehensive view. Metagenomics serves as an overview of the genetic diversity present in a given environment and can be used to identify previously unculturable microorganisms and characterize the genes possibly related to specific bioremediation activities, like pollutant degradation [39]. The actual protein expression patterns can be revealed by proteomics, revealing which genes are being translated into proteins under specific environmental conditions. The ability to analyze proteomes can help identify key enzymes involved in pollutant degradation, or stress responses, and thus is important to understanding microbial function in bioremediation.

Table 2. Integration of Molecular Techniques for Environmental Solutions.

Aspect	Combining Metagenomics, Proteomics, and Microarrays	Big Data and Bioinformatics in Environmental Microbiology	Systems Biology Approaches for Bioremediation
Definition	Integration of multiple molecular techniques to obtain a holistic understanding of microbial systems and their functionality in environmental contexts.	Utilization of computational tools to analyze and interpret large-scale datasets from molecular studies in environmental microbiology.	Application of multidisciplinary approaches to model and predict microbial interactions and functions for effective bioremediation strategies.
Objective	To link genetic potential (metagenomics), protein expression (proteomics), and real-time gene activity (microarrays) for a comprehensive microbial profile.	To process, analyze, and visualize complex datasets to uncover patterns, interactions, and correlations in microbial communities.	To integrate genomic, proteomic, and environmental data to create predictive models for optimizing bioremediation under varying environmental conditions.
Key Methodologies	- Sequencing (NGS, WGS) and functional annotation - Protein quantification using LC-MS/MS - DNA microarrays for gene expression analysis	- Data storage in databases (MG-RAST, EMBL-EBI) - Machine learning for pattern recognition - Statistical tools for network analysis	- Development of metabolic and regulatory network models - Flux balance analysis (FBA) - Dynamic system simulations
Applications in Environmental Solutions	- Identifying pollutant-degrading microbes - Correlating functional genes with protein activity - Tracking real-time microbial responses to pollutants	- Predicting the impact of environmental changes on microbial ecosystems - Mapping pollutant degradation pathways - Identifying microbial biomarkers	- Designing optimized microbial consortia for specific pollutants - Predicting system-wide effects of environmental interventions
Challenges	- Data standardization for cross-platform integration - Handling vast and diverse datasets - Limited compatibility of some molecular techniques	- Managing high computational demands - Ensuring data accuracy and reproducibility - Difficulty in interpreting highly complex datasets	- Complexity of modeling dynamic environmental systems - Limited availability of curated data for accurate predictions

(Table 2) cont.....

Aspect	Combining Metagenomics, Proteomics, and Microarrays	Big Data and Bioinformatics in Environmental Microbiology	Systems Biology Approaches for Bioremediation
Future Innovations	- Development of hybrid platforms integrating sequencing, proteomics, and microarrays - Real-time, field-deployable molecular tools	- Advanced AI-driven analytics for improved prediction - Cloud-based platforms for global collaboration and data sharing	- Enhanced predictive models incorporating climate and ecosystem variables - Inclusion of synthetic biology for engineering robust microbial solutions

DNA Microarrays are used for the high-throughput analysis of gene expression, which provides detailed information on how environmental factors influence microbial communities. Gene expression patterns in response to pollutants, nutrient availability, or other stressors can be monitored on microarrays, and the bioremediation process can be tracked in real-time. Taken together, these molecular techniques allow strains of microbe capable of degrading pollutants to be identified, the pathways they use to do so to be tracked, and the specific proteins involved in these processes to be determined [40]. This integration is particularly useful in designing the optimized bioremediation strategy applicable for large-scale environmental restoration.

Big Data and Bioinformatics in Environmental Microbiology

With the amount of data created by metagenomic, proteomic, and microarray-based analyses, there is a pressing need for these advanced tools for data analysis. Environmental microbiology has inevitably depended on big data technologies and bioinformatics to manage, analyze, and interpret complex datasets. The integration of different data types of genomic sequences, protein profiles, and gene expression data, through these technologies, is possible and helps in a better comprehension of microbial communities. Environmental microbiology has become data intensive with the growth of high throughput sequencing, mass spectrometry, and microarray platforms [41]. Environmental studies generate large volumes of data and big data technologies, including cloud computing, distributed data storage, and high performance computing, are needed to manage the data. These tools are capable of processing and analyzing data from hundreds or thousands of samples at the same time, allowing large-scale studies of microbial communities in polluted environments.

The analysis and interpretation of complex molecular data are critical to bioinformatics. Bioinformatics tools are exploited to assemble genomic sequences, annotate genes, and predict microbial functions in metagenomics.

Bioinformatics tools are used in proteomics to identify proteins, study their function, and predict interactions. On DNA microarrays, bioinformatics is applied to the analysis of gene expression data, identification of differentially expressed genes, and association of these with environmental variables [42]. Environmental microbiologists can use big data and bioinformatics to enhance the understanding of microbial communities' functional potential, identify biomarkers for pollution monitoring, and optimize bioremediation strategies based on real-time data analysis.

Systems Biology Approaches for Bioremediation

By applying experimental data from genomics, transcriptomics, proteomics, and metabolomics to construct highly detailed models of biological systems, systems biology approaches are used. Systems biology is particularly strong for investigations of microbial community complexity through interactions within the microbial community or between microbes and their environment in environmental microbiology. This integrated approach is desirable for bioremediation in which microbial activity is governed by several environmental variables and pollutants [43]. Computational models are built that simulate the behavior of microbial communities when treated to changes in environmental conditions, using systems biology. By predicting the response of microbial communities to different pollutants, nutrients, or environmental stressors, these models can help scientists determine the most favorable conditions for bioremediation.

Integrating genomic, proteomic, and metabolic data at the systems level allows us to identify key metabolic pathways responsible for pollutant degradation and environmental detoxification. This information can be used to elucidate how microorganisms degrade specific pollutants, to quantify the rate-limiting steps in degradation processes, and to design microbial strains or consortia with improved bioremediation abilities. Finally, systems biology is used to facilitate an understanding of how multiple microorganisms interact in a consortium. Cooperation, competition, or cross-feeding play an important part in the efficiency of the bioremediation processes, and these interactions. Modeling such interactions enables researchers to design microbial consortia, in which each partner enhances the others' activity so that they can degrade pollutants more effectively. Bioremediation strategies must be adapted to meet the (highly variable) conditions of environmental systems [44]. These systems biology approaches provide means to develop personalized bioremediation protocols, and to adapt microbial communities to specific types of contamination or environmental stressors. Therefore, systems biology handles bioremediation projects by constantly monitoring microbial responses and subsequently adjusting the conditions as required.

In the future of environmental microbiology, molecular techniques, such as metagenomics, proteomics or DNA microarrays, will integrate to provide big data analytics and systems biology. By providing integrated microbial diversity, functionality, and metabolic approaches, these approaches provide a comprehensive framework for the development of effective bioremediation strategies [45]. With pollution and climate change becoming larger and larger challenges to the environment, these advanced technologies will be critical in helping us develop solutions for sustainability. Molecular techniques and computational tools taken together with environmental scientists and engineers provide a better understanding of microbial ecosystems and of environmental restoration options for more efficient and targeted bioremediation.

CHALLENGES AND FUTURE DIRECTIONS

A major contribution of molecular techniques such as metagenomics, proteomics and DNA microarrays in environmental research, is their potential to integrate microbial communities and their role in environmental processes, for example, bioremediation. However, there are a host of challenges to overcome in order to actually realize the full benefits of these technologies.

Technical and Analytical Challenges in Molecular Techniques

There has been great interest in developing molecular techniques to study environmental microbiology as environmental samples are complex and highly variable. While microbial diversity is typically high in environmental samples, including from polluted or harsh environments, microbial communities are highly genetically and functionally heterogeneous. However, because the microbial world is so diverse, it is difficult to properly capture and analyze the microbial DNA, proteins, and gene expression profiles. Additionally, it was technically difficult to extract high-quality DNA and proteins from environmental samples due to the presence of the sample inhibitors that adversely affect downstream analyses [46]. Finally, the computational requirements for the processing and analysis of the large volumes of data produced from these technologies represent a serious challenge. While bioinformatics tools continue to enhance, they still struggle to correctly annotate sequences, predict protein functions, and model the complexity of microbial interactions. Additionally, there is still much work to be done with data integration from different high throughput platforms (*e.g.* metagenomics, proteomics, or microarray) because of the mismatch between data types. To make sure spouses' data from a variety of sources are effectively integrated and interpreted, this requires an advanced computational framework and expertise [47].

Emerging Trends and Innovations in Microbial Research

However, if the challenges described can be conquered, then new exciting emerging trends and innovations in microbial research may help to make molecular techniques more applicable in environmental study. Such an innovation is in the development of single-cell genomics, which studies individual microbial cells. The problem with traditional methods of studying microorganisms is that the functional potential is something that is difficult to study, so this technology can more accurately provide insight into the functional potential of microorganisms that are difficult to study. In addition, we are seeing the increasing use of Next-Generation Sequencing (NGS) technologies that will provide greater and faster sequencing of microbial diversity in complex environments. These improvements will more precisely, more accurately, and more fully report the metabolic strengths of microbial communities [48]. The second promising trend is AI and ML efforts to analyze large-scale microbial data. The patterns and correlations in microbial communities can be identified using AI and ML algorithms, which in turn will produce more accurate predictions of microbial behavior under exposure to environmental stressors. Such understanding could inform the development of predictive models for bioremediation processes as well as provide optimization for intervention plan strategies for pollution management. Additionally, enhanced pollutant degrading ability by engineered microorganisms is a promising innovation in synthetic biology that could improve bioremediation. Researchers are using molecular techniques and synthetic biology to learn how to design microorganisms to degrade particular contaminants more efficiently [41, 42].

FUTURE PERSPECTIVES IN ENVIRONMENTAL APPLICATIONS

The future of molecular techniques in environmental microbiology is bright, with the potential to provide much needed insight into our global environmental challenges. Inaddition, bioremediation is one area where these technologies will play a huge role. Rapid and accurate assessment of microbial diversity, function, and activity in polluted environments will provide the tools to design more efficient and specific bioremediation strategies. Through the application of advanced molecular tools, researchers can find the most suitable microbial strains and consortia to deal with different kinds of contaminations, and get rid of environmental conditions to increase bioremediation. Finally, as molecular technologies continue to advance, they will increasingly be used in bioremediation, but also in climate change mitigation and sustainable agriculture, environmental monitoring and elsewhere. Monitoring microbial communities in natural ecosystems should, in turn, provide important insights into ecosystem health and the effects of environmental stressors (such as climate change or habitat destruction) [48]. These technologies may also be developed for the promotion of sustainable agricultural practices involving the use of beneficial microbial communities that improve soil fertility and crop yield while minimizing dependency on chemical fertilizers and pesticides. Overall, the integration of molecular techniques, big data analytics, and systems biology will lead to the introduction of a more sustainable and data-oriented approach to environmental management [43]. Combined, these technologies together with policy and community engagement efforts can help prompt much more effective mitigation strategies for environmental degradation, as well as improvements in ecosystem services and the promotion of a healthier planet for future generations.

CONCLUSION

Recent advances in molecular techniques, such as metagenomics, proteomics, and DNA microarrays, have greatly increased our understanding of microbial communities and what they can do in terms of solving ecological problems. These methods contribute to detailed and comprehensive views on microbial diversity, functionality and interactions, especially in polluted environments. By taking advantage of these technologies, we gain important information into microbial mechanisms involved in pollutant degradation and bioremediation, and can thus engineer more efficient and tailored strategies for environmental cleanup. Metagenomics and proteomics taken together allow microbial community genetic and functional exploration on an unprecedented scale while revealing the contribution these communities make to environmental remediation. However, DNA microarrays allow real-time monitoring of microbial activity and detection of specific gene expression correlated with pollutant breakdown. Taken together, these techniques provide a systems approach to studying and controlling microbial systems for environmental purposes. However, as with the deployment of these technologies, there are challenges, especially in data integration and interpretation, which are highlighted in this chapter. As the field evolves, these hurdles will be overcome by innovation in data analysis, big data integration and bioinformatics. However, the advancements discussed in this chapter help researchers and practitioners to use microbial systems to mitigate environmental degradation. As these molecular techniques continue to evolve, new opportunities for advancing the management of the environment and developing sustainable solutions for a cleaner, greener planet exist now, and they will evolve further.

REFERENCES

[1]   Suman J, Rakshit A, Ogireddy SD, Singh S, Gupta C, Chandrakala J. Microbiome as a key player in sustainable agriculture and human health. Review. 2022; 2.
[http://dx.doi.org/10.3389/fsoil.2022.821589]

[2]   Ghose M, Parab A S, Manohar C S, Mohanan D, Toraskar A. Unraveling the role of bacterial communities in mangrove habitats under the urban influence, using a next-generation sequencing approach. J Sea Res 2024; 198: 102469.
[http://dx.doi.org/10.1016/j.seares.2024.102469]

[3]   Arevalo P, VanInsberghe D, Elsherbini J, Gore J, Polz M F. A reverse ecology approach based on a biological definition of microbial populations. Cell 2019; 178(4): 820-834.e14.
[http://dx.doi.org/10.1016/j.cell.2019.06.033]

[4]   Becsei Á, *et al.* Time-series sewage metagenomics distinguishes seasonal, human-derived and environmental microbial communities potentially allowing source-attributed surveillance. Nat Commun 2024; 15(1): 7551.
[http://dx.doi.org/10.1038/s41467-024-51957-8]

[5]   Zhang S, Li X, Wu J, *et al.* Molecular methods for pathogenic bacteria detection and recent advances in wastewater analysis. Water 2021; 13(24): 3551.
[http://dx.doi.org/10.3390/w13243551]

[6]   Jain H. From pollution to progress: groundbreaking advances in clean technology unveiled. Innov Green Dev 2024; 3(2): 100143.
[http://dx.doi.org/10.1016/j.igd.2024.100143]

[7]   Jain H. Data analytics enabled by the Internet of Things and artificial intelligence for the management of Earth's resources. In: Kumar D, Tewary T, Shekhar S, Eds. Data analytics and artificial intelligence for earth resource management. Elsevier 2025; pp. 19-36.
[http://dx.doi.org/10.1016/B978-0-443-23595-5.00002-4]

[8]   Jain H, Dhupper R, Shrivastava A, Kumar D, Kumari M. Leveraging machine learning algorithms for improved disaster preparedness and response through accurate weather pattern and natural disaster prediction. Review. 2023; 11.
[http://dx.doi.org/10.3389/fenvs.2023.1194918]

[9]   Jain H, Dhupper R, Shrivastava A, Kumari M. Enhancing groundwater remediation efficiency through advanced membrane and nano-enabled processes: a comparative study. Groundw Sustain Dev 2023; 23: 100975.
[http://dx.doi.org/10.1016/j.gsd.2023.100975]

[10]   Aplakidou E, *et al.* Visualizing metagenomic and metatranscriptomic data: a comprehensive review. Comput Struct Biotechnol J 2024; 23: 2011-33.
[http://dx.doi.org/10.1016/j.csbj.2024.04.060]

[11]   Trends in biological processes in industrial wastewater treatment,. M. P. Shah, ed.: IOP Publishing 2024.
[http://dx.doi.org/10.1088/978-0-7503-5678-7]

[12]   Shah MP. Emerging innovative trends in the application of biological processes for industrial wastewater treatment. Elsevier 2024.

[13]   Shah MP. Application of sewage sludge in industrial wastewater treatment. John Wiley & Sons 2024.
[http://dx.doi.org/10.1002/9781119857396]

[14]   Shah MP, Shah N. Environmental approach to remediate refractory pollutants from industrial wastewater treatment plant. Elsevier 2024.

[15]   Gobena S, Admassu B, Kinde MZ, Gessese AT. Proteomics and its current application in biomedical area: Concise review. ScientificWorldJournal 2024; 2024: 1-13.
[http://dx.doi.org/10.1155/2024/4454744] [PMID: 38404932]

[16] Gangola S, *et al.* Exploring microbial diversity responses in agricultural fields: A comparative analysis under pesticide stress and non-stress conditions. Orig Res. 2023; 14.

[17] Takizawa N, Kaida N, Torigoe S, *et al.* Identification and characterization of genes encoding polycyclic aromatic hydrocarbon dioxygenase and polycyclic aromatic hydrocarbon dihydrodiol dehydrogenase in Pseudomonas putida OUS82. J Bacteriol 1994; 176(8): 2444-9.
[http://dx.doi.org/10.1128/jb.176.8.2444-2449.1994] [PMID: 8157615]

[18] Methé BA, Nelson KE, Eisen JA, *et al.* Genome of Geobacter sulfurreducens: metal reduction in subsurface environments. Science 2003; 302(5652): 1967-9.
[http://dx.doi.org/10.1126/science.1088727] [PMID: 14671304]

[19] Solis M I V, Abraham P E, Chourey K, Swift C M, Löffler F E, Hettich R L. Targeted detection of *Dehalococcoides mccartyi* microbial protein biomarkers as indicators of reductive dechlorination activity in contaminated groundwater. Sci Rep 2019; 9(1): 10604.
[http://dx.doi.org/10.1038/s41598-019-46901-6]

[20] Johnson GR, Olsen RH. Nucleotide sequence analysis of genes encoding a toluene/benzene--monooxygenase from *Pseudomonas sp.* strain JS150. Appl Environ Microbiol 1995; 61(9): 3336-46.
[http://dx.doi.org/10.1128/aem.61.9.3336-3346.1995] [PMID: 7574644]

[21] An X, Chen Y, Chen G, Feng L, Zhang Q. Integrated metagenomic and metaproteomic analyses reveal potential degradation mechanism of azo dye-Direct Black G by thermophilic microflora. Ecotoxicol Environ Saf 2020; 196: 110557.
[http://dx.doi.org/10.1016/j.ecoenv.2020.110557]

[22] Thakur M, Medintz IL, Walper SA. Enzymatic bioremediation of organophosphate compounds-progress and remaining challenges. Front Bioeng Biotechnol 2019; 7: 289.
[http://dx.doi.org/10.3389/fbioe.2019.00289] [PMID: 31781549]

[23] Wright CL, Lehtovirta-Morley LE. Nitrification and beyond: metabolic versatility of ammonia oxidising archaea. ISME J. 2023; 17: pp. (9)1358-68.

[24] Gregson BH, Metodieva G, Metodiev MV, Golyshin PN, McKew BA. Protein expression in the obligate hydrocarbon-degrading psychrophile *Oleispira antarctica* RB-8 during alkane degradation and cold tolerance. Environ Microbiol 2020; 22(5): 1870-83.
[http://dx.doi.org/10.1111/1462-2920.14956] [PMID: 32090431]

[25] Snyder-Mackler N, Lea AJ. Functional genomic insights into the environmental determinants of mammalian fitness. Curr Opin Genet Dev 2018; 53: 105-12.
[http://dx.doi.org/10.1016/j.gde.2018.08.001] [PMID: 30142491]

[26] Tsakou F, Jersie-Christensen R, Jenssen H, Mojsoska B. The role of proteomics in bacterial response to antibiotics. Pharmaceuticals (Basel) 2020; 13(9): 214.
[http://dx.doi.org/10.3390/ph13090214] [PMID: 32867221]

[27] Knapik K, Bagi A, Krolicka A, Baussant T. Metatranscriptomic analysis of oil-exposed seawater bacterial communities archived by an Environmental Sample Processor (ESP). Microorganisms 2020; 8(5): 744.
[http://dx.doi.org/10.3390/microorganisms8050744] [PMID: 32429288]

[28] Li L, Meng D, Yin H, Zhang T, Liu Y. Genome-resolved metagenomics provides insights into the ecological roles of the keystone taxa in heavy-metal-contaminated soils. Front Microbiol 2023; 14: 1203164.
[http://dx.doi.org/10.3389/fmicb.2023.1203164] [PMID: 37547692]

[29] Pratap D, Ranjith NK. Metagenomics — a technological drift in bioremediation. Adv Bioremediat Wastewater Pollut Soil. Rijeka: IntechOpen 2015; p. Ch. 4.

[30] Jadhav S, Bhave M, Palombo E A. Methods used for the detection and subtyping of *Listeria monocytogenes*. J Microbiol Methods 2012; 88(3): 327-41.
[http://dx.doi.org/10.1016/j.mimet.2012.01.002]

[31]   Douterelo I, Boxall J B, Deines P, Sekar R, Fish K E, Biggs C A. Methodological approaches for studying the microbial ecology of drinking water distribution systems. Water Res 2014; 65: 134-56.
[http://dx.doi.org/10.1016/j.watres.2014.07.008]

[32]   Ali AA, Altemimi AB, Alhelfi N, Ibrahim SA. Application of biosensors for detection of pathogenic food bacteria: A review. Biosensors (Basel) 2020; 10(6): 58.
[http://dx.doi.org/10.3390/bios10060058] [PMID: 32486225]

[33]   Brown DC, Turner RJ. Assessing microbial monitoring methods for challenging environmental strains and cultures. Microbiol Res (Pavia) 2022; 13(2): 235-57.
[http://dx.doi.org/10.3390/microbiolres13020020]

[34]   Zhou J, Thompson D K. Challenges in applying microarrays to environmental studies. Curr Opin Biotechnol 2002; 13(3): 204-7.
[http://dx.doi.org/10.1016/S0958-1669(02)00319-1]

[35]   Aparna GM, Tetala KKR. Recent progress in development and application of dna, protein, peptide, glycan, antibody, and aptamer microarrays. Biomolecules 2023; 13(4): 602.
[http://dx.doi.org/10.3390/biom13040602] [PMID: 37189350]

[36]   Gong J, Yang C. Advances in the methods for studying gut microbiota and their relevance to the research of dietary fiber functions. Food Res Int 2012; 48(2): 916-29.

[37]   Negi A, Shukla A, Jaiswar A, Shrinet J, Jasrotia RS. Applications and challenges of microarray and RNA-sequencing. In: Singh DB, Pathak RK, Eds. Bioinformatics. Academic Press 2022; pp. 91-103.
[http://dx.doi.org/10.1016/B978-0-323-89775-4.00016-X]

[38]   Karunakaran E, *et al.* Integrating molecular microbial methods to improve faecal pollution management in rivers with designated bathing waters. Sci Total Environ 2024; 912: 168565.
[http://dx.doi.org/10.1016/j.scitotenv.2023.168565]

[39]   Bikel S, *et al.* Combining metagenomics, metatranscriptomics and viromics to explore novel microbial interactions: towards a systems-level understanding of human microbiome. Comput Struct Biotechnol J 2015; 13: 390-401.
[http://dx.doi.org/10.1016/j.csbj.2015.06.001]

[40]   Simpson J B, *et al.* Metagenomics combined with activity-based proteomics point to gut bacterial enzymes that reactivate mycophenolate. Gut Microbes 2022; 14(1): 2107289.
[http://dx.doi.org/10.1080/19490976.2022.2107289]

[41]   Hallin S. Environmental microbiology going computational—Predictive ecology and unpredicted discoveries. Environ Microbiol 2023; 25(1): 111-4.
[http://dx.doi.org/10.1111/1462-2920.16232] [PMID: 36181387]

[42]   Gupta A, Kumar S, Kumar A. Big data in bioinformatics and computational biology: Basic insights. Methods Mol Biol 2024; 2719: 153-66.
[http://dx.doi.org/10.1007/978-1-0716-3461-5_9] [PMID: 37803117]

[43]   Saranya S, Thamanna L, Chellapandi P J S M. Unveiling the potential of systems biology in biotechnology and biomedical research. , 2024; 4(4): 1217-38.

[44]   Dash D M, Osborne W J J C. A systematic review on the implementation of advanced and evolutionary biotechnological tools for efficient bioremediation of organophosphorus pesticides. , 2023; 313: 137506.
[http://dx.doi.org/10.1016/j.chemosphere.2022.137506]

[45]   de Lorenzo VJCb. Systems biology approaches to bioremediation. 2008; 19: pp. (6)579-89.
[http://dx.doi.org/10.1016/j.copbio.2008.10.004]

[46]   Walter W, Pfarr N, Meggendorfer M, Jost P, Haferlach T, Weichert W. Next-generation diagnostics for precision oncology: Preanalytical considerations, technical challenges, and available technologies. Semin Cancer Biol 2022; 84: 3-15.

[http://dx.doi.org/10.1016/j.semcancer.2020.10.015] [PMID: 33171257]

[47]   Feng W, *et al.* Molecular diagnosis of COVID-19: challenges and research needs. 2020; 92: pp. (15)10196-209.

[48]   dos Reis GA, *et al.* Comprehensive review of microbial inoculants: agricultural applications, technology trends in patents, and regulatory frameworks 2024; 16(19): 8720.

CHAPTER 10

Integrative Approaches in Microbial Biosensing: Towards Efficient Environmental Monitoring

Vidiksha Singla[1] and **Geetansh Sharma**[1,*]

[1] *School of Bioengineering and Food Technology, Shoolini University, Solan, Himachal Pradesh-173229, India*

Abstract: Microbial biosensors have gained significant attention as effective tools for *in situ* environmental monitoring due to their portability, cost-efficiency, and user-friendly design. These biosensors, based on the biological responses of microorganisms, have proven particularly valuable in detecting and quantifying pollutants, including heavy metals, pesticides, and emerging contaminants. Recent advancements in genetic engineering have enabled the development of microbial biosensors with increased specificity and sensitivity by integrating reporter genes with regulatory elements that respond dose-dependently to target chemicals. Such modifications allow for targeted detection, increased accuracy, and expanded application range. This chapter reviews current trends in microbial biosensor technology, with particular emphasis placed on toxicity assessment using microbial biosensors, which provide critical insights into ecotoxicity in water, soil, and air, offering a less costly and rapid substitute to traditional bioassays. The integration of transducers, including electrochemical, optical, and microbial fuel cells, further expands their functionality, allowing for versatile monitoring of pollutants in complex environments. We explore the latest advancements in microbial biosensor applications for environmental, food, and biomedical fields and discuss the technical and societal challenges impeding their widespread adoption. Through highlighting these advancements, this chapter underscores the role of microbial biosensors in enabling sustainable, efficient environmental monitoring, as well as their potential for broader application as we continue to refine and expand their capabilities.

Keywords: Advancements, Ecotoxicity, Genetic engineering, Microbial biosensors, Transducers.

INTRODUCTION

The fast pace of industrialization and the high use of chemicals in agriculture have emitted many toxic compounds into the environment, affecting soil, water, and

---

* **Corresponding author Geetansh Sharma:** School of Bioengineering and Food Technology, Shoolini University, Solan, Himachal Pradesh-173229, India; E-mail: geetanshsharma.gs@gmail.com

Harshita Jain & Maulin P. Shah (Eds.)

air. This contamination poses significant environmental challenges and health risks to animals and humans. Monitoring and detecting pollutants are essential for evaluating their harmful impacts on ecosystems and living organisms. Traditional chemical analysis methods, such as Inductively Coupled Plasma-Mass Spectrometry (ICP-MS), High-Performance Liquid Chromatography (HPLC), and gas Chromatography (GC) are widely used for their sensitivity and accuracy in analyzing environmental samples. However, to better understand the bioavailability of environmental pollutants, biosensors have been developed. These biosensors utilize biological elements to detect pollutants and offer advantages such as low cost, energy efficiency, and the capability for *in situ* real-time monitoring [1, 2].

According to the International Union of Pure and Applied Chemistry (IUPAC), a biosensor is defined as: A device that uses specific biochemical reactions mediated by, immune systems, isolated enzymes, organelles, tissues, or whole cells to detect chemical compounds usually by electrical, thermal, or optical signals The development of biosensors began in the 1950s. In 1956, Professor Leland C. Clark created the first oxygen electrode, establishing a foundation for future advancements in biosensor technology. The breakthrough came in 1962 when Clark and Lyons introduced the first true biosensor. They combined an enzyme, glucose oxidase, with an electrochemical sensor to measure glucose levels. This innovation marked the beginning of the widespread use of biosensors for various quantitative assessments [3].

Biosensors, which utilize the sensitivity and selectivity of biological components (such as biomolecules, organelles or whole cells) coupled with signal transducers, offer significant advantages over conventional methods. These advantages include high specificity, ease of use, fast response times, continuous real-time signal monitoring, and low cost. These detectors are able to identify specific chemical agents or mixtures of environmental relevance like industrial effluents, pesticides, and insecticides, among others, and biological activities such as genotoxicity, immunotoxicity, and endocrine activities. Classical analytical methods often require extensive sample pretreatment and are relied on labour-intensive work. In contrast, biosensors present substantial advantages in identifying and measuring the amounts of specific compounds directly in air or water. Conventional methods usually calculate the total concentration of possible toxic chemicals but do not account for an accurate assessment of bioavailability, including toxic impact on a living organism. Biosensors may be used to augment such classical methods by detecting both the bioavailable as well as unavailable forms of contaminants thus providing a more comprehensive understanding of environmental toxicity [1, 4].

The basic structure of a biosensor consists of sensing elements and signalling elements. Biological sensing elements, including DNA (aptamers), proteins (enzymes), antibodies, and whole cells (bacteria), can detect environmental pollutants such as organic compounds and heavy metals [5, 6]. When these sensing elements detect pollutants, they trigger signalling elements to produce various signals, such as fluorescence, luminescence, colour changes, pH changes, or electrical signals, which can be measured or detected by operators [2]. Technological advancements have enabled the development of genetically engineered biosensors. Additionally, nowadays biosensors are being utilized for biosafety purposes as well. This chapter reviews the various types, working principles, advantages, disadvantages, and future perspectives of microbial biosensors.

MICROBIAL BIOSENSORS

Microbial biosensors are analytical devices that utilize microorganisms or their metabolites to detect and quantify specific substances. According to the American Society for Microbiology (ASM), microbial biosensors, are engineered to signal and sense the existence of specific compounds. These biosensors leverage the natural responses of bacteria to environmental stimuli, often using genetic engineering to produce measurable signals in response to target molecules. Karube *et al.* investigated the first microbial biosensor in 1977. Compared to enzymatic biosensors, microbial biosensors exhibit lower sensitivity to inhibitory substances, greater accuracy in measuring temperature and pH, lower cost, and longer lifespan. These sensors can be categorized based on signal type, microbe type, sensor type, analyte specificity, microbial response, and more. The first microbial biosensor used a batch electrochemical type of MFC system where the anaerobic bacterium, *Clostridium butyricum*, was immobilized in a gel membrane on top of a platinum rode [7, 8]. Fig. (**1**). gives a graphical representation of how a microbial biosensor works.

An ideal biosensor should have two critical characteristics: sensitivity (the ability to detect low concentrations of analytes) and specificity (the ability to differentiate between various analytes through the aid of bio-recognition elements) [9]. The sensitivity of biosensors is not only dependent upon the sample's chemical complexity, including the quantity and type of analytes in it, but also dependent upon the physiological state of the DNA, enzymes or cells at the quantification moment [rs have gained significant attention as effective tools for *in situ* environmental monitoring due to their portability, cost-efficiency, and user-friendly design. These biosensors, based on the biological responses of microorganisms, have proven particularly valuoorganisms as a mechanism (Fig. **1**). The mechanism relies on the metabolism of living cells by using the total capacity for substrate assimilation by microorganisms in place of respiratory metabolic activity, such as in the case discussed by Xu *et al.*, 2011.

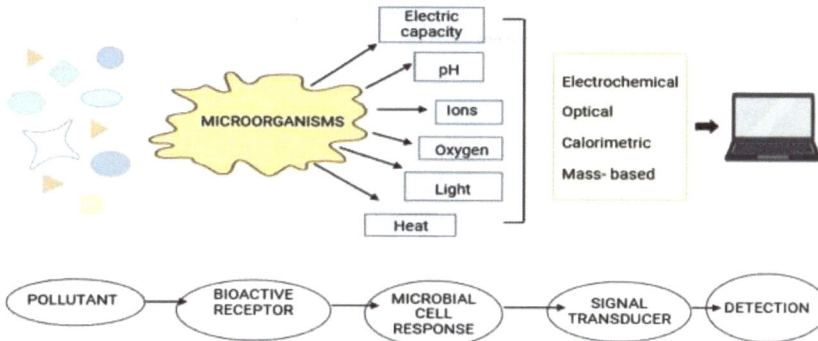

Fig. (1). Working of a microbial biosensor.

One of the significant advantages of microbial biosensors is their adaptability. It is relatively easy to modify these organisms to consume and degrade novel substrates under particular conditions of cultivation. Advances in molecular biology and recombinant DNA technologies have further opened up the possibility for customizing microorganisms. This allows for the enhancement of existing enzyme activities or the expression of foreign enzymes and proteins in host cells [10].

IMMOBILIZATION

For effective transduction of the biochemical response into the physical signal, the recognition element of the microbial cell should be immobilized onto the transducer. Immobilization is also the basis of making microbial biosensors, relying on the proximity between biomaterials and the transducer. Unfortunately, however, different characteristics of microbes and transducers make the proper selection of immobilization techniques difficult. Both viable and nonviable microbial cells can be used in the immobilization technique, having their requirements. Major considerations are, that the immobilization of the microbes should take place on the transducer surface in an efficient and stable manner, using reaction conditions and reagents that are not toxic to viable cells, and operation should be easy in a sterile environment [1, 11]. Immobilization techniques can be broadly categorized into two types, namely passive and active. Various microorganisms naturally tend to grow by attaching to the surfaces, which can be utilized for passive immobilization on different carriers. However, active immobilization techniques involve the use of chemical attachment, gel entrapment, and flocculant agents. Active immobilization can further be divided into two major categories, named chemical and physical [11]. The simplest method for microbial immobilization is physical adsorption as these do not form covalent bonds, causing minimal disruption to their native structure and function, making them ideal for viable cells.

There are two major physical methods. Entrapment is usually obtained by keeping cells in the vicinity of the transducer surface with filter membranes or dialysis, or by chemical/biological polymers and gels like polyvinyl alcohol, carrageenan, polyacrylamide, *etc*. However, one of the major drawbacks of entrapment is that there is extra diffusion resistance due to the material, which results in lowering the sensitivity and detection limits. Chemical methods for immobilizing microbes include cross-linking and covalent binding. In covalent binding, stable covalent bonds are formed between functional groups on components of the cell wall. Cells can be cross-linked directly onto a removable support membrane or on the surface of the transducer, which can then be put on the respective transducer. Cross-linking is especially favoured if the viability of the cells is compromised, and detection involves only intracellular enzymes [10].

SELECTION OF MICRO-ORGANISMS

Selecting an appropriate microorganism for detecting pollutants and their effects in the environment, and integrating it with a suitable transducer, is crucial in developing an environmental biosensor [12]. The chosen microorganism must be robust and capable of detecting specific pollutants at low concentrations to ensure cost-effective detection [13, 14]. Most microbes used in biosensors are single-celled and simple unicellular/ multicellular structures. Bacteria and yeast are the most commonly used microorganisms (Table **1**).

Bacteria

According to Caroline Ajo-Franklin, Ph.D., a biosciences professor at Rice University, "Bacteria rely on their natural sensors to understand what's going on around them". These sensors not only can detect changes but also make the bacteria respond. Researchers now try to exploit these natural responses to create microbial biosensors that can communicate environmental conditions back to humans. Typically, a bacterial cell is genetically engineered to recognize a specific molecule and produce a measurable signal in response. This process involves a compound binding to a receptor on the bacterium, initiating a transcriptional cascade and resulting in the synthesis of proteins with specific purposes [15].

In environmental applications, whole-cell biosensors can monitor nutrient levels and organic compounds in soil to help in crop management. Moreover, bacterial biosensors can be developed to detect other bacteria. For example, a biosensor was recently designed with the bacterium *Acinetobacter baylyi* that could detect cancer cells' DNA both *in vitro* and in a mouse model. Bacterial biosensors are preferred because of their high specificity, sensitivity, versatility, and capability to provide monitoring in real-time [16].

Table 1. Various microorganisms that have been used for detection.

S.NO	Analyte	Micro-organism	Transducer	Detection limit/ Range of detection (ROD)	Reference
1.	Paraoxon	*P. putida* JS444	Clark oxygen electrode	55ppb	[19]
2.	Cd^{2+}	*C. vulgaris*	Conductometric	1ppb	[20]
3.	Cephalosporins	*P. aeruginosa*	Potentiometric	ROD: $0.1f?"11mM$	[21]
4.	Cu^{2+}	*C.* sp.	Voltametric	54nM	[22]
5.	Glucose	*Bacteria consortium*	MFC	25mg/L	[23]
6.	Lysine	*E. coli*	Fluorescent	3 ug/ml	[24]
7.	Tetracyclines, 4-epimer derivatives	*Bacillus cereus*	Bioluminescence	25 ng/g	[25]
8.	Cadmium	Recombinant *E. coli*	Amperometry	25 nM	[26]
9.	Trichloroethylene	*P. aeruginosa* JI104	Chloride ion selective electrode	0.2, 0.06 I1/4M	[27]

Yeast

Many laboratory species of yeast, such as *Saccharomyces cerevisiae*, are classified as Generally Regarded As Safe (GRAS) by the US FDA and have significantly contributed to advances in genomics, transcriptomics, and metabolomics. One of its pioneering applications in environmental science is measuring Biological Oxygen Demand (BOD) in wastewater. Traditional BOD detection methods take about five days, whereas yeast-based biosensors can provide reliable measurements in minutes [17]. Yeast cells, being eukaryotes, share more similarities with human cells than bacterial cells do, making them preferred for toxicity and drug assessments. Additionally, these biosensors can be engineered to detect environmental pollutants and disease markers [18].

Yeast biosensors have also been used to detect the estrogenic activity of pollutants. This involves a plasmid vector with the reporter gene *lacZ* under an inducible promoter with estrogen response elements, integrating the human estrogen receptor gene into the yeast genome. When pollutants interact with the yeast biosensor, the estrogen response elements activate the reporter gene. Additionally, yeast biosensors can detect metals in the environment using specific promoters like the CUP1 promoter, which responds to copper presence [17].

CLASSIFICATION BASED ON WORKING

Optical Biosensors

Optical biosensors are devices that employ optical principles such as bioluminescence, fluorescence, and colorimetry to convert biochemical interactions into output signals. They work based on detecting variations in optical properties like UV-vis absorption, bioluminescence, chemiluminescence, reflectance, and fluorescence, that happen when the biocatalyst interacts with the target analyte. These biosensors have advantages in their compactness, flexibility, immunity to electrical noise, small probe size, and non-destructive nature [10, 28].

Bioluminescence, or light emission by living microorganisms, is an important application in real-time process monitoring. Bioluminescent microbial biosensors have been extensively investigated in the monitoring of bioavailable metals. For instance, *Ralstonia eutropha* AE2515 was engineered by fusing the cnrYXH regulatory genes with the bioluminescent luxCDABE reporter system to produce a whole-cell biosensor for the detection of bioavailable concentrations of Ni^{2+} and Co^{2+} in soil [10].

Fluorescence spectroscopy is a very sensitive analytical technique that can detect very low analyte concentrations, with the emission intensity being directly proportional to concentration. Fluorescent materials and Green Fluorescent Protein (GFP) are widely used in the construction of fluorescent biosensors. Fluorescent biosensors can be classified into two categories: *in vivo* and *in vitro*. *in vivo* types of fluorescent microbial biosensors produce fluorescent compounds such as GFP without further additives, whereas *in vitro* types involve alterations in the environment that modify light intensities. For example, Fiorentino *et al.* (2009) designed a microbial biosensor for measuring aromatic aldehydes through genetically engineered E. coli containing GFP in an aqueous system [1, 29]. Colorimetric microbial biosensors produce colored products that can be quantified and calibrated to corresponding analyte concentrations [30]. Lei *et al.* (2006a) demonstrated a highly sensitive biosensor for microbial toxin detection based on the color response of living fish cells. When cells are exposed to toxins from microbial pathogens, a visible color change occurs, which is proportional to the concentration of the toxin [31].

Electrochemical Biosensors

Electrochemical methods are most used in preparing microbial biosensors that include techniques like potentiometry, amperometry, conductometry, microbial fuel cells, and voltammetry. These biosensors are sensitive, compatible with advanced technologies for micro-fabrication, portable, and draw minimal power [1].

Amperometric microbial biosensors function at a set potential with respect to a reference electrode, sensing the current from reduction or oxidation on the electrode surface. Sugars are essential for all media, so detection sensors for sugars are in high demand. Designs range from modified Clark and microfabricated oxygen electrodes using *S. cerevisiae* and *E. coli* K12 mutants to graphite electrodes with *G. oxydans* and hexacyanoferrate (III) as mediators of redox reaction [10]. Conventional potentiometric microbial biosensors use gas-sensing or ion-selective electrodes coated with microbes that are immobilized. These sensors detect potential changes due to ion accumulation or depletion by the microbes. They offer a wide detection range but require a stable reference electrode. Recently, a potentiometric biosensor was established with a pH electrode that was modified by permeabilized *P. aeruginosa* for the quick detection of cephalosporin antibiotics. Rapid, selective detection of these antibiotic agents was done in by Kumar *et al.* (2008) and Hassan *et al.* (2016). Conductometric microbial biosensors are based on enzymatic reactions where the consumption and production of charged species produce changes in the ionic composition of the sample [1, 32].

These biosensors have many advantages: they employ thin-film electrodes, amenable to miniaturization and mass production, do not require reference electrodes, are insensitive to light, have low power consumption, and can recognize a wide variety of compounds.

Recently, there was the development of a new type of biosensor using SOB that detects water toxicity. SOB oxidizes inorganic sulfur to sulfuric acid; this changes the ionic composition, hence water toxicity can be detected by conductometric biosensors. Voltammetry is an electrochemical analysis technique that is fairly versatile as it measures both current and potential with high sensitivity due to low noise levels. Microbial fuel cells convert chemical energy into electricity through microbial metabolism, whereas changes in electrical current directly indicate the changes in the quality of water. MFCs were recently applied as biosensors to detect acetate with the help of *Geobacter sulfurreducens*. The current by this bacterium was proportional to different acetate concentration levels in the influent [1, 33].

Mass-based Biosensors

Mass-based biosensors measure changes in mass to identify specific analytes. They have high sensitivity and can monitor the process in real-time. These biosensors use a transducer to convert mass changes into electrical signals. The most common types include:

1. Quartz Crystal Microbalance (QCM): This sensor works on the principle of detecting changes in the frequency of a quartz crystal resonator as a function of mass. When the target molecule binds to the sensor surface, the frequency is altered because of added mass.
2. Surface Acoustic Wave (SAW) Sensors: These sensors use the measurement of the velocity or amplitude of acoustic waves that travel along the surface of a piezoelectric material to detect mass changes. The properties of the wave are changed by the binding of analytes to the sensor surface, which can be measured.

Both SAW and QCM sensors are very efficient in terms of their high sensitivity and real-time delivery, making them useful in all types of analytical applications [34, 35].

Calorimetric Biosensors

Calorimetric microbial biosensors detect specific analytes by measuring the heat change from biochemical reactions, that is directly proportional to the concentration of the respective analyte. The main calorimetric techniques include:

1. Isothermal Calorimetry: Measures heat changes at a constant temperature, providing real-time data on microbial activity and growth. For example, Isothermal Microcalorimetry (IMC) has been used to monitor microbial activity in contaminated soil by measuring heat flow from microbial metabolism [36].
2. Adiabatic Calorimetry: Ensures no heat exchange with the environment, making it highly accurate for measuring enthalpy changes. This technique can study the thermodynamics of microbial reactions, such as the heat produced by *Saccharomyces cerevisiae* during glucose breakdown, providing precise measurements of reaction enthalpy and enzyme activity [37].
3. Heat Conduction Calorimetry: Involves heat transfer from the reaction vessel to the environment. This method has been used to detect pathogens in water samples by measuring the metabolic heat produced by *Escherichia coli*, offering a rapid and sensitive approach for pathogen detection [38, 39] (Table **2**).

Table 2. Overview of different biosensors.

S.NO	Biosensor Type	Transducer	Advantages	Disadvantages	Reference
1.	Optical	Bioluminescence Fluorescence Colorimetric	High sensitivity and specificity, non-invasive, real-time monitoring, multi-channel sensing, electrically passive.	Complex instrumentation, light interference.	[1, 40]
2.	Electrochemical	Amperometry Potentiometry Conductometry Voltammetry Microbial fuel cells	High-sensitivity, cost-effective, miniaturization, compatible with modern microfabrication technologies.	Limited selectivity, maintenance, and interferences related to electrochemical reactions.	[41]
3.	Mass-based	Quartz Crystal Microbalance (QCM) Surface Acoustic Wave (SAW) sensors	Direct measurement, versatility, simplicity.	Complex setup, environmental sensitivity.	[42]
4.	Calorimetric	Isothermal Adiabatic Heat Conduction	Label-free detection, high throughput, continuous monitoring.	Less sensitivity, and heat management.	[43]

ENVIRONMENTAL MONITORING

Soil

Soil is vital in human life, as it provides water and nutrients to plants, which are essential in our food supply chain. Soil pollution, caused both by natural processes and man's activities, reduces productivity in the soil through adverse effects on its physical, chemical, and biological properties. This includes natural sources such as volcanic eruptions and rocks weathering, which are major contributors. Others include industrial processes, agricultural activities, improper spills, and waste disposal. Soil pollution affects the human body and the environment since it causes contamination of ecosystems, water quality, and productivity of agricultural systems. Several guidelines regarding soil pollution detection and its control have been released by institutions like the European Environment Agency (EEA) and the United States Environmental Protection Agency (EPA). Some of the most recommended guidelines include SSL or the set soil screening limits, usage control for fertilizers, pesticides, and so forth [44, 45].

Soil Organic Matter (SOM) is crucial for the reduction of soil pollution since it adsorbs pollutants like heavy metals and organic contaminants, thus reducing their mobility and bioavailability. High SOM levels improve the structure of the soil, reduce erosion, and retain pollutants within certain areas. Monitoring SOM is important for maintaining the health of the soil [8]. Mishra and Saini.,2024, designed a fluorescence-based microbial biosensor with Pseudomonas fluorescens for soil humic substance analysis. This biosensor responds by giving a fluorescent signal as a function of the humic substance concentration. Therefore, it provides a fast method of determining the levels of SOM. An enzyme-based electrochemical biosensor with Escherichia coli has been designed to measure the levels of soil respiration as carbon dioxide production from samples of soil. This method gives real-time data regarding microbial activity and SOM content, thus helping to evaluate soil health and organic matter decomposition [45]. Fig. (**2**). provides a graphical representation of the environmental monitoring parameters.

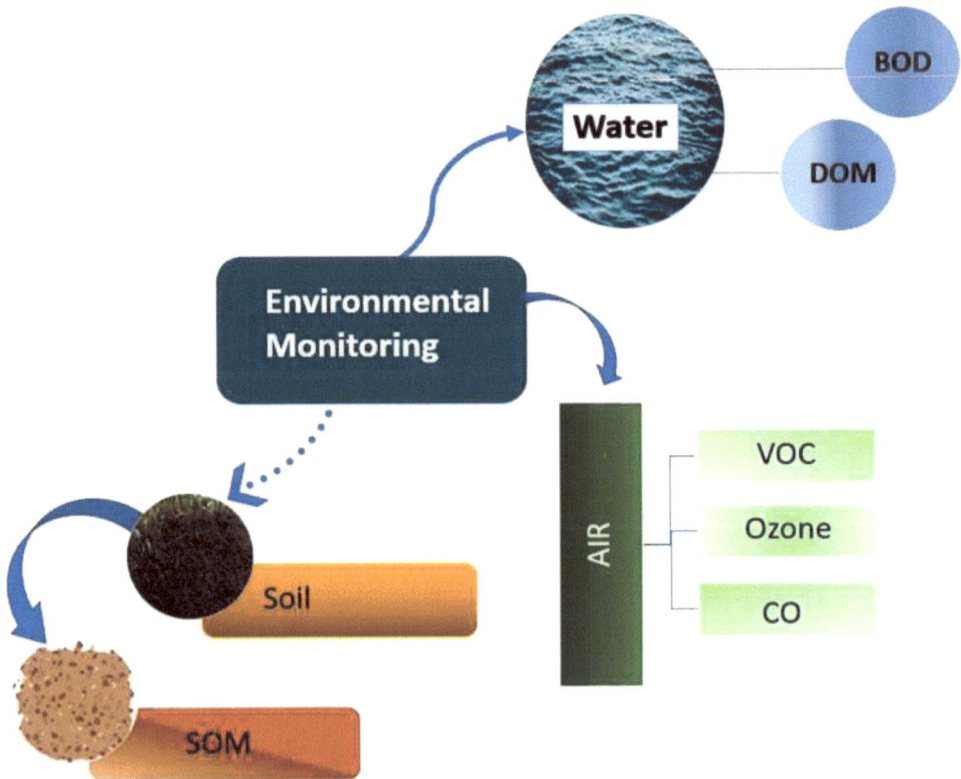

Fig. (2). Detection parameters used for environmental monitoring.

Water

Water pollution is one of the most critical worldwide issues that require continuous monitoring of water resource policies in an effort to tackle the problems effectively. All factors including climate, geology, soil, vegetation, rainfall, groundwater, human activities, and flow conditions have an influence on water quality. However, the major sources that affect water quality are point sources from industries and municipalities. Pollutants like herbicides, pesticides, industrial chemicals, and pathogens pose significant risks to water quality. Several organizations, including the United States Environmental Protection Agency (EPA), the World Health Organization (WHO), and the European Environment Agency (EEA), have established guidelines and regulations to address these concerns. This includes Guidelines for Drinking Water Quality (GDWQ), National Primary Drinking Water Regulations (NPDWRs), and the Water Framework Directive (WFD) [46].

Biochemical Oxygen Demand (BOD) and Dissolved Organic Matter (DOM) are crucial indicators for assessing water pollution. BOD measures the oxygen required by bacteria to decompose organic matter in water, with high levels indicating a significant presence of organic pollutants that can deplete oxygen and harm aquatic life [47]. DOM consists of organic molecules from decomposed plant and animal material, playing a vital role in the carbon cycle and influencing water quality by affecting nutrient availability and pollutant behavior [48]. Elevated DOM levels can increase microbial activity, subsequently raising BOD levels and further depleting oxygen in the water, thus highlighting the interconnectedness of these factors in maintaining aquatic ecosystem health. In 2011, scientists from New Zealand Webber and his colleagues tested various bacterial strains for fast and accurate BOD sensing using their product MICREDOX assay, pointing out that *Arthrobacter globiformis* could be applied as a workable strain for measurement of BOD [49]. Ivandini and Einaga reported on an oxygen-sensitive gold-modified boron-doped diamond electrode for amperometric measurement of BOD utilizing *Rhodotorula* sp. Yeast. Other than the heterotrophic microbes, there are also effective measures for estimating biodegradable DOM by using eukaryotic microbes like fungi-containing yeasts that can assimilate organic matter which is otherwise hard to degrade. In 2014, Chen *et al.* investigated the rapid biodegradation of several insoluble fine fibers in effluent, choosing fungi which are marine-derived, *Penicillium janthinellum* P1 and *Pestalotiopsis* sp. J63 as potential biorecognition elements for measuring biodegradable DOMs [8, 50].

Air

Although natural events, such as volcanoes and fires, can emit pollutants into the environment, human activities remain the main source of air pollution. Air pollutants are substances that can damage humans, animals, vegetation, or materials. These include ozone (O_3), nitrogen oxides (NO_x), Carbon Monoxide (CO), particulate matter (PM2.5 and PM10), Volatile Organic Compounds (VOCs), sulfur dioxide (SO_2), and heavy metals. They differ in their chemical components, their reactivity, sources of emission, degradation time, and even their ability to travel considerable distances. Air pollution affects human health both acutely and chronically, affecting various organs and systems in the body. Health effects span from minor respiratory irritation to chronic conditions such as chronic respiratory diseases, lung cancer, heart diseases, acute respiratory infections in children, and exacerbation of previously existing heart and lung disease, including asthma attacks and chronic bronchitis in adults. The United States Environmental Protection Agency (EPA) manages air pollution control through the Clean Air Act (CAA), which includes the National Ambient Air Quality Standards (NAAQS), National Emission Standards for Hazardous Air Pollutants (NESHAPs), and New Source Performance Standards (NSPS) [51]. Moreover, the World Health Organization (WHO) has developed several air quality guidelines to manage and reduce air pollution around the world [52]. With increasing organic pollution, organic matter accumulates at the bottom of water bodies, shifting the environment from aerobic to anaerobic as decomposition progresses.

This process releases methane gas from the water's surface, which makes methane measurement a good indicator of organic pollution levels. In 2015, Zarei and Farahbakhsh from Iran developed an optical methane biosensor based on methanotrophic bacteria and gold nanoparticles for detecting methane through a gas flow system [8, 53] (Fig. **2**).

GENETICALLY ENGINEERED BIOSENSORS AND BIOSECURITY

Recombinant DNA technology has been utilized to modify microorganisms, incorporating natural regulatory genes encoding the transcriptional regulators and with promoters and bio-recognition genes. This creates the basis for biosensors capable of detecting targeted compounds. Sayler and Ripp (2000) were pioneers in demonstrating that genetically modified microorganisms really could be used practically as bioremediation components [54]. In non-specific biosensors, the toxicity of a target compound is determined by monitoring a decrease in the activity of a reporter protein through the integration of a reporter gene with a constitutive promoter. More specific biosensors operate on transcriptional regulators that are activated by specific target chemicals. They interact with a constitutive promoter to generate a dose-dependent signal from a reporter gene [55]. The recombinant microbial biosensors are developed from genes linking pollutant-activatable, which consists of the transcriptional regulator as well as a promoter/operator element linked to a promoterless reporter gene. Recombinant genes are capable of being inserted either in microorganisms as a plasmid or being inserted directly into their chromosome. When the target chemical or specific physiological condition is present, the transcriptional regulator activates the promoter, which results in the expression of the reporter gene and subsequently the generation of a measurable signal. The specificity and sensitivity of these biosensors are mainly based on how effectively the regulatory gene responds to its intended target [56]. Commonly used reporter genes include lux/luc, which are genes encoding bacterial or firefly luciferase, gfp, encoding green fluorescent protein, and lacZ encoding Iý-galactosidase.

Gennaro *et al.* (2011) developed a system, making use of two genetically engineered strains of E. coli that was used to monitor benzene pollution in environmental air samples coming from an oil refinery, exemplifying the practical application of microbial biosensors in detecting environmental pollutants [57]. Considering the high risks associated with endocrine-disrupting chemicals to human health and aquatic life, Cevenini *et al.* (2017) developed a sensitive, portable cell-based device. The cell-based device used a new yeast-estrogen screen (nanoYES) and a mini-camera as a light detector to identify such compounds. The scientists genetically engineered *Saccharomyces cerevisiae* cells with a yeast codon-optimized variant of NanoLuc luciferase (yNLucP) and prepared ready-to-use 3D-printed cartridges with immobilized cells. This portable analytical platform allows a quick and quantitative estimation of the total estrogenic activity of 17Iý-estradiol in samples with a detection limit of 0.08 nM. This highly sensitive yeast biosensor proved effective in water samples spiked with a variety of chemicals having estrogen-like activity. Therefore, this biosensor can be widely used for on-site monitoring of endocrine disruptors as well as other micropollutants in aquatic systems. Microbial biosensors have in recent times been identified with the potential to improve biosecurity [58].

The World Health Organization defines biosecurity as the state of being free from the accidental or intentional release of biological agents or toxins. It embraces the prevention of misuse of microorganisms and biotechnology, as well as controlling infectious disease outbreaks. Microbial biosensors have been used effectively in detecting waterborne pathogens, preventing bioterrorism, and ensuring food safety. Anthrax is essentially an herbivore disease. Human infection occurs through contact with infected animals or products. *Bacillus anthracis* is the causative agent, a Gram-positive, spore-forming bacterium. It exists in both vegetative and spore forms. The spores are highly resistant and can survive extreme conditions for decades. Traditional detection of *B. anthracis* is by culturing samples, but biosensors made from microbes offer faster alternatives. In 2010, Guo *et al.* presented a phage display-based biosensor to meet this need. Phage display is based on the insertion of a foreign gene into a phage genome, which results in displaying the obtained protein on the phage surface [59]. This system is cheap and easy to engineer. The authors employed one chain variable fragment (scFv) derived from monoclonal antibodies specific to B. anthracis in order to create a bifunctional biosensor. The scFv was joined with the phage protein pIII for coexpression and linked with a gold-binding peptide that used the phage protein pVIII. This engineered phage could detect B. anthracis specifically and sensitively with the help of the scFv and bind gold nanoparticles for signal amplification. This allowed the detection by observation [60, 61].

CHALLENGES

With regard to these microbial-derived biosensors, innovative developments are seen because of their efficient monitoring capacities in terms of environmental contaminant detection. In most remote or isolated areas where the transportation of test samples to laboratories is not very feasible, creating robust whole-cell biosensors for such detection, especially of toxic pollutants, emerging contaminants, and heavy metals, is an extremely beneficial process. Although numerous kinds of microbial biosensors have been developed, very few have been applied to *in situ* monitoring. Most of these microbial biosensors have a concentration detection limit below 0.1 I1/4M and delayed response times because reporter gene expression requires some time to produce the response signal, thereby hindering real-time monitoring. Researchers are improving the response times of these biosensors as well as sensitivity and specificity by using more sensitive promoters, refining host strains, designing strains capable of producing modifying enzymes, employing proteins bound to periplasm or surface-expressed, and monitoring the physiology of bacteria [1].

Maintenance of activity in complex environments and viability of cells that may not supply sufficient nutrients or carry inhibitory compounds represents another problem, which might lead to an underestimation of toxicants. Viability issues are probably less severe for disposable biosensors aimed at short-term on-site monitoring but certainly represent critical concerns for continuous-use online biosensors. Different species and strains exhibit variability in activity, and activity variations occur even within the same species or strain based on physiological status during and following preservation. Hence, careful choice of appropriate microorganisms and very strict control over cell physiology is necessary for the biosensor development process. Ideally, a biosensor must retain its activity under the target field conditions, especially in harsh environments that include extreme temperatures, the existence of complex organic solvents, and highly alkaline or acidic pH [56].

Public concerns about releasing genetically modified microbial biosensors in the environment may even restrain further use. However, the potential risk is in reality pretty low, mostly because many biosensor cells are bound to matrices or kept within compartments, whereas environmental samples are applied in biosensors. The practical advantage can be much greater compared with this potential risk.

CONCLUSION

Microorganisms are highly suitable as biosensing elements for constructing biosensors that detect environmental pollutants. During the past two decades, microbial-derived biosensors have gained a lot of attention from researchers due to advancements in environmental detection and biomedical diagnostics. The most commonly used microorganisms in these biosensors are bacteria and yeast, which can be classified into four major types based on their working principles. Each type has its advantages and disadvantages and is used for various toxicity measurements and monitoring environmental pollution by measuring parameters, such as Dissolved Organic Matter (DOM), Biochemical Oxygen Demand (BOD), and Suspended Organic Matter (SOM). Microbial biosensors are preferred over traditional chemical biosensors due to their high sensitivity and ease of use.

Recently, these biosensors have also been applied to the field of biosecurity for concerns regarding the use of microorganisms as bioterrorist agents, like in the case of anthrax. It is therefore essential to be knowledgeable about the operating principle of these biosensors to effectively analyze the information gathered. Nanotechnology has also led to the incorporation of nanostructured materials in biosensing because of their great biocompatibility, amplified electron transfer characteristics, and higher surface area.

Genetically engineered biosensors have improved specificity for analyte detection. It utilizes advancements in recombinant DNA technology, enabling such biosensors to be targeted for high precision at detecting particular target compounds. The pollutant-responsive genes can be combined with reporter genes to develop very sensitive and specific biosensors.

However, there are a number of limitations that must be addressed before these microbial biosensors can be broadly used for real-time environmental applications and commercialization. Some of the main limitations include increasing sensitivity, selectivity, and maintainable viability of cells along with functionality under non-native environmental conditions. Thus, further developments in microbial biosensor technologies are required.

REFERENCES

[1] Hassan SHA, Van Ginkel SW, Hussein MAM, Abskharon R, Oh SE. Toxicity assessment using different bioassays and microbial biosensors. Environ Int 2016; 92-93: 106-18.
[http://dx.doi.org/10.1016/j.envint.2016.03.003] [PMID: 27071051]

[2] Huang CW, Lin C, Nguyen MK, Hussain A, Bui XT, Ngo HH. A review of biosensor for environmental monitoring: principle, application, and corresponding achievement of sustainable development goals. Bioengineered 2023; 14(1): 58-80.
[http://dx.doi.org/10.1080/21655979.2022.2095089] [PMID: 37377408]

[3] Palchetti I, Mascini M. Biosensor technology: A brief history. 2010; pp. 15-23.
[http://dx.doi.org/10.1007/978-90-481-3606-3_2]

[4] Batzias F, Siontorou CG. A novel system for environmental monitoring through a cooperative/synergistic scheme between bioindicators and biosensors. J Environ Manage 2007; 82(2): 221-39.
[http://dx.doi.org/10.1016/j.jenvman.2005.12.023] [PMID: 16569474]

[5] Shah M, Shah N. Environmental approach to remediate refractory pollutants from industrial wastewater treatment plant 2024. Available from: https://books.google.com/books?hl=en&lr=&id=eAvmEAAAQBAJ&oi=fnd&pg=PP1&dq=Environmental+Approach+to+Remediate+Refractory+Pollutants+from+Industrial&ots=LbgQjREMND&sig=Wd5OEbDzrzaXAynMurbBCr5bu6I

[6] Shah M. Application of sewage sludge in industrial wastewater treatment 2024. Available from: https://books.google.com/books?hl=en&lr=&id=M6X7EAAAQBAJ&oi=fnd&pg=PA1&dq=Application+of+Sewage+Sludge+in+Industrial+Wastewater+Treatment&ots=-WsC0IdrBX&sig=mEoy5cLyvDqcBKBfjhB_4qq35fQ
[http://dx.doi.org/10.1002/9781119857396]

[7] "Bacterial Biosensors: The Future of Analyte Detection".

[8] H. Nakamura. "Current status of water environment and their microbial biosensor techniques – Part II: Recent trends in microbial biosensor development". Anal Bioanal Chem 2018; 410(17): 3967-89.
[http://dx.doi.org/10.1007/s00216-018-1080-0] [PMID: 29736704]

[9] Eltzov E, Marks RS. Whole-cell aquatic biosensors. Anal Bioanal Chem 2011; 400(4): 895-913.
[http://dx.doi.org/10.1007/s00216-010-4084-y] [PMID: 20835820]

[10] Capin J, Chabert E, Zuñiga A, Bonnet J. Microbial biosensors for diagnostics, surveillance and epidemiology: today's achievements and tomorrow's prospects. Microb Biotechnol 2024; 17(11): e70047.
[http://dx.doi.org/10.1111/1751-7915.70047]

[11] Xu X, Ying Y. Microbial biosensors for environmental monitoring and food analysis. Food Rev Int 2011; 27(3): 300-29.
[http://dx.doi.org/10.1080/87559129.2011.563393]

[12] Bacterial Biosensors: The Future of Analyte Detection.

[13] Jain H. From pollution to progress: Groundbreaking advances in clean technology unveiled. Innovation and Green Development 2024; 3(2): 100143.
[http://dx.doi.org/10.1016/j.igd.2024.100143]

[14] Bhardwaj LK, *et al.* COVID-19 and the interplay with antibacterial drug resistance Available from: https://services.igi-global.com/resolvedoi/resolve.aspx?doi=10.4018/979-8-3693-4139-1.ch010
[http://dx.doi.org/10.4018/979-8-3693-4139-1.ch010]

[15] Murphy L. Biosensors and bioelectrochemistry. Curr Opin Chem Biol 2006; 10(2): 177-84.

[http://dx.doi.org/10.1016/j.cbpa.2006.02.023] [PMID: 16516536]

[16] Mishra P, Saini P. Microbial biosensors: Design, types and applications. Bioprospecting of microbial resources for agriculture, environment and bio-chemical industry. Cham: Springer Nature Switzerland 2024; pp. 153-61.
[http://dx.doi.org/10.1007/978-3-031-63844-2_9]

[17] Dhakal S, Macreadie I. The use of yeast in biosensing. Microorganisms 2022; 10(9): 1772.
[http://dx.doi.org/10.3390/microorganisms10091772] [PMID: 36144374]

[18] Walmsley RM, Keenan P. The eukaryote alternative: Advantages of using yeasts in place of bacteria in microbial biosensor development. Biotechnol Bioprocess Eng; BBE 2000; 5(6): 387-94.
[http://dx.doi.org/10.1007/BF02931936]

[19] Lei Y, Mulchandani P, Chen W, Mulchandani A. Direct determination of *p*-nitrophenyl substituent organophosphorus nerve agents using a recombinant *Pseudomonas putida* JS444-modified Clark oxygen electrode. J Agric Food Chem 2005; 53(3): 524-7.
[http://dx.doi.org/10.1021/jf048943t] [PMID: 15686397]

[20] Guedri H, Durrieu C. A self-assembled monolayers based conductometric algal whole cell biosensor for water monitoring. Mikrochim Acta 2008; 163(3-4): 179-84.
[http://dx.doi.org/10.1007/s00604-008-0017-2]

[21] Kumar S, Kundu S, Pakshirajan K, Dasu VV. Cephalosporins determination with a novel microbial biosensor based on permeabilized *Pseudomonas aeruginosa* whole cells. Appl Biochem Biotechnol 2008; 151(2-3): 653-64.
[http://dx.doi.org/10.1007/s12010-008-8280-6] [PMID: 18551255]

[22] Alpat S, Alpat S, Cadirci B, Yasa I, Telefoncu A. A novel microbial biosensor based on *Circinella* sp. modified carbon paste electrode and its voltammetric application. Sens Actuators B Chem 2008; 134(1): 175-81.
[http://dx.doi.org/10.1016/j.snb.2008.04.044]

[23] Gangwar R, Ray D, Rao K T, *et al.* Plasma functionalized carbon interfaces for biosensor application: toward the real-time detection of *Escherichia coli* O157: H7. ACS Omega 2022; 7(24): 21025-34.
[http://dx.doi.org/10.1021/acsomega.2c01802]

[24] Chalova VI, Zabala-DA-az IB, Woodward CL, Ricke SC. Development of a whole cell green fluorescent sensor for lysine quantification. World J Microbiol Biotechnol 2008; 24(3): 353-9.
[http://dx.doi.org/10.1007/s11274-007-9479-3]

[25] Virolainen NE, Pikkemaat MG, Elferink JWA, Karp MT. Rapid detection of tetracyclines and their 4-epimer derivatives from poultry meat with bioluminescent biosensor bacteria. J Agric Food Chem 2008; 56(23): 11065-70.
[http://dx.doi.org/10.1021/jf801797z] [PMID: 18998699]

[26] Biran I, Babai R, Levcov K, Rishpon J, Ron EZ. Online and *in situ* monitoring of environmental pollutants: electrochemical biosensing of cadmium. Environ Microbiol 2000; 2(3): 285-90.
[http://dx.doi.org/10.1046/j.1462-2920.2000.00103.x] [PMID: 11200429]

[27] Han T, Sasaki S, Yano K, *et al.* Flow injection microbial trichloroethylene sensor. Talanta 2002; 57(2): 271-6.
[http://dx.doi.org/10.1016/S0039-9140(02)00027-9] [PMID: 18968627]

[28] Vogrinc D, Vodovnik M, Marinsek-Logar R. Microbial biosensors for environmental monitoring. Acta Agric Slov 2015; 106(2)
[http://dx.doi.org/10.14720/aas.2015.106.2.1]

[29] Fiorentino G, Ronca R, Bartolucci S. A novel E. coli biosensor for detecting aromatic aldehydes based on a responsive inducible archaeal promoter fused to the green fluorescent protein. Appl Microbiol Biotechnol 2009; 82(1): 67-77.
[http://dx.doi.org/10.1007/s00253-008-1771-0] [PMID: 18998120]

[30]   Su L, Jia W, Hou C, Lei Y. Microbial biosensors: A review. Biosens Bioelectron 2011; 26(5): 1788-99.
[http://dx.doi.org/10.1016/j.bios.2010.09.005] [PMID: 20951023]

[31]   Lei Y, Chen W, Mulchandani A. Microbial biosensors. Anal Chim Acta 2006; 568(1-2): 200-10.
[http://dx.doi.org/10.1016/j.aca.2005.11.065] [PMID: 17761261]

[32]   Shideler S, Bookout T, Qasim A, *et al.* Biosensor-guided detection of outer membrane-specific antimicrobial activity against *Pseudomonas aeruginosa* from fungal cultures and medicinal plant extracts. Microbiol Spectr 2023; 11(6): e01536-23.
[http://dx.doi.org/10.1128/spectrum.01536-23]

[33]   Tront JM, Fortner JD, PlAtze M, Hughes JB, Puzrin AM. Microbial fuel cell biosensor for *in situ* assessment of microbial activity. Biosens Bioelectron 2008; 24(4): 586-90.
[http://dx.doi.org/10.1016/j.bios.2008.06.006] [PMID: 18621521]

[34]   Tetyana P, Morgan Shumbula P, Njengele-Tetyana Z. Biosensors: Design, development and applications. Nanopores. IntechOpen 2021.
[http://dx.doi.org/10.5772/intechopen.97576]

[35]   What are biosensors: definition, examples, types and applications?

[36]   Braissant O, Wirz D, GApfert B, Daniels AU. Use of isothermal microcalorimetry to monitor microbial activities. FEMS Microbiol Lett 2010; 303(1): 1-8.
[http://dx.doi.org/10.1111/j.1574-6968.2009.01819.x] [PMID: 19895644]

[37]   Fessas D, Schiraldi A. Isothermal calorimetry and microbial growth: beyond modeling. J Therm Anal Calorim 2017; 130(1): 567-72.
[http://dx.doi.org/10.1007/s10973-017-6515-x]

[38]   Morazzoni C, Sirel M, Allesina S, *et al.* Proof of concept: real-time viability and metabolic profiling of probiotics with isothermal microcalorimetry. Front Microbiol 2024; 15: 1391688.
[http://dx.doi.org/10.3389/fmicb.2024.1391688]

[39]   Isothermal microcalorimetry to monitor bacteria growth - TA instruments.

[40]   Advantages of Biosensors | Disadvantages of Biosensors.

[41]   Wang K, Lin X, Zhang M, Li Y, Luo C, Wu J. Review of electrochemical biosensors for food safety detection. Biosensors (Basel) 2022; 12(11): 959.
[http://dx.doi.org/10.3390/bios12110959] [PMID: 36354467]

[42]   Zhou Q A, Zheng C, Zhu L, Wang J. A review on rapid detection of modified quartz crystal microbalance sensors for food: contamination, flavour and adulteration. TrAC Trends Anal Chem 2022; 157: 116805.
[http://dx.doi.org/10.1016/j.trac.2022.116805]

[43]   Lee W, Lee , Koh . Development and applications of chip calorimeters as novel biosensors. Nanobiosensors in Disease Diagnosis 2012; (Apr): 17.
[http://dx.doi.org/10.2147/NDD.S26438]

[44]   A complete guide to soil pollution: causes, effects, and solutions | Atlas Scientific.

[45]   A complete guide to soil pollution: causes, effects, and solutions | Atlas Scientific. Available from: https://greencoast.org/prevention-of-soil-pollution/

[46]   Guidelines for drinking-water quality, 4th edition, incorporating the 1st addendum.

[47]   Biochemical Oxygen Demand (BOD) and Water | U.S. Geological Survey.

[48]   5.2 Dissolved Oxygen and Biochemical Oxygen Demand | Monitoring & Assessment | US EPA.

[49]   Ivandini TA, Saepudin E, Wardah H, Harmesa N, Dewangga N, Einaga Y. Development of a biochemical oxygen demand sensor using gold-modified boron doped diamond electrodes. Anal Chem

2012; 84(22): 9825-32.
[http://dx.doi.org/10.1021/ac302090y] [PMID: 23088708]

[50]   Chen H, Wang M, Shen Y, Yao S. Optimization of two-species whole-cell immobilization system constructed with marine-derived fungi and its biological degradation ability. Chin J Chem Eng 2014; 22(2): 187-92.
[http://dx.doi.org/10.1016/S1004-9541(14)60024-0]

[51]   Regulatory and Guidance Information by Topic: Air | US EPA.

[52]   Kampa M, Castanas E. Human health effects of air pollution. Environ Pollut 2008; 151(2): 362-7.
[http://dx.doi.org/10.1016/j.envpol.2007.06.012] [PMID: 17646040]

[53]   Poma N, Bonini A, Vivaldi F, *et al.* Biosensing systems for the detection and quantification of methane gas. Appl Microbiol Biotechnol 2023; 107(18): 5627-34.
[http://dx.doi.org/10.1007/s00253-023-12629-7] [PMID: 37486352]

[54]   Sayler GS, Ripp S. Field applications of genetically engineered microorganisms for bioremediation processes. Curr Opin Biotechnol 2000; 11(3): 286-9.
[http://dx.doi.org/10.1016/S0958-1669(00)00097-5] [PMID: 10851144]

[55]   Bilal M, Iqbal HMN. Microbial-derived biosensors for monitoring environmental contaminants: Recent advances and future outlook. Process Saf Environ Prot 2019; 124: 8-17.
[http://dx.doi.org/10.1016/j.psep.2019.01.032]

[56]   Shin HJ. Genetically engineered microbial biosensors for *in situ* monitoring of environmental pollution. Appl Microbiol Biotechnol 2011; 89(4): 867-77.
[http://dx.doi.org/10.1007/s00253-010-2990-8] [PMID: 21063700]

[57]   Di Gennaro P, Bruzzese N, Anderlini D, *et al.* Development of microbial engineered whole-cell systems for environmental benzene determination. Ecotoxicol Environ Saf 2011; 74(3): 542-9.
[http://dx.doi.org/10.1016/j.ecoenv.2010.08.006] [PMID: 20980054]

[58]   Cevenini L, Lopreside A, Calabretta MM, *et al.* A novel bioluminescent NanoLuc yeast-estrogen screen biosensor (nanoYES) with a compact wireless camera for effect-based detection of endocrine-disrupting chemicals. Anal Bioanal Chem 2018; 410(4): 1237-46.
[http://dx.doi.org/10.1007/s00216-017-0661-7] [PMID: 28965124]

[59]   Guo Y, Liang X, Zhou Y, *et al.* Construction of bifunctional phage display for biological analysis and immunoassay. Anal Biochem 2010; 396(1): 155-7.
[http://dx.doi.org/10.1016/j.ab.2009.08.026] [PMID: 19699710]

[60]   Hammond E, Li CP, Ferro V. Development of a colorimetric assay for heparanase activity suitable for kinetic analysis and inhibitor screening. Anal Biochem 2010; 396(1): 112-6.
[http://dx.doi.org/10.1016/j.ab.2009.09.007] [PMID: 19748475]

[61]   Wang DB, Cui MM, Li M, Zhang XE. Biosensors for the detection of *Bacillus anthracis*. Acc Chem Res 2021; 54(24): 4451-61.
[http://dx.doi.org/10.1021/acs.accounts.1c00407] [PMID: 34846836]

CHAPTER 11

Case Studies and Practical Applications of Microbial Technologies

Shivali Pal[1], Harshita Jain[1,*], Anamika Shrivastava[1], Maya Kumari[2] and Renu Dhupper[1]

[1] *Amity Institute of Environmental Sciences, Amity University, Noida, Gautam Budh Nagar, Uttar Pradesh-201313, India*

[2] *Amity School of Natural Resources and Sustainable Development, Amity University Uttar Pradesh, Noida, Uttar Pradesh, India*

Abstract: This chapter studies the transformative usefulness of microbial technologies in real-world considering their application in various environments. Case studies on microbial bioremediation of micropollutants, heavy metals, petroleum-based pollutants, and pesticides are discussed in context illustrating their effectiveness in attaining pollution control. The chapter also discusses the application of microbial waste to energy systems as shown by the potential contribution to the production of sustainable energy and waste management. Besides, it investigates microbial biosensors' development and deployment for environmental monitoring and its precision in detecting and minimizing pollution. This chapter draws together theoretical concepts and practical applications to emphasize the central importance of microbial technologies for environmental restoration and sustainable practices.

Keywords: Biosensors, Environmental monitoring, Microbial bioremediation, Waste-to-energy.

INTRODUCTION

Modern environmental science has relied on microbial technologies as a cornerstone. These are sustainable solutions to some of the most pressing ecological challenges of our time. The contribution of microorganisms in biogeochemical cycles, pollutant degradation, resource recovery, and energy production is essential owing to their diverse metabolic potential. The use of these technologies in innovative microbial tools has had a great impact on pollution control, waste management, and environmental monitoring [1]. The use of microorganisms has been termed microbial technologies and associated metabolic

---

* **Corresponding author Harshita Jain:** Amity Institute of Environmental Sciences, Amity University, Noida, Gautam Budh Nagar, Uttar Pradesh- 201313, India; E-mail: hjain@amity.edu

Harshita Jain & Maulin P. Shah (Eds.)

processes are applied as a solution to overcome environmental challenges. These are technologies that tap into the microbial tendency to degrade or to transform or detoxify, harmful substances in the environment. Different microbial systems are used in many different contexts including emission monitoring, remediation of soils, tokenizer, and treatment of wastewater [2, 3]. The exceptional adaptability and efficiency of microbes are dictated by the phenomenal diversity of microbial species, each with a uniquely matched suite of species that can metabolize specific compounds under diverse ecological conditions. For example, microbes that frequently use *Pseudomonas* and *Bacillus* species with capacity for hydrocarbon degradation and heavy metal detoxification, respectively, are extensively used for bioremediation. All the same, the anaerobic digestion process relies on methanogenic archaea converting organic waste into biogas - a renewable energy source. Such microbial processes have become understood and applied, offering environmental management the opportunity to reduce reliance on chemical treatments and transition to more eco-friendly alternatives [4].

The great versatility of microbial technologies means that they can be used to address wide regimes of environmental issues. Microbial consortia are used in wastewater treatment to degrade organic matter and reduce nitrogenous compounds improving water quality for reuse or discharge. The bioaugmentation and biostimulation techniques in contaminated soils capitalize on microbial activity to degrade hydrocarbons and reduce the toxicity of heavy metals. Real-time monitoring of pollutants has also been facilitated with the development of microbial biosensors for decision-making [5]. Furthermore, microbial interventions have become more necessary as a consequence of the growing problem of micropollutants including pharmaceuticals and personal care products. These persistent compounds are often hard to remove by conventional treatment systems, with these compounds accumulating in aquatic environments. However, utilizing microbial technologies by deploying species that can degrade these micropollutants into nontoxic metabolites to protect ecosystems and public health, is a promising solution.

The practical implications of microbial technologies can be better understood with case studies. Empirical evidence is provided for how they work, what they work on, and what scaling up might look like in different settings. The focus of this chapter is on the application of detailed case studies that not only validate the theoretical underpinnings of microbial applications but also the adaptability of microbial applications to real-world challenges. Various contexts have been included in these case studies ranging from micropollutants for bioremediation, to heavy metals, petroleum-based pollutants, and pesticides. The specific microbial species used, environments optimized, and outcomes obtained are provided for each case [6]. For example, the contaminated soils with petroleum hydrocarbons can bioremediate and the same applies to the hydrocarbonoclastic bacteria's role in restoring soil health. Like this, microbial strategies for heavy metal detoxification, such as the use of sulfate-reducing bacteria in mine tailings, show us how microbes can immobilize or transform any toxic metals into less harmful forms. In these examples, we demonstrate the potential of microbial technologies to manage pollution at its source and minimize impacts on the secondary environment [7].

There are very few areas as promising as waste-to-energy systems, where microbial processes are used to convert organic waste into renewable energy. Microbial biodegradation of organic matter under anaerobic conditions by a consortium of microorganisms is well established and involves the production of biogas from a mixture of methane and carbon dioxide [8 - 10]. Besides mitigating waste accumulation, this process can also act as a sustainable energy source that can lessen the need for fossil fuels. Case studies of waste-to-energy systems show how microbial technologies are being adapted to maximize energy yields and increase the number of feedstocks that are amenable to use. Microbial technologies can convert diverse waste streams such as municipal solid waste into agricultural residues that turn into valuable energy products in an efficient manner. In this frontier, the integration of microbial fuel cells that directly convert organic matter into electricity represents a promising avenue toward a sustainable energy generation pathway while tackling waste management problems [11].

Microbial biosensors represent another critical application of microbial technologies, providing tools for the precise detection and quantification of environmental pollutants. These biosensors utilize microbial cells or enzymes as biological recognition elements, coupled with transducers that convert biochemical signals into measurable outputs. Their high specificity, rapid response times, and ability to operate under diverse environmental conditions make them indispensable for monitoring water, soil, and air quality. Real-world applications of microbial biosensors have demonstrated their utility in detecting heavy metals, nitrates, and organic pollutants at trace levels. For example, biosensors based on *Escherichia coli* have been used to monitor arsenic contamination in drinking water, providing accurate and cost-effective solutions for resource-limited settings. By enabling early detection of pollutants, microbial biosensors play a crucial role in preventing environmental degradation and protecting public health [12].

While microbial technologies have demonstrated remarkable potential, their widespread adoption is not without challenges. Factors such as the complexity of microbial ecosystems, the variability of environmental conditions, and the need for robust monitoring and control systems can affect the scalability of these technologies. However, advancements in synthetic biology, omics technologies, and bioprocess engineering are addressing these challenges by enabling the design of more efficient and resilient microbial systems. The scope of microbial technologies aligns closely with the United Nations Sustainable Development Goals (SDGs), particularly those related to clean water and sanitation (SDG 6), affordable and clean energy (SDG 7), and responsible consumption and production (SDG 12). By providing sustainable solutions for waste management, pollution control, and energy production, microbial technologies contribute to the broader agenda of environmental sustainability and social well-being [13].

MICROBIAL BIOREMEDIATION CASE STUDIES

Microbial bioremediation is a process in which pollutant compounds are removed from a contaminated environment or neutralized by using microorganisms capable of transforming them into less toxic, or non-toxic forms. This process deforms from the intrinsic metabolic properties of microbes, which classifies it as an environment-friendly approach more than the chemical or physical approach to remediation. Table **1** highlights specific case studies in microbial bioremediation, focusing on four key areas: options include bioremediation of micropollutants, treatment options for heavy metals, the breakdown of hydrocarbons, and microbial methods for pesticide removal.

Table 1. Microbial Bioremediation Case Studies: Applications and Outcomes.

Category	Pollutant Type	Microbial Species/Involved Mechanism	Case Study/Example	Outcome
Bioremediation of Micropollutants	Pharmaceuticals and personal care products (PPCPs).	*Pseudomonas putida, Comamonas testosteroni.*	Removal of ibuprofen and triclosan from wastewater in municipal treatment facilities.	Reduction of PPCP concentrations by up to 95% within treatment cycles.
Heavy Metals Detoxification	Lead (Pb), Cadmium (Cd), Arsenic (As).	Sulfate-reducing bacteria (*Desulfovibrio vulgaris*), *Bacillus* spp.	Bioremediation of mine tailings in a heavy metal-contaminated site.	Immobilization of heavy metals *via* precipitation as metal sulfides; significant reduction in toxicity.
Degradation of Petroleum-Based Pollutants	Hydrocarbons from oil spills.	*Alcanivorax borkumensis, Pseudomonas aeruginosa.*	Cleanup of Deepwater Horizon oil spill using hydrocarbonoclastic bacteria.	Breakdown of aliphatic and aromatic hydrocarbons; significant restoration of marine ecosystems.

(Table 1) cont.....

Category	Pollutant Type	Microbial Species/Involved Mechanism	Case Study/Example	Outcome
Pesticide Remediation	Organophosphates, DDT, Carbamates	*Flavobacterium sp., Rhodococcus erythropolis*	Remediation of pesticide-contaminated agricultural soils.	Degradation of organophosphates and DDT residues; improvement in soil health and crop productivity.

Bioremediation of Micropollutants

Pharmaceuticals and Personal Care Products (PPCPs), and Endocrine-Disrupting Compounds (EDCs) are significant environmental pollutants because of high resistance, bioaccumulation, and threats to biological and health systems. Most of these pollutants come from municipal wastewater discharge, agricultural leaching, and industrial effluents. Micropollutants are commonly found in water in concentrations as low as ng/L to µg/L yet they can pollute water and exert toxic effects that affect the life of organisms and aquatic systems and promote hormonal change in wildlife and breeding of antibiotic-resistant germs in bacteria. Many of the conventional technologies used to eliminate micropollutants from the wastewater, including activated sludge processes and advanced oxidation, are less effective for this purpose [14]. These systems are mainly used for dealing with the volume of organic matter and nutrients but many micropollutants are refractory and do not decompose or transform under these systems. For instance, substances such as ibuprofen, triclosan, or synthetic hormones may remain relatively unaffected by the commonly used water treatment processes, and after being discharged into natural waters, they may be ingested by human beings who then experience potentially adverse consequences. Microbial bioremediation has the potential to complement the limitations observed with the conventional treatment modalities. Some species of microbes relevant to biodegradation have enzymatic pathways that can break down organic micropollutants into less toxic functionalities (Fig. 1). For example, researchers have looked at *Pseudomonas putida* as a means to degrade ibuprofen, a monooxygenase enzyme, into pollutant soluble intermediates that are non-toxic. In the same context, *Comamonas testosteroni* proves effectiveness in degrading triclosan, which is an antibacterial used in many products, through dechlorination and hydroxylation pathways [15]. These microbes may naturally inhabitate treatment systems or can be introduced through bioaugmentation when selected strains with superior pollutant removal capabilities are introduced.

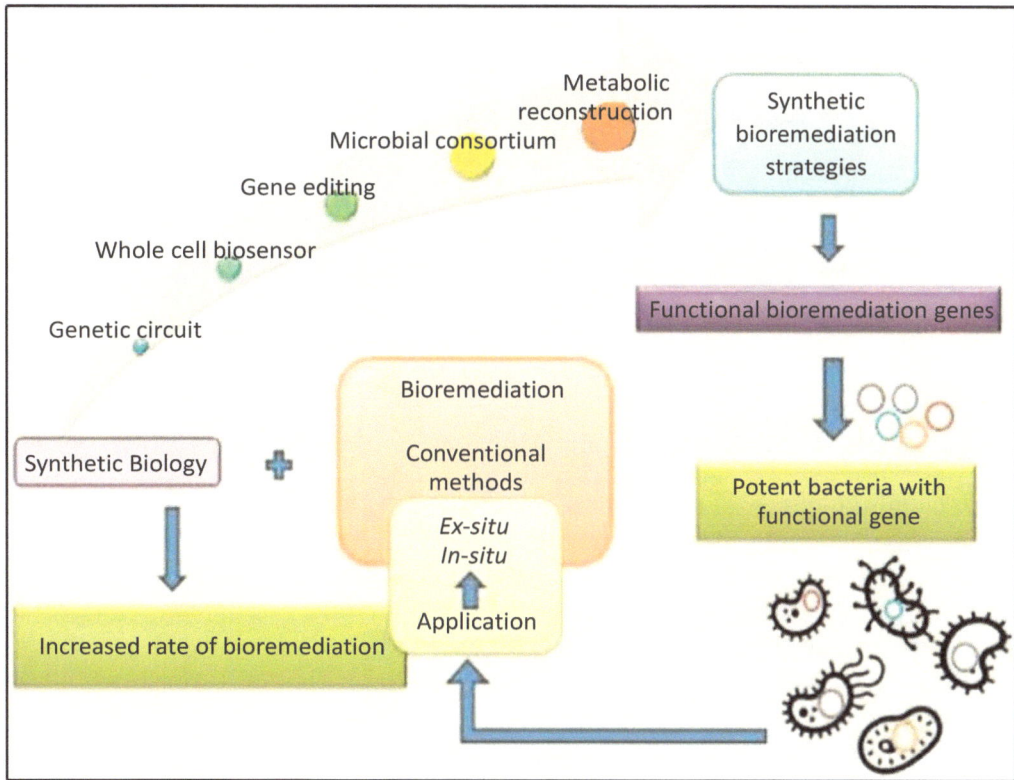

Fig. (1). Bioremediation strategies for removing micropollutants [16].

Various cases have been presented and discussed in order to bring examples of microbial bioremediation as an effective working process. For example, risk minimization strategies of municipal wastewater treatment plants in Europe and North America have incorporated PPCPs using bioaugmentation, particularly microbial strains, such as *Pseudomonas* spp. and *Comamonas* spp. An illustrative study indicated that, when bioaugmentation was used in secondary treatment stages, the concentrations of ibuprofen and triclosan in effluent water were reduced by 95% [17]. For instance, a treatment facility in Germany used a biofilm reactor with *Pseudomonas* species for the removal of synthetic hormones, and the reactor was shown to remove them to over 90 percent. These outcomes show that microbial bioremediation can greatly improve the capability of the elimination of micropollutants, thereby contributing to a safer discharge of effluents into natural water bodies. The use of microbial bioremediation on micropollutants holds a number of advantages. This process presents a cost-effective and eco-friendly solution that taps into naturally occurring biological factors and hence does not necessitate expensive chemical treatment or high energy using advanced oxidation processes [18]. Further, complex microbial systems can be well-tuned to capture a specific pollutant and to achieve the desired biodegradation level through the use of genetically modified microbial strains or by bringing changes in pH, temperature, or nutrient level of the polluted medium. Future work aims toenhance the stability and expandability of microbial systems into large-scale applications. With the help of synthetic biology practices, the current generation of microbes is being developed for better degradative traits, and on the other hand, the omics tools play a vital role in finding new microbial pathways for degradation of refractive pollutants. Furthermore, the combination of microbial bioremediation with other treatment technologies for example, membrane bioreactors or constructed wetlands as well as other treatment technologies could increase overall removal efficiency.

Heavy Metals' Detoxification Using Microbial Pathways

Lead, cadmium, arsenic, and other heavy metals are toxic and non-biodegradable pollutants threatening the environment and human health. The fact that they are non-biodegradable allows their use in different systems, including soil and water, and infiltrate the food chain [9]. Long-term effects of these metals are adverse, affecting the neurological system, and kidneys, and leading to cancer. In addition, the toxicity of their effects impairs microbial and plant diversity and alters ecosystem services. The above shows that mitigation of heavy metal contamination is, therefore, a daunting environmental task. Chemical precipitation, ion exchange, and adsorption are the usual techniques for dealing with HM contamination. However, costs such as high power consumption, rate of secondary waste production, and high costs characteristic of these techniques restrict their application [19]. Further, these techniques may not be appropriate to remediate low concentrations of heavy metal or to immobilize heavy metals in a way that they do not leach back into the environment thus compromising on the recontamination potential. Such limitations underscore the need for much-improved approaches to site remediation.

Microbial detoxification has therefore been developed as a new technique based on the potential of microorganisms to convert toxic heavy metals into non-toxic forms [8]. Some microbes for instance have evolved their own metabolic pathways, which enable them to accept, process, and eliminate the heavy metal in question. Such a mechanism is most efficient with Sulfate-Reducing Bacteria (SRB) like *Desulfovibrio vulgaris*. Some depend on sulfate as their terminal electron acceptor in the process of anaerobic respiration, giving off hydrogen sulfide (H_2S). The H_2S then forms a reaction with soluble heavy metal ions, which form insoluble metal sulfides like PbS and CdS. These metal sulfides are stable and non-reactive and substantially decrease the solubility and transportability of heavy metals in the environment [21]. The second microbial mechanism that is applied in heavy metal removal is biosorption and bioaccumulation. Metal-tolerant bacteria including *Bacillus* spp. are capable of sorbing heavy metal ions by virtue of functional groups including carboxyl, hydroxyl, and amino groups. This process known as biosorption fixes the heavy metals in place to mean that the metals can not come into contact with other organisms or perhaps leech into water sources. Microbes also accumulate heavy metal ions either by bioaccumulation within cells in intracellular compartments. These processes effectively remove unhealthy heavy metals while also helping to concentrate them so they can be reclaimed commercially Fig. (2) [22].

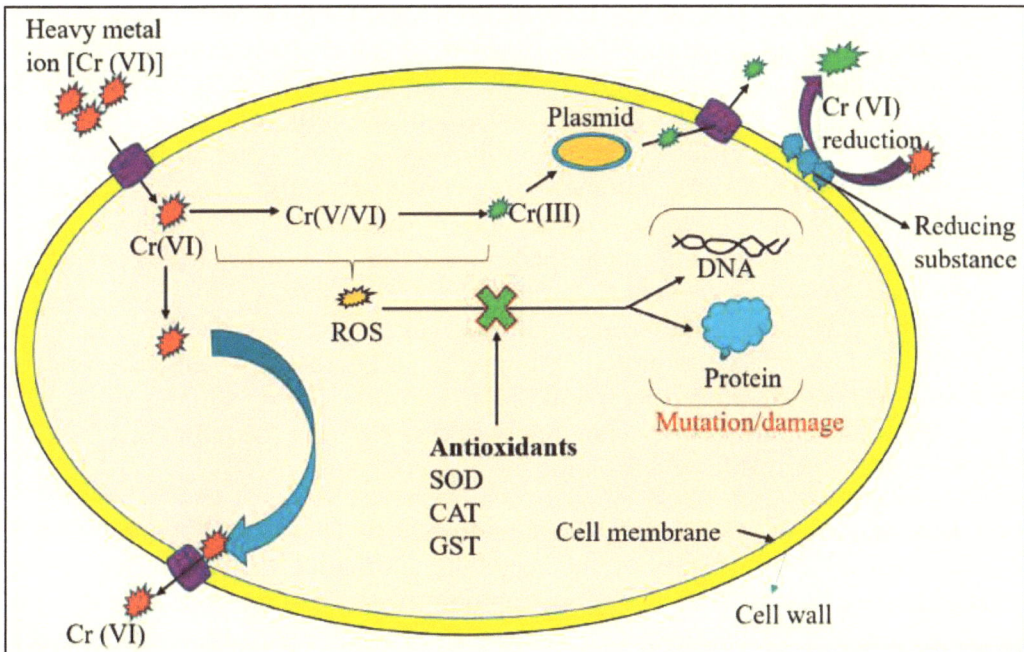

Fig. (2). Remediation of heavy metals using bacteria and the mechanism of heavy metal reduction in bacterial cells [20].

Some of the case studies have revealed how microbial pathways can be useful in practice. For instance, in the environments influenced by mine tailings where the microbial populations show high densities of SRB the latter has been known to immobilise the dangerous heavy metals. Likewise, numerous wastewater treatment plants have used bioaugmentation with MTB to treat industrial effluents with lead and cadmium and the result showed a high percentage reduction of metals [23]. These successes can be taken as evidence supporting the viability of microbial detoxification as an inexpensive and environmentally friendly solution to the incident of heavy metals. Microbial detoxification is a feasible solution to the problem of heavy metal contamination; it overcomes the drawbacks of other methods and long-term positive effects on the environment are expected. As future research investigates further, microbial strategies can be thought to become even more significant in the protection of ecosystems and human individuals.

Degradation of Petroleum-Based Pollutants

Crude oil, gasoline, and other hydrocarbon products arise as major pollutants as evidenced by the oil spill examples. Such occurrences lead to mass destruction of the environment; aquatic and terrestrial bio-systems, animals, and crops. Hydrocarbons are in general, toxic, and persistent and can accumulate in organisms leading to long-term hazards of raising ecological and health impacts. The existing means of dealing with oil pollution include sweeping and the use of chemical emulsifying agents but these methods present drawbacks including high costs for cleanup, only partial cleanup, and secondary pollution. In this regard, the biodegradation of hydrocarbons by microbes is the most potentially viable and eco-friendly method [24]. Hydrocarbonoclastic Bacteria (HCB) are some microbial species that have developed the ability to derive energy and carbon from hydrocarbons. Among such species, the following deserve mentioning: *Alcanivorax borkumensis* and *Pseudomonas aeruginosa*, as they are directly involved in petroleum pollution degradation. These microbes also have enzymes that can metabolize large saturated carboxylic hydrocarbons, aliphatic compound hydrocarbons, and aromatic compound hydrocarbons all into simpler substances like carbon dioxide and water. They not only disintoxicate the pollutants but they are also naturally capable of redeeming harmed ecosystems [25].

One of the most novel utilizations of microbial hydrocarbon degradation was claimed during the mitigation of the Deepwater Horizon oil spill that occurred in the Gulf of Mexico in 2010. This catastrophic spill was estimated to have dumped about 4.9 million barrels of crude oil into the marine habitat negatively affecting most of the species. In response, the bioremediation approaches and techniques were applied to solve this problem and attempted to use the native bio-populations of hydrocarbon-degrading bacteria which include *Alcanivorax borkumensis*. Such microbes benefited from the oil-polluted water, nurtured the process of breaking down the hydrocarbons, and helped to rejuvenate the marine environment. It shows excellent reduction in both aliphatic and aromatic hydrocarbon concentrations proving that microbial intervention works well in large-scale oil spill situations [27]. In addition to the immediate effects of utilizing microbes within the oil industry for bioremediation of oil spills, microbial degradation of petroleum-based pollutants has been utilized in other ways including in treating industrial effluent and soil remediation. For example, bioaugmentation of *Pseudomonas* spp. in petroleum-contaminated soils effectively biodegraded the soil-borne Polycyclic Aromatic Hydrocarbons (PAHs) for a better fit of the agro-resource equivalent. In a similar way, microbial consortiums in refinery wastewater treatment systems have been effective in reducing the hydrocarbon content to comport with required environmental discharge limits. Microbial degradation in these two case studies has provided good results, making microbial-based solutions efficient in petroleum pollution remediation [28]. However, the efficiency with which these microbes degrade the liquid hydrocarbons depends on the type and concentration of the hydrocarbon in question, environmental characteristics such as temperature and oxygen availability, and the availability of nutrient resources such as nitrogen and phosphorus (Fig. **3**). Biostimulation and bioaugmentation of these conditions can further improve the extent of microbial remediation.

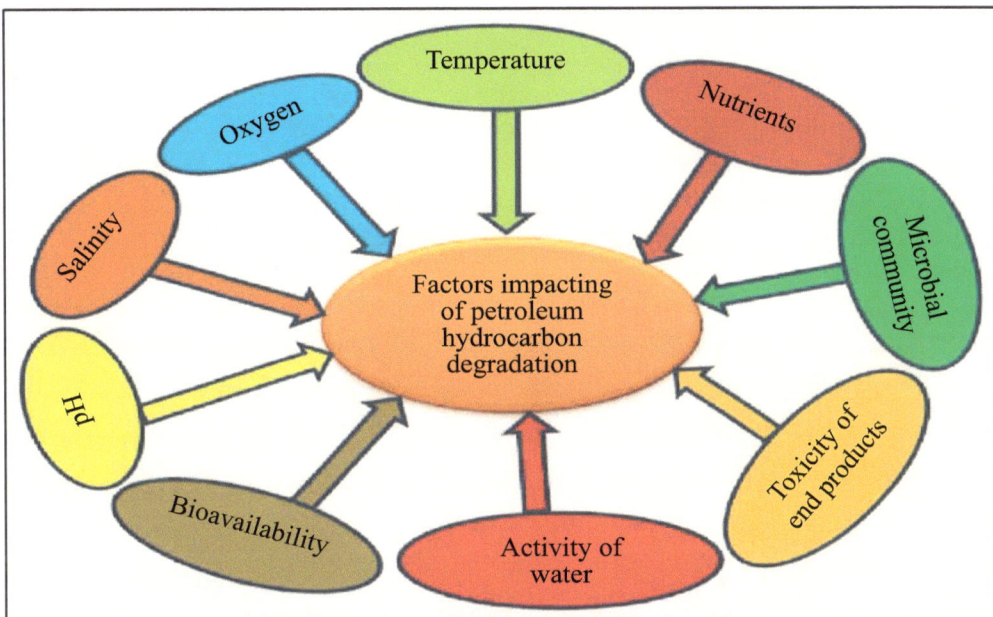

Fig. (3). Schematic representation of factors that influence microbial remediation [26].

Microbial Strategies for Pesticide Remediation

Pesticides are commonly used in modern farming techniques to combat pests and extend crop yields, containing lasting residues that prove to be very hazardous to the environment and the health of the community. It becomes acutely evident that even common pesticide classes, such as organophosphate, DDT, and carbamate damage the soil microbial community, contaminate water sources, and enter the food chain because of their persistence in water and soil. These residues not only impose a negative impact on the quality of the soil but also prolong the risks to biological diversity and human well-being, including endocrine disruption and carcinogenicity (Fig. **4**) [29]. Due to some drawbacks of conventional remediation technologies, microbial techniques have been presented as an efficient method of detoxifying environments contaminated with pesticides. Some microbial strains have genetically developed their ability to degrade pesticide products into harmless products or products that are less harmful to human beings. Of these genera, *Flavobacterium* sp. and *Rhodococcus erythropolis* are worth mentioning due to their versatility in the breakdown of various pesticide classes. These microbes employ enzymes like phosphatases, dehydrogenases, and hydrolases that help in the biodegradation of the pesticide molecules by breaking chemical bonds and converting them into simpler molecules such as CO_2, water, and organic acidic products. This process not only purifies the pesticides but also prevents the harmful effects of such bad products on the environment, restoring ecological health [30].

An example was undertaken that proves that microbial remediation is effective when used to treat pesticide-contaminated soils. Samples of soil contaminated with organophosphates and DDT were bioaugmented with *Flavobacterium* sp. and *Rhodococcus Erythropolis*. At some point, the microbial communities resulted in the disinfection and degradation of pesticide components reducing the concentration by more than 90%. Most importantly, this remediation also subjected the soil to a positive change in nutrients with the microbial breakdown of pesticides, releasing such nutrients as phosphorus and nitrogen back into the soil [31]. This improvement in the quality of the soil they cultivated made it safe to cultivate crops making food security better. Microbial pesticide remediation has also been successfully used in water treatment systems. For example, bioreactors that contained *Flavobacterium* sp. have been applied for the treatment of pesticide-contaminated agricultural effluent and yielded significant pollutant removal efficiencies. These systems have the mandate to further protect the aquatic ecosystems while at the same time helping to uphold water quality standards, which can also help serve public health protections. Several factors could influence the extent of microbial remediation, these include: the type of pesticides used, environmental conditions, and co-factors or nutrients. Some of the bioremediation techniques, that have been used in order to improve the efficiency of the process include biostimulation, which is the enhancement of the soil microbial activities through the addition of nutrients, and bioaugmentation, which involves the introduction of pesticide-degrading microbial strains in the soil. Moreover, microbial activity employing more than one strain of the bacteria with the specific degradative enzymes has been described to have reduced efficiency in the degradation of the complex multinary pesticide mixtures [32].

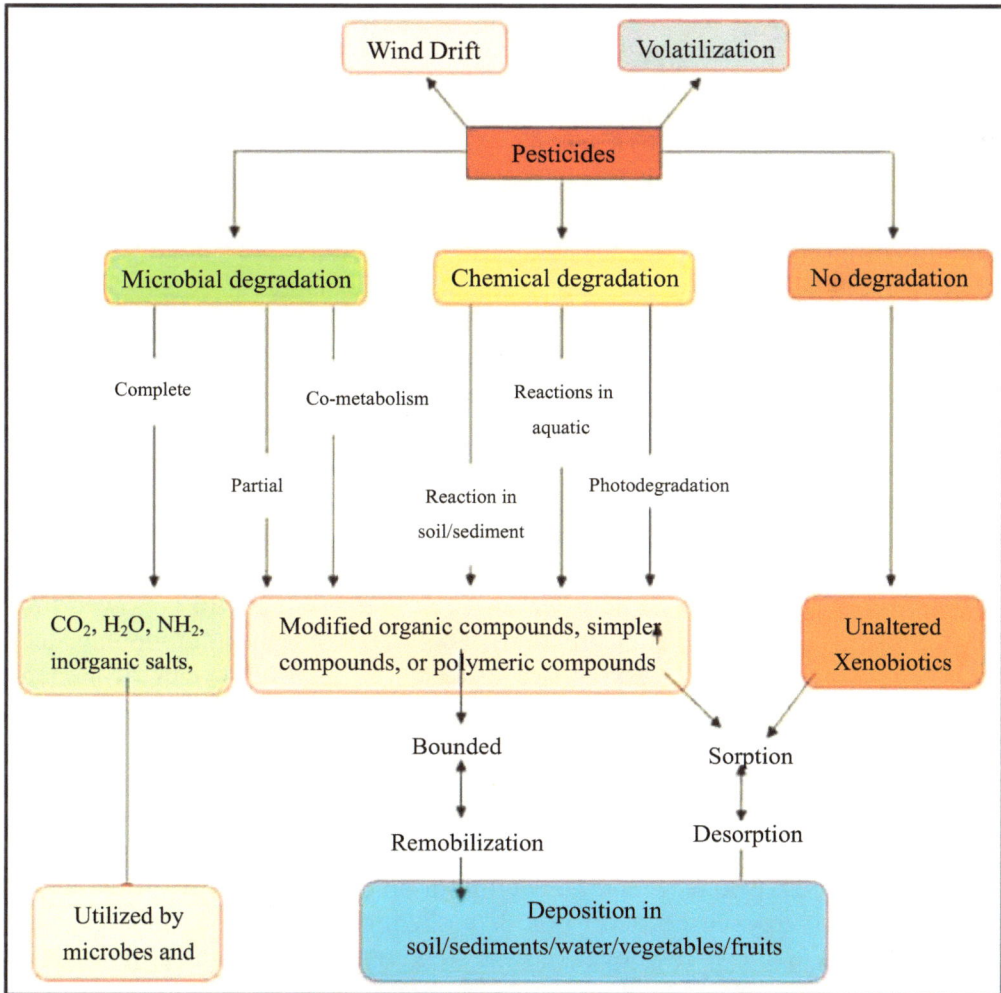

Fig. (4). Possible fate of pesticides in the environment [33].

MICROBIAL WASTE-TO-ENERGY SYSTEMS

Microbial technologies play a pivotal role in converting organic waste into bioenergy, offering sustainable solutions to both waste management and renewable energy production. Two significant approaches in this domain are Microbial Fuel Cells (MFCs) and Anaerobic Digestion (AD). Case studies in microbial bioenergy conversion demonstrate the transformative potential of microbial fuel cells and anaerobic digestion. These technologies not only offer effective waste treatment solutions but also contribute to energy sustainability by recovering valuable bioenergy [34]. By harnessing the power of microbes, these systems address environmental challenges while creating opportunities for decentralized energy systems and circular economy models. Microbial fuel cells utilize the metabolic processes of microbes to generate electricity directly from organic waste. In an MFC, electrogenic bacteria such as *Geobacter sulfurreducens* and *Shewanella oneidensis* oxidize organic substrates, releasing electrons that flow to an anode, and creating an electric current [35].

Municipal wastewater treatment presents a dual challenge: ineffectively handling a large volume of effluents charged with organic matter and reducing operational energy. Currently used treatment processes include Activated sludge systems that are efficient in polishing pollutants but are energy-demanding. This has put focus on emergent, eco-friendly technologies such as Microbial Fuel Cells (MFCs), which are capable of both wastewater treatment and electricity generation. The practical application of the technology in the context of the organization's operation takes into consideration the following characteristics of MFC technology: In one case study, a municipal wastewater treatment facility incorporated the use of MFC technology in its operations with a view of making the processes more efficient and sustainable [36]. The system used *Geobacter sulfurreducens*, a famous electrogenic microorganism, that degrades the organic contaminants found in the wastewater. *G. sulfurreducens* is well-endowed in the transfer of electrons derived from the oxidation of organic compounds to an electrode to enable the direct conversion of chemical energy into electrical energy. Accordingly, the present work found that the MFC system was able to produce a maximum power density of 150 mW/m². Though the amount might not be enough to provide the whole energy needed by the plant, it helped to an extent to reduce the plant's energy consumption hence improving the overall cost of operations. As it stands, the MFC also facilitated an 80% removal of the COD of the wastewater. COD is one of the most important measurements involving the concentration of oxygen needed to oxidize chemical compounds in wastewater. COD reduction to a very low level shows effective removal of the pollutants, and compliance with environmental discharge requirements [37].

Solid waste disposal especially food waste is mostly dumped in municipal waste bins and ends up in landfills as an additional source of greenhouse gases. Microbial Fuel Cells (MFCs) make this possible by using food waste to produce electricity and at the same time reduce the volumetric waste fraction. A case study where an MFC was inserted was where food waste slurry from a local restaurant was used. Organotrophic bacteria in the slurry further reduced the likelihood because they used the organic compounds to create a peak voltage of 0.8V. In addition to electricity generation, the system attained a 70% reduction in the volume of waste emphasizing MFCs as a viable local waste treatment technology [38]. This technology can be of most help in restaurants or small community centers and manage their waste while using it to generate electricity. Anaerobic digestion on the other hand deals with organic waste which is efficient in large-scale firms. AD is dependent on a group of bacteria that comprise hydrolytic bacteria, acidogenic bacteria, acetogenic bacteria, and methanogenic bacteria to decompose organic substances in anaerobic conditions. The process results in biogas-which is a blend of methane and carbon dioxide and is usable for electricity, and heat or can be upgraded to vehicle grade fuel. Besides, it results in the production of digestate, a material that is rich in nutrients and can be used as a fertiliser, due to its benefits to sustainable farming. Whereas MFCs yield comparatively small amounts of electricity, AD systems are well adapted to larger centralised waste treatment plants due to the increased energy production. MFCs and AD systems combined are promising technologies for reclaiming value-added products from organic waste and increasing energy recovery with reduced waste output in various applications [39].

One dairy farm utilized an AD system to control the large amount of cattle dung and crop residues produced on the farm. The system was found to be efficient, making 500 m³ of biogas per day, with 65% methane content. The biogas was used mainly to produce electricity and heat so that a high amount of energy was being produced internally hence minimizing the importation of energy as well as reducing greenhouse gas emissions. The generation of its electricity also helped cut out expenses and enabled the farm to be ecological through the decrease of carbon emissions. The AD process also 'produced' digestate, the residual, high nutrient content output of the finished anaerobic digestion of organic material [40]. Digestate can be used as an organic manure to improve soil quality and reduce the application of chemical fertilisers, which are more energy-demanding in their manufacture. This closed-loop system can be described as a circular model of agriculture wherein waste byproducts generated from the process were used as resource inputs; the biogas for power production, and the digestate for enhancing the fertility of farmland. This paper presents an example of how AD technology can be utilized to promote sustainable agriculture through effectively using waste products. Besides disposal problems, such as the emission of gases that can harm the environment and humans, biogas production enhances the generation of renewable energy, which creates economic and environmental benefits for the farm [41]. On the same note, the use of digestate as a fertilizer is environmentally friendly in agriculture since it replaces the use of chemical fertilizers and hence has an enhanced impact on soil health. This example shows how AD systems can be a remarkable solution to deal with waste management, energy generation, and nutrient recovery on farms that create value for farm sectors as well as improve the environment [42].

Another case study described the installation of an AD system at a landfill facility to deal with landfill leachate which consists of significant levels of organic substance. The liquid that flows through the waste material, commonly referred to as leachate is often extremely poisonous and rare to treat *via* standard techniques. However, by inoculating methanogenic bacteria including *Methanosaeta* and *Methanosarcina*, the organic load in the leachate was well converted into biogas. Hence these microbes are very important in the anaerobic digestion process and are responsible for the reduction of complex organic matter into methane (CH_4) and carbon dioxide (CO_2) that forms biogas [43]. The biogas produced by this process was channeled to a combined heat and power plant to generate electricity and heat. This not only ensured the availability of renewable energy to support operations at the landfill but also made sure that the landfill had a minimized environmental impact by reusing the energy generated from waste. The generation of biogas from landfill leachate reduced the toxic effect of the leachate since the process aids in the reduction of toxicity hence making the leachate have the least adverse effects on the neighbourhood environment. The present work is an example of the applicability of AD technology in the treatment of landfill leachate and at the same time, hinders an alternative method to produce renewable energy [44]. The utilisation of methane production and energy recovery is also environmentally friendly where a waste management problem turns out to be an advantage with the added benefit of producing energy. It helps in minimising landfill pollution and in the reduction of Green House Gas effect through the use of biogas generated for energy uses instead of being emitted as methane.

MICROBIAL BIOSENSORS AND ENVIRONMENTAL MONITORING

Pollution biosensors are highly effective in detecting different pollutants with the help of microbial activity. These biosensors are derived from the capacity of certain microbial communities to have a physiological response to the pollutant and, in doing so, generate an observable signal, possibly a change in color, luminescence, or electrical potential [45]. These make them extremely useful in environmental management as they afford real-time, sensitive, and economical techniques for measuring air, water, and soil quality. Some of the major areas in which microbial biosensors have been used in the detection of environmental pollutants in air, water, and soil are also presented in Table **2**. It has features like the kind of pollutants to be sensed, the microorganisms that are used in the biosensors, and some case studies on the usage of the biosensors in environmental management. The selected case studies exemplify the applicability of microbial biosensors for continuous, economical, and highly selective detection of pollutants–hence improving pollution control and environmentally friendly technologies and practices [45].

In air quality assessment, microbial biosensors are able to identify Volatile Organic Compounds (VOCs), greenhouse gases, and other chemicals. For instance, biosensors based on *Pseudomonas putida* bacteria can identify VOCs like benzene and toluene; equally, biosensors that incorporate *Saccharomyces cerevisiae* can be used in IAQ to detect formaldehyde. These biosensors have relatively high sensitivity with a detection range of 1ppm, proving useful in the continuous monitoring of the quality of the air in industrial and or indoor spaces [46]. The present-day water quality monitoring offers more prospects through the use of microbial biosensors that are uniquely programmed to identify several pollutants including heavy metals, organic and microbial pollutants as well as pathogens. *Escherichia coli* biosensors have recently been used that are constructed through genetic engineering for the identification of mercury and arsenic ions in water ingredients resulting in luminescent indications. In one of the case studies, a microbial biosensor developed with *Shewanella Oneidensis* was inserted into a river to detect the levels of heavy metal pollutants. The data generated was quite useful for timely action and underlined the effectiveness of microbial biosensors for water quality monitoring [47].

Table 2. Applications of Microbial Biosensors in Pollution Detection

Application	Pollutants Detected	Microorganism Used	Case Study/Example
Air Quality Monitoring	Volatile Organic Compounds (VOCs), Formaldehyde	*Pseudomonas putida, Saccharomyces cerevisiae*	VOCs like benzene and toluene in industrial emissions and formaldehyde in indoor air monitoring.
Water Quality Monitoring	Heavy Metals (Mercury, Arsenic), Organic Pollutants	*Escherichia coli, Shewanella oneidensis*	Detection of heavy metals in a river using *Shewanella oneidensis*; luminescent response to pollutants.
Soil Quality Monitoring	Pesticides (Organophosphates, Carbamates), Hydrocarbons	*Acinetobacter sp., Rhodococcus erythropolis*	Monitoring pesticide degradation in agricultural soils and hydrocarbon degradation in oil-contaminated soil.

Another area where microbial biosensor technology proves advantageous is in soil quality monitoring, in relation to the identification of soil pollutants including pesticides, hydrocarbons, and nutrient levels. For example, *Acinetobacter* sp has been applied in biosensors for monitoring the rate of organophosphates and carbamates degradation in agricultural soils to support the decision-making process in pest control. Similarly, in yet another case, a *Rhodococcus erythropolis* biosensor was used for tracking the degradation of hydrocarbon in the oil-laden soil. Through the biosensor, the rate at which bioremediation was being achieved was monitored effectively hence determining the recovery rate of the soil [48]. From the above-mentioned case studies, we can understand the importance of microbial biosensors in the detection and estimation of various pollution indices in the environment. Some of them include high selectivity, resolution, and sensitivity, portability in data output and low cost of microbial biosensors have made them suitable for monitoring of environment and pollution control. They help to identify a large number of different impurities, which allows time for interventions and contributes to the environmentally friendly approach.

INTEGRATING LESSONS FROM CASE STUDIES

The only major drawback voiced by members of the audience was the absence of practical examples of microbial technologies when studying numerous case studies of microbial technologies for environmental pollution detection, bioremediation, and generating bioenergy. To this end, through synthesizing the findings of these cases, researchers and practitioners in this field could enhance known solutions and develop better novel ones that will effectively address these environmental problems. The following conclusions along with the potential contributions to future research and development in this field are as follows.

Key Insights from Case Studies

Various microbial bioremediation and biosensor case studies have confirmed that microbial technologies are real and effective. For example, Microbial Fuel Cells (MFCs) can demonstrate the possibility of bioelectricity production from wastewater microbial biosensors for real-time and effective detection of depraved air, water, and soil. This application also supports the importance of microorganisms in the mitigation of pollution and the production of energy. Among the strengths of microbial technologies, the system can effectively adapt to any form of environmental pollution [49]. From petroleum hydrocarbons to heavy metals to pesticides, microbes can be genetically modified or naturally chosen to effectively attack certain pollutants in certain locales. Large microbial systems can also be seen in the case of enhanced biological phosphorus removal, in which wastewater treatment involves, bioremediation of contaminated soils and volatile organic compounds detection. Microbial-based solutions are generally more economical and environmentally friendly than conventional methods. The ability to generate electricity from the microbial fuel cell on waste material or to use biosensors for low-cost environmental detection is solely an important benefit of microbial technologies economically and friendly to the environment. Such systems provide a direction for responsible environmental management as they minimize the use of chemicals as well as expensive complex structures [3, 4].

Future Directions

According to the present and future analyses, the improvement of microbial functions is achievable through the refinement of genetic engineering methods. Additional optimization of these systems can occur through genetic modification of microbes to break down a greater number and types of pollutants at a faster rate, or to identify pollutants at even ppb levels. Application design of microbes with particular pathways for specific pollutants is likely to be a major one in the future due to synthetic biology [50]. The opportunity to use microbial biosensors with IoT technology shows much potential for monitoring the environment in real-time. When microbial sensors are connected with Cloud computing, the data obtained can be transmitted as well as analyzed in real-time, allowing quick action against pollution incidents. This integration will open the way for smart monitoring networks, which will permit the constant survey of the scale of the problem for widespread application. Microbial technologies are promising, however, scaling these systems from the laboratory-scale to full industrial and field applications is not easy. Further research will look at increasing the yield of microbial fuel cells, increasing the efficiency of bioremediation as well as biosensor networks for large-scale wastes and pollutants [6, 13, 19]. To enable a transition from laboratory to large-scale manufacturing and use, industry links will have to come into play.

Implications for Research and Industry

Thus, interdisciplinary cooperation of microbiology, engineering, and environmental science will remain crucial to the further development of microbial technologies. They should consider addressing challenges that include improvement of microbial efficiency as well as biosensor stability, and integrating them into real environments for practical use. Intersectoral cooperation between universities and researchers, pragmatic industries, and governments will be considered essential for the development of breakthrough technologies. To industries, implementing the bio-technologies involving microbes is an opportunity for organizations to minimize their impact on the environment, by optimizing their production processes at the same time [24, 36]. The application of microbial systems in the development of future industrial waste treatment, environmental monitoring, and integrated energy production processes is seen. This shift calls for innovation in the development of these technologies besides pushing governments and their regulatory agencies to provide support to these technologies to make them safe and effective [50, 51]. These industries could particularly benefit from such innovations; microbial solutions are more sustainable and cost-effective than conventional methods applied in agriculture, water, and energy industries.

CONCLUSION

In this chapter, different and useful aspects and successful examples of microbial technologies in environmental management have been discussed in the context of microbial bioremediation, waste to energy, and biosensors for the detection of pollution. From these case studies, it is clear that microbes have the potential to help solve numerous environmental problems, including the degradation of hazardous chemicals like heavy metals and pesticides, as well as carrying out the production of electricity through microbial fuel cells. The use of microbial technologies in field-level complementary applications stands to benefit from practical, attractive features such as relatively low cost, renewability, and versatility with regard to pollutants. That said, problems persist in how these technologies may be applied at a commercial scale and fine-tuned for various climates. The future of microbial technologies is in the further development of genetic engineering, SMART technologies, and IoT integration, and in the designing of technology systems that have the ability to increase in scale to meet the growing global environmental demands. With these trends in mind, microbial technologies represent the next frontier and significant possibilities for solving pollution, waste, and energy challenges throughout the rest of this century. In light of these case studies, major conclusions and implications for future research and adoption of these technologies are presented. As academia, industry, and policymakers' cooperation progresses, microbial approaches are destined to become the driving force in developing a world free of pollution.

REFERENCES

[1]　Mahmud K, Missaoui A, Lee K, Ghimire B, Presley H W, Makaju S. Rhizosphere microbiome manipulation for sustainable crop production. Curr Plant Biol 2021; 27: 100210.
[http://dx.doi.org/10.1016/j.cpb.2021.100210]

[2]　Shah MP. Emerging innovative trends in the application of biological processes for industrial wastewater treatment. Elsevier 2024.

[3]　Shah MP, Shah N. Environmental approach to remediate refractory pollutants from industrial wastewater treatment plant. Elsevier 2024.

[4]　Yin Q, He K, Collins G, De Vrieze J, Wu G. Microbial strategies driving low concentration substrate degradation for sustainable remediation solutions. npj Clean Water 2024; 7(1): 52.
[http://dx.doi.org/10.1038/s41545-024-00348-z]

[5]　Santomartino R, *et al.* Toward sustainable space exploration: a roadmap for harnessing the power of microorganisms. Nat Commun 2023; 14(1): 1391.
[http://dx.doi.org/10.1038/s41467-023-37070-2]

[6]　Chaudhary S, Sindhu S S, Dhanker R, Kumari A. Microbes-mediated sulphur cycling in soil: impact on soil fertility, crop production and environmental sustainability. Microbiol Res 2023; 271: 127340.
[http://dx.doi.org/10.1016/j.micres.2023.127340]

[7]　Koshlaf E, S Ball A. Soil bioremediation approaches for petroleum hydrocarbon polluted environments. AIMS Microbiol 2017; 3(1): 25-49.
[http://dx.doi.org/10.3934/microbiol.2017.1.25] [PMID: 31294147]

[8]　Agarwal S, Jain H. Investigating the potential of floral waste as a vermicompost and dual-functional biosorbent for sustainable environmental management. Water, Air, & Soil Pollut 2024; 235(5): 322-2.
[http://dx.doi.org/10.1007/s11270-024-07144-y]

[9]　Jain H. From pollution to progress: groundbreaking advances in clean technology unveiled. Innov

Green Dev 2024; 3(2): 100143-3.
[http://dx.doi.org/10.1016/j.igd.2024.100143]

[10]   Jain H, Dhupper R, Shrivastava A, Kumar D, Kumari M. Leveraging machine learning algorithms for improved disaster preparedness and response through accurate weather pattern and natural disaster prediction. 2023.
[http://dx.doi.org/10.3389/fenvs.2023.1194918]

[11]   Prasanna Kumar D J, Mishra R K, Chinnam S, Binnal P, Dwivedi N. A comprehensive study on anaerobic digestion of organic solid waste: a review on configurations, operating parameters, techno-economic analysis and current trends. Biotechnol Notes 2024; 5: 33-49.
[http://dx.doi.org/10.1016/j.biotno.2024.02.001]

[12]   Moraskie M, *et al.* Microbial whole-cell biosensors: current applications, challenges, and future perspectives. Biosens Bioelectron 2021; 191: 113359-9.
[http://dx.doi.org/10.1016/j.bios.2021.113359]

[13]   Sagova-Mareckova M, *et al.* Expanding ecological assessment by integrating microorganisms into routine freshwater biomonitoring. Water Res 2021; 191: 116767-7.
[http://dx.doi.org/10.1016/j.watres.2020.116767]

[14]   Osuoha J O, Anyanwu B O, Ejileugha C. Pharmaceuticals and personal care products as emerging contaminants: need for combined treatment strategy. J Hazard Mater Adv 2023; 9: 100206-6.
[http://dx.doi.org/10.1016/j.hazadv.2022.100206]

[15]   Ebele A J, Abou-Elwafa Abdallah M, Harrad S. Pharmaceuticals and Personal Care Products (PPCPs) in the freshwater aquatic environment. Emerg Contam 2017; 3(1): 1-16.
[http://dx.doi.org/10.1016/j.emcon.2016.12.004]

[16]   Qamar SA, *et al.* 19 - Biodegradation of micropollutants. In: Iqbal HMN, Bilal M, Nguyen TA, Yasin G, Eds. Biodegradation and biodeterioration at the nanoscale. Elsevier 2022; pp. 477-507.
[http://dx.doi.org/10.1016/B978-0-12-823970-4.00018-X]

[17]   Singh D, Goswami R K, Agrawal K, Chaturvedi V, Verma P. Bio-inspired remediation of wastewater: a contemporary approach for environmental clean-up. Curr Res Green Sustain Chem 2022; 5: 100261-1.
[http://dx.doi.org/10.1016/j.crgsc.2022.100261]

[18]   Vaksmaa A, Guerrero-Cruz S, Ghosh P, Zeghal E, Hernando-Morales V, Niemann H. Role of fungi in bioremediation of emerging pollutants. Rev. 2023; 10.
[http://dx.doi.org/10.3389/fmars.2023.1070905]

[19]   Jain H. Exploring the emergence of sustainable practices in healthcare research and application as a path to a healthier future. In: Prabhakar PK, Leal Filho W, Eds. Preserving health, preserving earth: The path to sustainable healthcare. Cham: Springer Nature Switzerland 2024; pp. 121-37.
[http://dx.doi.org/10.1007/978-3-031-60545-1_7]

[20]   Zhou B, Zhang T, Wang F. Microbial-based heavy metal bioremediation: toxicity and eco-friendly approaches to heavy metal decontamination. Appl Sci (Basel) 2023; 13(14): 8439.
[http://dx.doi.org/10.3390/app13148439]

[21]   Singh SB, Carroll-Portillo A, Lin HC. *"Desulfovibrio* in the Gut: The Enemy within?"* (in eng). Microorganisms 2023; 11(7): 1772.
[http://dx.doi.org/10.3390/microorganisms11071772] [PMID: 37512944]

[22]   Kushkevych I, Kovářová A, Dordevic D, *et al.* Distribution of sulfate-reducing bacteria in the environment: Cryopreservation techniques and their potential storage application. Processes (Basel) 2021; 9(10): 1843.
[http://dx.doi.org/10.3390/pr9101843]

[23]   Tran TTT, Kannoorpatti K, Padovan A, Thennadil S. Sulphate-reducing bacteria's response to extreme ph environments and the effect of their activities on microbial corrosion. Appl Sci (Basel) 2021; 11(5):

2201.
[http://dx.doi.org/10.3390/app11052201]

[24] Hassanshahian M, Amirinejad N, Askarinejad Behzadi M. Crude oil pollution and biodegradation at the Persian Gulf: A comprehensive and review study. J Environ Health Sci Eng 2020; 18(2): 1415-35.
[http://dx.doi.org/10.1007/s40201-020-00557-x] [PMID: 33312652]

[25] Ambaye T G, *et al.* Remediation of soil polluted with petroleum hydrocarbons and its reuse for agriculture: recent progress, challenges, and perspectives. Chemosphere 2022; 293: 133572-2.
[http://dx.doi.org/10.1016/j.chemosphere.2022.133572]

[26] Adedeji JA, Tetteh EK, Opoku Amankwa M, *et al.* Microbial bioremediation and biodegradation of petroleum products: A mini review. Appl Sci (Basel) 2022; 12(23): 12212.
[http://dx.doi.org/10.3390/app122312212]

[27] Kostka JE, Prakash O, Overholt WA, *et al.* Hydrocarbon-degrading bacteria and the bacterial community response in gulf of Mexico beach sands impacted by the deepwater horizon oil spill. Appl Environ Microbiol 2011; 77(22): 7962-74.
[http://dx.doi.org/10.1128/AEM.05402-11] [PMID: 21948834]

[28] Kimes NE, Callaghan AV, Suflita JM, Morris PJ. Microbial transformation of the Deepwater Horizon oil spill past, present, and future perspectives. Front Microbiol 2014; 5: 603.
[http://dx.doi.org/10.3389/fmicb.2014.00603] [PMID: 25477866]

[29] Kaur R, *et al.* Pesticides: An alarming detrimental to health and environment. Sci Total Environ 2024; 915: 170113-3.
[http://dx.doi.org/10.1016/j.scitotenv.2024.170113]

[30] Ahmad MF, Ahmad FA, Alsayegh AA, *et al.* Pesticides impacts on human health and the environment with their mechanisms of action and possible countermeasures. Heliyon 2024; 10(7): e29128.
[http://dx.doi.org/10.1016/j.heliyon.2024.e29128] [PMID: 38623208]

[31] Pathak VM, *et al.* Current status of pesticide effects on environment, human health and it's eco-friendly management as bioremediation: a comprehensive review. Rev. 2022; 13.
[http://dx.doi.org/10.3389/fmicb.2022.962619]

[32] Tudi M, Daniel Ruan H, Wang L, *et al.* Agriculture development, pesticide application and its impact on the environment. Int J Environ Res Public Health 2021; 18(3): 1112.
[http://dx.doi.org/10.3390/ijerph18031112] [PMID: 33513796]

[33] Tarfeen N, Nisa KU, Hamid B, *et al.* Microbial remediation: A promising tool for reclamation of contaminated sites with special emphasis on heavy metal and pesticide pollution: a review Processes (Basel) 2022; 10(7): 1358.
[http://dx.doi.org/10.3390/pr10071358]

[34] Obileke K, Onyeaka H, Meyer E L, Nwokolo N. Microbial fuel cells, a renewable energy technology for bio-electricity generation: A mini-review. Electrochem Commun 2021; 125: 107003-3.
[http://dx.doi.org/10.1016/j.elecom.2021.107003]

[35] Hoang A T, *et al.* Microbial fuel cells for bioelectricity production from waste as sustainable prospect of future energy sector. Chemosphere 2022; 287: 132285-5.
[http://dx.doi.org/10.1016/j.chemosphere.2021.132285]

[36] Apollon W. An overview of microbial fuel cell technology for sustainable electricity production. Membranes (Basel) 2023; 13(11): 884.
[http://dx.doi.org/10.3390/membranes13110884] [PMID: 37999370]

[37] Moradian J M, Fang Z, Yong Y-C. Recent advances on biomass-fueled microbial fuel cell. Bioresour Bioprocess 2021; 8(1): 14-4.
[http://dx.doi.org/10.1186/s40643-021-00365-7]

[38] Kakkar S, Dharavat N, Sudabattula S K. Transforming food waste into energy: a comprehensive review. Results Eng 2024; 24: 103376-6.

[http://dx.doi.org/10.1016/j.rineng.2024.103376]

[39]   Bhatia L, Jha H, Sarkar T, Sarangi PK. Food waste utilization for reducing carbon footprints towards sustainable and cleaner environment: A review Int J Environ Res Public Health 2023; 20(3): 2318.
[http://dx.doi.org/10.3390/ijerph20032318] [PMID: 36767685]

[40]   Soha T, Papp L, Csontos C, Munkácsy B. The importance of high crop residue demand on biogas plant site selection, scaling and feedstock allocation – a regional scale concept in a Hungarian study area. Renew Sustain Energy Rev 2021; 141: 110822-2.
[http://dx.doi.org/10.1016/j.rser.2021.110822]

[41]   Pochwatka P, Kowalczyk-Juśko A, Sołowiej P, Wawrzyniak A, Dach J. Biogas plant exploitation in a middle-sized dairy farm in poland: Energetic and economic aspects. Energies 2020; 13(22): 6058.
[http://dx.doi.org/10.3390/en13226058]

[42]   Li Y, Zhao J, Achinas S, Zhang Z, Krooneman J, Euverink G J W. The biomethanation of cow manure in a continuous anaerobic digester can be boosted *via* a bioaugmentation culture containing Bathyarchaeota. Sci Total Environ 2020; 745: 141042-2.
[http://dx.doi.org/10.1016/j.scitotenv.2020.141042]

[43]   El-Saadony M T, *et al.* Hazardous wastes and management strategies of landfill leachates: a comprehensive review. Environ Technol Innov 2023; 31: 103150-0.
[http://dx.doi.org/10.1016/j.eti.2023.103150]

[44]   Khan O, Mufazzal S, Sherwani A F, Khan Z A, Parvez M, Idrisi M J. Experimental investigation and multi-performance optimization of the leachate recirculation based sustainable landfills using Taguchi approach and an integrated MCDM method. Sci Rep 2023; 13(1): 19102-2.
[http://dx.doi.org/10.1038/s41598-023-45885-8]

[45]   Huang CW, Lin C, Nguyen MK, Hussain A, Bui XT, Ngo HH. A review of biosensor for environmental monitoring: principle, application, and corresponding achievement of sustainable development goals. Bioengineered 2023; 14(1): 58-80.
[http://dx.doi.org/10.1080/21655979.2022.2095089] [PMID: 37377408]

[46]   Tecon R, Van der Meer JR. Bacterial biosensors for measuring availability of environmental pollutants. Sensors (Basel) 2008; 8(7): 4062-80.
[http://dx.doi.org/10.3390/s8074062] [PMID: 27879922]

[47]   Kunze-Szikszay N, Euler M, Perl T. Identification of volatile compounds from bacteria by spectrometric methods in medicine diagnostic and other areas: current state and perspectives. Appl Microbiol Biotechnol 2021; 105(16-17): 6245-55.
[http://dx.doi.org/10.1007/s00253-021-11469-7] [PMID: 34415392]

[48]   Lindquist HDA. Microbial biosensors for recreational and source waters. J Microbiol Methods 2020; 177: 106059.
[http://dx.doi.org/10.1016/j.mimet.2020.106059] [PMID: 32946871]

[49]   Cui Y, Lai B, Tang X. Microbial fuel cell-based biosensors. Biosensors (Basel) 2019; 9(3): 92.
[http://dx.doi.org/10.3390/bios9030092] [PMID: 31340591]

[50]   Yaashikaa P R, Devi M K, Kumar P S. Engineering microbes for enhancing the degradation of environmental pollutants: a detailed review on synthetic biology. Environ Res 2022; 214: 113868.
[http://dx.doi.org/10.1016/j.envres.2022.113868]

[51]   Nadarajah K, Abdul Rahman NSN. The microbial connection to sustainable agriculture. Plants 2023; 12(12): 2307.
[http://dx.doi.org/10.3390/plants12122307] [PMID: 37375932]

CHAPTER 12

Revolutionizing Microbial Nanotechnology: A Green Approach to Sustainable Energy Production

Jaya Dayal[1,*]

[1] *Department of Microbiology, S.S. Jain Subodh P.G. College, Jaipur, India*

Abstract: Green nanotechnology is a new interdisciplinary field that aims to utilize eco-friendly materials and methods in the generation of different applications. Among the most innovative in this area is to employ microorganisms for the manufacturing and design of nanomaterials at a large scale to be utilized in energy, environmental clean-up, and green engineering applications. In this chapter, we look more into the microbial role in green nanotechnology, especially when it deals with energy production, storage, and conversion. Here, this chapter discusses the existence of various microorganisms, including bacteria, fungi, yeast, and algae, that can easily synthesize a wide range of nanomaterials under ambient environmental conditions. It is considered a greener technology compared to traditional chemical methods that use toxic reagents, consume high energy, and produce hazardous byproducts.

Embedding these biosynthesized nanomaterials into energy-adaptive systems has demonstrated a promise to improve system efficiency and reduce carbon footprints to achieve sustainability. The important applications that have been discussed in this chapter are Microbial Fuel Cells (MFCs) for bioelectricity generation, Microbial Solar Cells (MSCs), which convert sunlight into electricity, and microbial-based hydrogen and biofuel production systems. The chapter also explores the potentials of microbial-assisted biogas production and carbon capture for CO_2 sequestration as an alternative strategy towards a circular economy.

The promise of microbial nanotechnology is considerable, but it faces issues like scaling, cost-effectiveness, and regulatory hurdles. While waste conversion and eco-friendly nanomaterial production are yet to make much progress, this area has the scope of spreading sustainable energy solutions. This chapter elaborates on microbial-based nanotechnology and how it acts as a potential means to innovate in green energy and the environment, focusing mainly on its mechanisms, current applications, and future prospects.

Keywords: Bioelectricity generation, Biofuel production, Green nanotechnology, Microbial nanomaterials, Sustainable energy.

---

* **Corresponding author Jaya Dayal:** Department of Microbiology, S.S. Jain Subodh P.G College, Jaipur, India; E-mail: jaya22subodh@gmail.com

Harshita Jain & Maulin P. Shah (Eds.)

INTRODUCTION

Over the last decade, a number of effective and ecologically friendly techniques for synthesizing nanomaterials have been developed. Conventionally, numerous physical and chemical techniques, including chemical vapor deposition, pyrolysis, laser ablation, aerosol technologies, lithography, and microemulsion synthesis, can be used to synthesize them. These technologies are costly due to limitations such as high energy, temperature, pH, and pressure demands [1]. Using toxic compounds like organic solvents, reducing and stabilizing agents, and producing harmful byproducts can cause toxicity and environmental issues. Green synthesis has recently emerged as a response to these limits, which restrict the ecosystem with its distinct features, broader applicability, and environmental sustainability.

Green nanomaterials are derived from a number of natural sources, including plants [2], algae [3], actinomycetes [4], bacteria [5], yeast [6], and fungi (Fig. (**1**). The ability of these organisms to synthesize inorganic substances on a nano- and microscale has resulted in the formation of a new research area. The efficiency of biosynthesis of "green" nanomaterials is determined by the medium used. Plant extracts are frequently utilized as an effective medium for large-scale biosynthesis in a variety of species [7]. However, the stability of plant-derived nanomaterials varies substantially according to the biochemical composition of plant extracts from the same species. Using microbial resources to synthesize nanomaterials is advantageous due to their ability to stabilize the particles [8]. The structural diversity and facile cultivability of microbes make them ideal for synthesizing green nanomaterials, making them potential nano factories. Microbial cells, such as algae, fungi, yeast, and bacteria, play an important role in the bioreduction of metal ions and bioactive components in the synthesis process of nanomaterials.

Nanotechnologies provide significant environmental benefits by encouraging clean and green practices. Nanotech-based microbial fuel cells have received interest due to their promise as clean and versatile energy sources. Nanotechnology can help establish new industries using cost-effective ways, hence promoting long-term economic development. Nanotechnology refers to materials, processes, and phenomena at the nanoscale. Nanotechnology and green nanotechnology can reduce the negative impact of energy generation, storage, and use [9, 10].

Microbially synthesized nanomaterials show high electrocatalytic activity and outstanding performance as active materials for various significant applications in green and sustainable energy generation. More specifically, the chapter aims to discuss the microbial synthesis of nanomaterials and the role and importance of nanotechnology in bioenergy generation. The challenges and prospects for further development in the use of nanotechnology to produce bioenergy are finally covered.

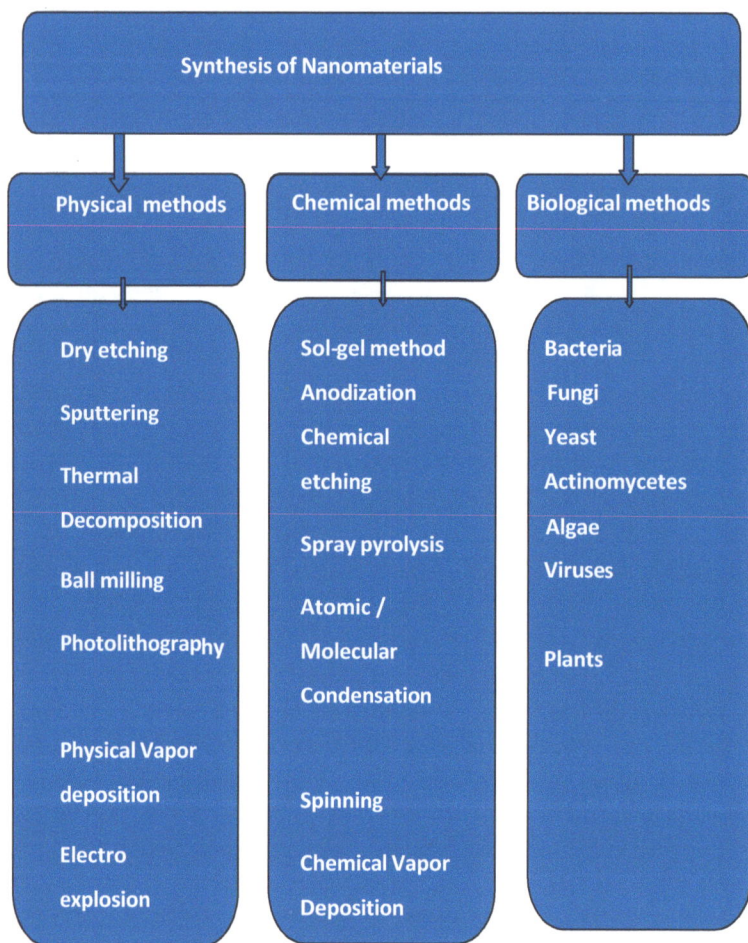

```
                    Synthesis of Nanomaterials

      Physical methods      Chemical methods      Biological methods

      Dry etching           Sol-gel method        Bacteria
                            Anodization            Fungi
      Sputtering            Chemical               Yeast
      Thermal               etching               Actinomycetes
      Decomposition                                Algae
                            Spray pyrolysis        Viruses
      Ball milling          Atomic /
      Photolithography      Molecular              Plants
                            Condensation

      Physical Vapor
      deposition            Spinning
      Electro               Chemical Vapor
      explosion             Deposition
```

Fig. (1). Different approaches of synthesis of nanomaterials.

THE ROLE OF MICROBES IN GREEN NANOTECHNOLOGY

Microorganisms are naturally capable of producing a variety of nanomaterials, such as metal nanomaterials, metal oxides, carbon nanotubes, and other nanostructures. They synthesize nanomaterials by internal and extracellular processes, including.

bioaccumulation, biomineralization, precipitation, and biosorption where metal ions are reduced to metal nanomaterials in the presence of cellular biomolecules, which are subsequently functionalized for specific uses [11, 12]. This biosynthesis of nanomaterials has several advantages over conventional chemical procedures, which usually include toxic chemicals, high temperatures, and significant energy use. Microbial synthesis of nanomaterials can be carried out in mild conditions, using renewable resources and producing little waste.

Types of Microbial Systems in Nanotechnology

The diversity of microorganisms, including bacteria, fungi, algae, and yeast, act as microbial nanofactories, and offer a wide range of pathways for synthesizing nanomaterials.

Bacteria

Bacteria are incredibly resilient to abiotic stressors and have the unique property to produce reduced metal ions due to their adaptability to environmental stress [13] Microorganisms are important nanofactories that can accumulate and detoxify heavy metals because they possess several reductase enzymes that can convert metal salts into metal nanomaterials. For example, some bacteria, such as *Pseudomonas stutzeri* (a facultative aerobe) and *P. aeruginosa* (a facultative anaerobe), can persist and endure in conditions with high levels of metals, particularly heavy metals.

It is commonly known that bacteria can produce inorganic compounds both inside and outside of their cells. In microorganisms, intracellular biosynthesis is strongly reliant on the negative charge of the cell wall. Electrostatic interactions transfer positively charged metal ions into negatively charged cell walls. Microorganisms use enzymes such as nitrate reductase to decrease ions and produce nanomaterials. Nanomaterials accumulate in the periplasmic area and are then transferred through the cell [14, 15].

Nanomaterials are synthesized extracellularly using nitrate reductase enzymes found in the cell wall or released into the growth medium. Nitrate reductase converts metal ions to metallic form [16]. The reduction of nanomaterials is also significantly aided by the presence of several components in microorganisms, including proteins, enzymes, and other biological molecules.

Fungi

A viable biogenic approach for creating various metal nanomaterials is the use of myco-nanotechnological techniques. Fungi secrete more proteins, increasing the productivity of nanomaterials. To survive metal stress and severe environments, fungi secrete enzymes and metabolites, including nitrate reductase, naphthoquinones, and anthraquinones, which act as reducing agents for nanomaterial production. Both extracellular and intracellular nanosynthesis are possible, as in bacteria.

Extracellular synthesis involves biosorption of metal ions on fungal filaments and reduction through secretion of enzymes, and metabolites. Intracellular synthesis involves the influx of metal salts within the fungal cell and bioreduction to metal ions. Fungi are more efficient than bacteria in producing nanomaterials due to their abundance of bioactive metabolites, high aggregation, and increased productivity. Their greater flexibility in different ecosystems, resilience to environmental stress, and diversity of forms (from unicellular structures to microfilaments) make them ideal for their role [17]. Moreover, fungal mycelia produce more biomass and have a larger surface area for interacting with metal salts, which promotes the production of nanomaterials. Therefore, it becomes more beneficial to explore and exploit fungi for the production of nanomaterials.

Yeast

Different yeast strains from various genera produce nanomaterials using unique processes and show notable differences in particle size, material composition, monodispersity, and other characteristics. One report stated that detoxification processes in yeast cells generate glutathione (GSH) and two types of metal-binding ligands: metallothioneins and phytochelatins. These substances stabilize the resulting complexes and affect the process for the synthesis of nanomaterials in most of the yeast species studied. A common defence mechanism used by yeast cells when exposed to toxic metals is the conversion of the ions into complex polymer molecules that are safe for the cell [18]. The ability of yeast cells to create semiconductor nanomaterials is very well-known.

Algae

Another innovative strategy is to synthesize nanomaterials using algae. The application of algae species in metal nanomaterial production depends on their structural and physical properties, as well as secreted biomolecules [19]. Algae release a variety of biomolecules, including carbohydrates, proteins, and secondary metabolites, which play a critical role in nanomaterial synthesis. Extracellular polysaccharides aid in reducing metal ions and stabilizing metal nanomaterials, while algal membrane proteins play a crucial role in synthesizing metal ions.

Flavonoids and terpenoids are examples of secondary metabolites that can stabilize and cap metal nanomaterials, changing their size, shape, and design [1, 20, 21]. Algal species can form nanostructures due to their skeletal and morphological diversity. Nanomaterials produced by algal species might be uneven, elongated, or spherical in shape and size [22].

MECHANISMS OF MICROBIAL NANOMATERIAL SYNTHESIS

Various processes are employed in the synthesis of microbial nanomaterials, with each method suited to specific types of nanomaterials and their production conditions.

Extracellular Synthesis

The process of extracellular biosynthesis includes the reduction of metal ions outside the cell, usually in the surrounding culture medium. Microorganisms form proteins, enzymes, and other biomolecules in which they help reduce metal ions and produce nanomaterial. These biomolecules function as a reducing and capping agent that transforms ionic precursors of metal ions into stabilized and capped nanomaterials [23].

Extracellular synthesis on nanomaterials has various advantages over other approaches. For example, because the nanomaterials originate outside the cell, they are easier to remove and purify than nanomaterials formed intracellularly. The easy execution of downstream processing reduces time, cost, and technical challenges, making this technique appropriate for large-scale production [24]. Because of its environmental sustainability, minimal toxicity, and scalability, extracellular synthesis is very useful in commercial and biomedical applications. In addition, this technique often produces nanomaterials free from cellular debris, which makes them suitable for a variety of applications, such as drug delivery, catalysis, and imaging.

Intracellular Synthesis

In intracellular synthesis, the generation of nanomaterial is dependent on absorbing and reducing metal ions inside the microbial cell. Often traveling past the cell membrane to the cytoplasmic region, these metal ions interact with a variety of intracellular components such as enzymes, proteins, and organelles. The reducing agent, biomolecules, and microbe enzymes reduce metal ions into nanomaterials within the cells. Typically, the nanomaterials fabricated through intracellular biosynthesis usually adhere to some constituents of the cell like cytoplasmic matrix, the cell membrane, or a particular organelle. Intracellular macromolecules present in microbial cells tend to hold onto the metal ions. Reductases and NADH (nicotinamide adenine dinucleotide) act as electron donors, aiding in the reduction and hence enabling metallic nanomaterials to form

[26, 27]. The basic mechanism of synthesis of nanomaterials is illustrated in Fig. (**2**).

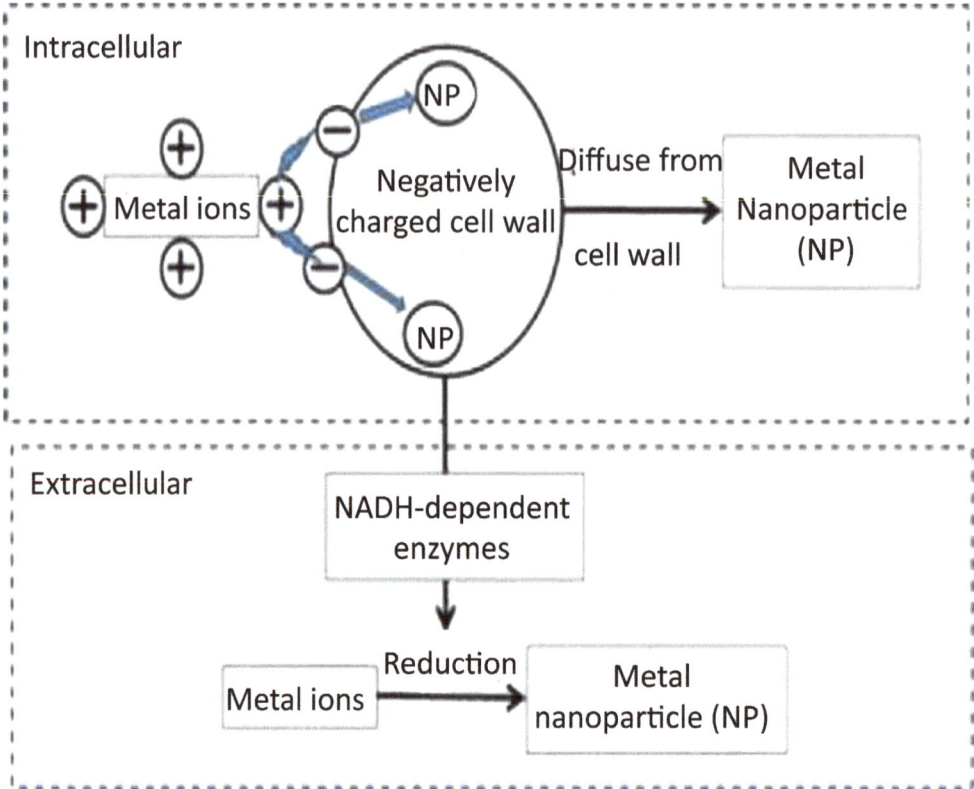

Fig. (2). Extracellular and intracellular synthesis of nanomaterials by bacteria [25].

Intracellular synthesis does, however, provide certain unique difficulties in relation to nanomaterial recovery. Cell lysis or mechanical disruption methods like enzymatic digestion, ultrasonication, or chemical treatments would have to be employed for the extraction since the nanomaterials remain inside the microbial cells. This additional process can be costly and time-consuming, especially when production levels are high. Moreover, comprehensive purification methods are essential to ensure that the nanomaterials are devoid of impurities and debris throughout the extraction process.

Despite its limitations, intracellular biosynthesis yields nanomaterials with outstanding crystallinity and biocompatibility, which is advantageous for applications requiring specific size, shape, and uniformity of nanomaterials.

MICROBIAL NANOTECHNOLOGY IN GREEN ENERGETICS

Microbial activities provide a sustainable method of energy generation and storage. Organic waste can be converted into bioelectricity *via* microbial metabolism, or nanomaterials can be used to improve the efficiency of energy conversion systems. Microbial nanotechnology has allowed for substantial advances in biofuel cells, microbial solar cells, and hydrogen production systems, with applications ranging from waste treatment to sustainable energy generation.

Microbial Fuel Cells (MFCs)

Microbial Fuel Cells (MFCs) are an innovative and eco-friendly source to provide methods for producing sustainable energy from organic waste. MFCs use the chemical bond energy of organic compounds to generate electricity directly using microbes, eliminating the need for intermediate processes and losses [28].

It has one or two separate chambers with a membrane in the MFC. When there are two chambers in the MFC, then the positive electrode is at the anode chamber while the negative electrode is found at the cathode chamber, which is separated by a PEM (Proton Exchange Membrane). PEM allows for proton transfer from the anode to the cathode. Organic or chemical waste is kept in the anode chamber. Bacteria are essential and efficient in breaking down organic compounds in the anode chamber, which produces electrons and protons [29]. An external circuit transfers electrons from the anode to the cathode, while the PEM transfers protons to the cathode. MFC performance and efficiency are affected by a variety of factors, including the organic material employed, bacteria that act as biocatalysts, internal resistance, membrane properties, electrode material characteristics and ion concentration [30].

Nanomaterial Integration

The incorporation of microbial biosynthesized nanomaterials, such as carbon nanotubes, graphene, and metal nanomaterials, into the MFC structure has been shown to improve their efficiency. For example, they offer a large surface area for the production of biofilms, which can speed up the biocatalytic activity of microorganisms. Nanoscale carbon or metal catalysts can enhance the pace of oxygen reduction reactions [31]. Nanomaterials with antibacterial properties, such as metal oxides, can extend air-cathode [32]. Nanomaterials are commonly utilized in MFC electrodes to increase active reaction sites. The membrane employed in MFCs must have high proton conductivity, low internal resistance, high energy recovery, and strong physical and chemical stability to avoid oxygen transport from the cathode to anode chambers. By adding nanomaterials to membrane architectures, the current approach aims to enhance membrane performance as well as their thermal and physical characteristics. Improvements have resulted in higher-performing MFCs for energy generation, wastewater treatment, and bio-sensing [33].

Microbial Solar Cells (MSCs)

Microbial Solar Cells (MSCs) or Biological Photovoltaics (BPV) are microbial bioelectrochemical systems that turn solar energy into electricity [34 - 37]. Unlike MFCs, phototropic organisms such as photosynthetic algae and cyanobacteria are utilized in MSCs to gather light, split water, and generate energy. The main steps involved in microbial solar cells are as follows: 1) production of organic matter by photosynthetic-active bacteria or algae; 2) anodic oxidation of organic matter by electrochemically active bacteria; and 3) cathodic reduction of oxygen. In MSCs, these microbes are used to produce organic materials in the anode chamber and provide electrical energy that can be harvested [38]. Recent research has focused on developing self-sustaining microbial solar cells that use light instead of pure organic matter as an energy source, as sunlight provides a limitless source of energy and the capacity to produce electricity even at night.

Quantum Dots and Carbon Nanotubes

Recent research suggests that quantum dots and carbon nanotubes can considerably improve the power output of MSCs. Quantum dots are capable of absorbing a wide range of light wavelengths, allowing for efficient energy conversion. Carbon nanotubes improve electron transport due to their superior conductivity and mechanical qualities, increasing the efficiency of energy conversion [39].

Microbial-Assisted Biogas Production

Microbial-assisted biogas production involves anaerobic digestion by microorganisms to transform organic waste materials into biogas, which is mainly made up of carbon dioxide, methane, water vapor, hydrogen sulfide (H_2S), hydrogen, and ammonia [40]. Anaerobic digestion is a multi-step process carried out by a consortium of microbes. It consists of four microbial sequential stages: hydrolysis, acidogenesis, acetogenesis, and methanogenesis. This gas is naturally produced in swamps, marshes, or urban waste disposal sites and must be confined before it may be used.

To economically utilize biogas, fermentation can be carried out in a relatively simple device called a digestion reservoir under regulated circumstances. Biogas is created through anaerobic fermentation of biomass, and the amount produced depends on the temperature of the digestion reservoir and the type of material employed. It possesses organic materials, and the proportions of gas are altered by them.

Biogas is a renewable energy source that may be used to heat, generate power, and fuel vehicles. Advances in microbial nanotechnology have enabled more efficient and cost-effective biogas production.

Nanomaterial Use in Biogas Systems

Nanomaterials have demonstrated promising results in anaerobic processes, specifically as electron donors, acceptors, and cofactors of key enzymes such as [Fe]- and [Ni-Fe]-hydrogenase [41]. Nanomaterials, for example, boost organic matter hydrolysis because of their high surface-area-to-volume ratio, allowing microbes to adhere to the active sites of molecules. This then activates the metabolic process that regulates the activity of hydrogenase enzymes and ferredoxins. Anaerobic digestion has used a variety of nanomaterials, including zero-valent metals, metal oxides, and carbon-based nanomaterials. Furthermore, the influence of nanomaterials on microbial populations during anaerobic digestion has also been investigated. ZVI nanomaterials have been shown to boost the population of methanogens in an anaerobic digester [42].

Biohydrogen Production

Hydrogen is a sustainable and renewable alternative energy source. It is an eco-friendly fuel that generates power and water, reducing the negative impact of fossil fuel consumption [43]. Hydrogen is the purest fuel and has the highest calorific value of all regularly used fuels, including coal and oil. It is produced as a byproduct through bacterial fermentation [44]. "Biohydrogen" is the dihydrogen gas (H^2) produced by microorganisms such as bacteria, algae, and archaea. Biohydrogen can be produced by many different biological processes, including biophotolysis, microbial electrolysis, and fermentation [45]. Biohydrogen synthesis uses several organic substrates and microorganisms [46]. Lignocellulosic biomass is a suitable substrate for large-scale biohydrogen generation due to its high cellulose content, renewable nature, and widespread availability.

While biomass to biohydrogen synthesis is the most sustainable technique for producing biofuels, there are some limitations that must be addressed before it can be commercialized on an industrial scale.

Application of Nanomaterial in Biohydrogen Production

Nanomaterials have unique features that can improve biomass-to-biohydrogen synthesis. Properties such as strong electro-conductivity, large surface area, and high surface-to-volume ratio can catalyze all stages of cellulose biohydrogen generation. Nanomaterials including Iron, Nickel, Copper, Gold, Silver, and Titanium have been shown to enhance biohydrogen production by multiple biological pathways, including bio-photolysis, dark fermentation, and photo-fermentation [47, 48]. Nanomaterials, particularly iron and nickel, serve as co-factors on hydrogenase and nitrogenase enzymes, leading to increased biohydrogen generation yield. Nanomaterials operate as oxygen scavengers during fermentation, reducing the possibility of oxidation-reduction. This approach creates an anaerobic environment for hydrogenase enzyme function, leading to increased biohydrogen output. Nanomaterials can improve the synthesis and stability of cellulase enzymes, leading to effective cellulose-t--sugar conversion processes [49]. Nanomaterials can improve a sustainable approach to producing high yields of biohydrogen on a practical scale.

Microbial CO_2 Sequestration and Conversion to Biofuel

Carbon dioxide (CO_2) is an essential component of plant life, and consequently animal and human life. The combustion of carbon-containing fuels produces CO_2 allowing humans to reside practically anywhere on Earth and power industrial production. However, the incomplete combustion of fossil fuels (coal, oil, and gas) used in electricity, transportation, and industrial operations contributes to the majority of CO_2 emissions [50]. CO_2 emissions are steadily increasing and are regarded as a major cause of climate change. Carbon capture and sequestration (CCS) systems can minimize CO_2 emissions from fixed resources (coal fires, cement plants, and other industries) [51, 52].

CO_2 includes carbon, which is a key component of biofuels. Therefore, CO_2 has the potential to be converted into biofuels and bioenergy, making it a viable option for the production of renewable energy. CO_2 is captured in a biological system before being turned into biofuels or bioenergy.

Microbial CO_2 sequestration systems use photosynthetic bacteria, algae, and other microorganisms to remove CO_2 from the atmosphere and convert it into biofuels. This technique frequently uses microorganisms such as cyanobacteria, algae, and acetogenic bacteria. They convert CO_2 into biofuels like ethanol, methane, and biodiesel. Cyanobacteria and microalgae use sunlight and CO_2 to create biomass that can be converted into biofuels. Non-photosynthetic microorganisms, including acetogens, use CO_2 as a carbon source to manufacture organic molecules in anaerobic conditions, known as chemoautotrophic fixation. In

Microbial Electrosynthesis (MESs), electrogenic microorganisms at the cathode of bioelectrochemical systems convert CO_2 into fuels such as acetate, methane, or ethanol [53, 54].

MESs are Bio-Electrochemical Systems (BESs) in which microbes act as biocatalysts to facilitate anodic oxidation or cathodic reduction processes. These microorganisms are electroactive, and their metabolic activity causes them to emit electrons to the anode or receive electrons from the cathode. The anode material used in MESs has a significant effect on both cost and system performance. Anode materials should normally be noncorrosive, electrochemically active in the desired oxidation reaction, physically and chemically stable, and biocompatible enough to allow electrogenic bacteria to form a biofilm. The cathode material in MESs varies according to the application [55]. The most popular cathode materials used to scale up MESs are carbon foam and carbon fibers, graphite plates, rods, carbon nanotube-based composites, carbon felt, titanium mesh, and stainless steel [56].

Nanomaterial Integration

The main electrochemical constraints for scaling up MES are electrolyte resistance, charge-transfer resistance due to slow reaction rates on electrodes, and resistance due to slow diffusion (mass transport). These limits lead to poor power densities and coulombic efficiency [57]. To solve the issue of insufficient power densities and coulombic efficiency, electrode and Proton Exchange Membrane (PEM) modification with Engineered Nanomaterials (ENMs) can be a viable solution. Conductivity, surface topology and chemistry, surface area, chemical and electrochemical stability, porosity, toxicity, and biocompatibility are all intrinsic properties of electrodes that influence their interactions with microbes. Modifying electrode materials or surfaces with ENMs has been the predominant approach to enhancing microorganism-electrode interactions [54]. Thus, initiatives are motivated to widely deploy the bio-recycling system, contributing to the circular bioeconomy of MES, where waste generated is recycled and turned into economic products.

SUSTAINABILITY AND ENVIRONMENTAL IMPACT

There are several sustainability and environmental advantages of using microbial processes in nanotechnology. Toxic chemicals, high temperatures, and hazardous byproducts are typically involved in conventional chemical synthesis techniques for nanomaterials. However, microbial synthesis is a more environmentally friendly method of producing nanomaterials since it may be done under ambient conditions and yields non-toxic byproducts.

Waste Utilization

Utilization of organic waste or the industrial byproduct as the feedstock for nanomaterial synthesis and energy production can be achieved through microbial nanotechnology. Microbial activity can also transform organic waste, such as agricultural or industrial effluent, into useful products like bioelectricity, biofuels, and nanomaterials, thereby contributing toward a circular economy.

Energy Efficiency

Microbial techniques for nanomaterial production are more energy efficient than standard chemical procedures. This is especially relevant given the increasing demand for renewable energy sources. The capacity to create nanomaterials at low temperatures and without the use of harmful chemicals minimizes the overall environmental impact of nanotechnology.

CHALLENGES

Despite the promising promise of microbial nanotechnology, many challenges remain to be addressed.

Limited Efficiency of Microbial Systems

Microorganisms used in nanotechnology may exhibit slow growth rates, restricted yields, and inadequate metabolic efficiency for large-scale bioenergetic applications [58].

Scalability Issues

Scaling up microbial nanotechnology from laboratory to industrial levels is a major problem because microbial cultures and nanomaterials have complex requirements. Moving from small-scale research to large-scale applications requires resolving issues with reactor design, nutrient delivery, and waste management [59, 60].

Toxicity of Nanomaterials

Since nanomaterials lower metabolic activity and cell viability, they may be harmful to microorganisms. As a result, microbial activity is less sustainable and effective. Biofilms, which are antibiotic-resistant microbial colonies covered in a protective extracellular matrix, have been demonstrated to be destroyed by certain nanomaterials [61 - 63].

Regulatory and Moral Issues

Microorganisms that have been genetically engineered and nanomaterials are areas of ethical and regulatory concern, particularly with respect to the long-term impacts and environmental release. These need to be considered in the safe and proper application of microbial nanotechnology in bioenergetics [63 - 65].

FUTURE TRENDS

Microbial systems lead green nanotechnology and the development of sustainable energy. Microorganisms present a unique opportunity to solve environmental issues in nanomaterial synthesis because of their capacity to biosynthesize a variety of nanomaterials under mild, eco-friendly conditions [66 - 69]. Future research work should focus on biocompatible nanomaterials to reduce toxicity, and cheaper production by adopting more efficient manufacturing processes, and optimizing microbial strains towards higher energy and nanomaterial synthesis. It is possible that the activity of microbes can be involved in energy systems like solar cells, microbial fuel cells, and biohydrogen production for further development of sustainable renewable energy technologies [70 - 72].

CONCLUSION

In conclusion, microbial nanotechnology stands at the forefront of revolutionizing sustainable energy production by harnessing the natural capabilities of microorganisms to synthesize eco-friendly nanomaterials. This green approach offers a promising alternative to conventional methods, significantly reducing environmental impact while enhancing the efficiency of energy systems. Through applications such as microbial fuel cells, microbial solar cells, and microbial-based hydrogen and biofuel production, this technology demonstrates immense potential in contributing to a circular economy and reducing carbon footprints. Despite the challenges related to scalability, cost-effectiveness, and regulatory frameworks, the field of microbial nanotechnology is poised for substantial growth, offering innovative solutions for clean energy and environmental remediation. As research advances, integrating microbial nanotechnology into mainstream energy production and environmental management systems could pave the way for a more sustainable and resilient future. This chapter underscores the importance of continued exploration and development in this interdisciplinary field, highlighting its pivotal role in driving green energy innovations and fostering environmental sustainability.

REFERENCES

[1]   Purohit J, Chattopadhyay A, Singh NK. Green synthesis of microbial nanoparticle: approaches to application. Nanotechnology in the life sciences. 2019; pp. 35-60.
[http://dx.doi.org/10.1007/978-3-030-16534-5_3]

[2]   Pantidos N, Horsfall LE. Biological synthesis of metallic nanoparticles by bacteria, fungi and plants. J Nanomed Nanotechnol 2014; 5(5)
[http://dx.doi.org/10.4172/2157-7439.1000233]

[3]   Aziz N, Faraz M, Sherwani MA, Fatma T, Prasad R. Illuminating the anticancerous efficacy of a new fungal chassis for silver nanoparticle synthesis. Front Chem 2019; 7: 65.
[http://dx.doi.org/10.3389/fchem.2019.00065] [PMID: 30800654]

[4]   Singh P, Kim YJ, Zhang D, Yang DC. Biological synthesis of nanoparticles from plants and microorganisms. Trends Biotechnol 2016; 34(7): 588-99.
[http://dx.doi.org/10.1016/j.tibtech.2016.02.006] [PMID: 26944794]

[5]   Prasad R. Advances and applications through fungal nanobiotechnology 2016.
[http://dx.doi.org/10.1007/978-3-319-42990-8]

[6]   Menon S, S R, S VK. A review on biogenic synthesis of gold nanoparticles, characterization, and its applications. Resource-Efficient Technologies 2017; 3(4): 516-27.
[http://dx.doi.org/10.1016/j.reffit.2017.08.002]

[7]   Dhillon GS, Brar SK, Kaur S, Verma M. Green approach for nanoparticle biosynthesis by fungi: current trends and applications. Crit Rev Biotechnol 2012; 32(1): 49-73.
[http://dx.doi.org/10.3109/07388551.2010.550568] [PMID: 21696293]

[8]   Selvakumar R, Seethalakshmi N, Thavamani P, Naidu R, Megharaj M. Recent advances in the synthesis of inorganic nano/microstructures using microbial biotemplates and their applications. RSC Advances 2014; 4(94): 52156-69.
[http://dx.doi.org/10.1039/C4RA07903E]

[9]   Ramazani A, Aghahosseini H. Engineered nanomaterial for catalysis industry. Elsevier eBooks. 2018; pp. 208-16.
[http://dx.doi.org/10.1016/B978-0-12-813351-4.00012-2]

[10]   Mahdavi Dehkharghani F, Ghahremanlou M, Zandi Z, Jalili M, Mozafari M, Mardani P. Future energy and therapeutic perspectives of green nano-technology: recent advances and challenges. Nano Micro Biosystems 2023; 2(1): 11-21.
[http://dx.doi.org/10.22034/nmbj.2023.385185.1013]

[11]   Singh M, Srivastava M, Kumar A, Pandey K. Role of plant growth promoting microorganisms in sustainable agriculture and nanotechnology. Amsterdam: Elsevier 2019.
[http://dx.doi.org/10.1016/C2018-0-01338-9]

[12]   Annamalai J, Ummalyma SB, Pandey A, Bhaskar T. Recent trends in microbial nanoparticle synthesis and potential application in environmental technology: a comprehensive review. Environ Sci Pollut Res Int 2021; 28(36): 49362-82.
[http://dx.doi.org/10.1007/s11356-021-15680-x] [PMID: 34331227]

[13]   Deobagkar D, Kulkarni R, Shaiwale N, Deobagkar D. Synthesis and extracellular accumulation of silver nanoparticles by employing radiation-resistant *Deinococcus radiodurans*, their characterization, and determination of bioactivity. Int J Nanomedicine 2015; 10: 963-74.
[http://dx.doi.org/10.2147/IJN.S72888] [PMID: 25673991]

[14]   Tiquia-Arashiro S, Rodrigues D. Nanoparticles synthesized by microorganisms. Extremophiles: Applications in Nanotechnology 2016; 1-51.
[http://dx.doi.org/10.1007/978-3-319-45215-9_1]

[15]   Hulkoti NI, Taranath TC. Biosynthesis of nanoparticles using microbes—A review. Colloids Surf B Biointerfaces 2014; 121: 474-83.

[http://dx.doi.org/10.1016/j.colsurfb.2014.05.027] [PMID: 25001188]

[16]   Mohd Yusof H, Mohamad R, Zaidan UH, Abdul Rahman NA. Microbial synthesis of zinc oxide nanoparticles and their potential application as an antimicrobial agent and a feed supplement in animal industry: a review. J Anim Sci Biotechnol 2019; 10(1): 57.
[http://dx.doi.org/10.1186/s40104-019-0368-z] [PMID: 31321032]

[17]   Velusamy P, Kumar GV, Jeyanthi V, Das J, Pachaiappan R. Bio-inspired green nanoparticles: Synthesis, mechanism, and antibacterial application. Toxicol Res 2016; 32(2): 95-102.
[http://dx.doi.org/10.5487/TR.2016.32.2.095] [PMID: 27123159]

[18]   Ghosh S, Ahmad R, Zeyaullah M, Khare SK. Microbial nano-factories: synthesis and biomedical applications. Front Chem 2021; 9: 626834.
[http://dx.doi.org/10.3389/fchem.2021.626834] [PMID: 33937188]

[19]   Siddiqi KS, Husen A. Fabrication of metal and metal oxide nanoparticles by algae and their toxic effects. Nanoscale Res Lett 2016; 11(1): 363.
[http://dx.doi.org/10.1186/s11671-016-1580-9] [PMID: 27530743]

[20]   Haroon Anwar S. A brief review on nanoparticles: types of platforms, biological synthesis and applications. Research & Reviews Journal of Material Sciences 2018; 6(2)
[http://dx.doi.org/10.4172/2321-6212.1000222]

[21]   Vasquez RD, Apostol JG, de Leon JD, *et al.* Polysaccharide-mediated green synthesis of silver nanoparticles from Sargassum siliquosum J.G. Agardh: Assessment of toxicity and hepatoprotective activity. OpenNano 2016; 1: 16-24.
[http://dx.doi.org/10.1016/j.onano.2016.03.001]

[22]   Patel V, Berthold D, Puranik P, Gantar M. Screening of cyanobacteria and microalgae for their ability to synthesize silver nanoparticles with antibacterial activity. Biotechnol Rep (Amst) 2015; 5: 112-9.
[http://dx.doi.org/10.1016/j.btre.2014.12.001] [PMID: 28626689]

[23]   Sharma D, Kanchi S, Bisetty K. Biogenic synthesis of nanoparticles: A review. Arab J Chem 2019; 12(8): 3576-600.
[http://dx.doi.org/10.1016/j.arabjc.2015.11.002]

[24]   Pasula RR, Lim S. Engineering nanoparticle synthesis using microbial factories. Eng Biol 2017; 1(1): 12-7.
[http://dx.doi.org/10.1049/enb.2017.0009]

[25]   Das RK, Pachapur VL, Lonappan L, *et al.* Biological synthesis of metallic nanoparticles: plants, animals and microbial aspects. Nanotechnology for Environmental Engineering 2017; 2(1): 18.
[http://dx.doi.org/10.1007/s41204-017-0029-4]

[26]   Lahiri D, Nag M, Sheikh HI, *et al.* Microbiologically-synthesized nanoparticles and their role in silencing the biofilm signaling cascade. Front Microbiol 2021; 12: 636588.
[http://dx.doi.org/10.3389/fmicb.2021.636588] [PMID: 33717030]

[27]   Zhang X, Yan S, Tyagi RD, Surampalli RY. Synthesis of nanoparticles by microorganisms and their application in enhancing microbiological reaction rates. Chemosphere 2011; 82(4): 489-94.
[http://dx.doi.org/10.1016/j.chemosphere.2010.10.023] [PMID: 21055786]

[28]   Yaqoob AA, Mohamad Ibrahim MN, Rafatullah M, Chua YS, Ahmad A, Umar K. Recent advances in anodes for microbial fuel cells: An overview. Materials (Basel) 2020; 13(9): 2078.
[http://dx.doi.org/10.3390/ma13092078] [PMID: 32369902]

[29]   Kumar R, Singh L, Zularisam AW, Hai FI. Microbial fuel cell is emerging as a versatile technology: a review on its possible applications, challenges and strategies to improve the performances. Int J Energy Res 2018; 42(2): 369-94.
[http://dx.doi.org/10.1002/er.3780]

[30]   Abd-Elrahman NK, Al-Harbi N, Basfer NM, Al-Hadeethi Y, Umar A, Akbar S. Applications of nanomaterials in microbial fuel cells: A review. Molecules 2022; 27(21): 7483.

[http://dx.doi.org/10.3390/molecules27217483] [PMID: 36364309]

[31]   Santoro C, Arbizzani C, Erable B, Ieropoulos I. Microbial fuel cells: From fundamentals to applications. A review. J Power Sources 2017; 356: 225-44.
[http://dx.doi.org/10.1016/j.jpowsour.2017.03.109] [PMID: 28717261]

[32]   Yang E, Chae KJ, Alayande AB, Kim KY, Kim IS. Concurrent performance improvement and biofouling mitigation in osmotic microbial fuel cells using a silver nanoparticle-polydopamine coated forward osmosis membrane. J Membr Sci 2016; 513: 217-25.
[http://dx.doi.org/10.1016/j.memsci.2016.04.028]

[33]   Mashkour M, Rahimnejad M, Raouf F, Navidjouy N. A review on the application of nanomaterials in improving microbial fuel cells. Biofuel Research Journal 2021; 8(2): 1400-16.
[http://dx.doi.org/10.18331/BRJ2021.8.2.5]

[34]   Reggente M, Politi S, Antonucci A, Tamburri E, Boghossian AA. Design of optimized PEDOT-based electrodes for enhancing performance of living photovoltaics based on phototropic bacteria. Adv Mater Technol 2020; 5(3): 1900931.
[http://dx.doi.org/10.1002/admt.201900931]

[35]   Mouhib M, Antonucci A, Reggente M, Amirjani A, Gillen AJ, Boghossian AA. Enhancing bioelectricity generation in microbial fuel cells and biophotovoltaics using nanomaterials. Nano Res 2019; 12(9): 2184-99.
[http://dx.doi.org/10.1007/s12274-019-2438-0]

[36]   Milano F, Punzi A, Ragni R, Trotta M, Farinola GM. Photonics and optoelectronics with bacteria: making materials from photosynthetic microorganisms. Adv Funct Mater 2019; 29(21): 1805521.
[http://dx.doi.org/10.1002/adfm.201805521]

[37]   Tseng CP, Silberg JJ, Bennett GN, Verduzco R. 100th anniversary of macromolecular science viewpoint: soft materials for microbial bioelectronics. ACS Macro Lett 2020; 9(11): 1590-603.
[http://dx.doi.org/10.1021/acsmacrolett.0c00573] [PMID: 35617074]

[38]   Gude V G, Kokabian Bahareh, Gadhamshett Venkataramana. Beneficial bioelectrochemical systems for energy, water, and biomass production. J Microb Biochem Technol. 2013.
[http://dx.doi.org/10.4172/1948-5948.S6-005]

[39]   Park SH, Bai SJ, Song YS. Improved performance of carbon nanotubes embedded photomicrobial solar cell. Nanotechnology 2020; 31(11): 115401.
[http://dx.doi.org/10.1088/1361-6528/ab5b2a] [PMID: 31766024]

[40]   Kulkarni MB, Ghanegaonkar PM. Biogas generation from floral waste using different techniques. Global Journal of Environmental Science and Management 2019; 5(1): 17-30.
[http://dx.doi.org/10.22034/gjesm.2019.01.02]

[41]   Abdelsalam E, Samer M, Attia YA, Abdel-Hadi MA, Hassan HE, Badr Y. Influence of zero valent iron nanoparticles and magnetic iron oxide nanoparticles on biogas and methane production from anaerobic digestion of manure. Energy 2017; 120: 842-53.
[http://dx.doi.org/10.1016/j.energy.2016.11.137]

[42]   Markandan K, Chai WS. Perspectives on nanomaterials and nanotechnology for sustainable bioenergy generation. Materials (Basel) 2022; 15(21): 7769.
[http://dx.doi.org/10.3390/ma15217769] [PMID: 36363361]

[43]   Poleto L, Souza P, Magrini FE, *et al.* Selection and identification of microorganisms present in the treatment of wastewater and activated sludge to produce biohydrogen from glycerol. Int J Hydrogen Energy 2016; 41(7): 4374-81.
[http://dx.doi.org/10.1016/j.ijhydene.2015.06.051]

[44]   Liu BF, Ren NQ, Tang J, *et al.* Bio-hydrogen production by mixed culture of photo- and dark-fermentation bacteria. Int J Hydrogen Energy 2010; 35(7): 2858-62.
[http://dx.doi.org/10.1016/j.ijhydene.2009.05.005]

[45] Srivastava N, Srivastava M, Mishra PK, *et al.* Advances in nanomaterials induced biohydrogen production using waste biomass. Bioresour Technol 2020; 307: 123094.
[http://dx.doi.org/10.1016/j.biortech.2020.123094] [PMID: 32249026]

[46] Nagarajan D, Lee DJ, Kondo A, Chang JS. Recent insights into biohydrogen production by microalgae – From biophotolysis to dark fermentation. Bioresour Technol 2017; 227: 373-87.
[http://dx.doi.org/10.1016/j.biortech.2016.12.104] [PMID: 28089136]

[47] Taherdanak M, Zilouei H, Karimi K. Investigating the effects of iron and nickel nanoparticles on dark hydrogen fermentation from starch using central composite design. Int J Hydrogen Energy 2015; 40(38): 12956-63.
[http://dx.doi.org/10.1016/j.ijhydene.2015.08.004]

[48] Lin R, Cheng J, Ding L, *et al.* Enhanced dark hydrogen fermentation by addition of ferric oxide nanoparticles using Enterobacter aerogenes. Bioresour Technol 2016; 207: 213-9.
[http://dx.doi.org/10.1016/j.biortech.2016.02.009] [PMID: 26890796]

[49] Srivastava N, Srivastava M, Manikanta A, Singh P, Ramteke PW, Mishra PK. Nanomaterials for biofuel production using lignocellulosic waste. Environ Chem Lett 2017; 15(2): 179-84.
[http://dx.doi.org/10.1007/s10311-017-0622-6]

[50] Dayal J, Singh A, Mathur A. Strategies of biosynthesis of nanoparticles and their potential applications. International Journal of Scientific Research and Reviews 2019; 8(1): 792-806.

[51] Yu X, Catanescu CO, Bird RE, *et al.* Trends in research and development for CO_2 capture and sequestration. ACS Omega 2023; 8(13): 11643-64.
[http://dx.doi.org/10.1021/acsomega.2c05070] [PMID: 37033841]

[52] Bhatia SK, Bhatia RK, Jeon JM, Kumar G, Yang YH. Carbon dioxide capture and bioenergy production using biological system – A review. Renew Sustain Energy Rev 2019; 110: 143-58.
[http://dx.doi.org/10.1016/j.rser.2019.04.070]

[53] Ibrahim I, Salehmin MNI, Balachandran K, *et al.* Role of microbial electrosynthesis system in CO_2 capture and conversion: a recent advancement toward cathode development. Front Microbiol 2023; 14: 1192187.
[http://dx.doi.org/10.3389/fmicb.2023.1192187] [PMID: 37520357]

[54] Thakur IS, Kumar M, Varjani SJ, Wu Y, Gnansounou E, Ravindran S. Sequestration and utilization of carbon dioxide by chemical and biological methods for biofuels and biomaterials by chemoautotrophs: Opportunities and challenges. Bioresour Technol 2018; 256: 478-90.
[http://dx.doi.org/10.1016/j.biortech.2018.02.039] [PMID: 29459105]

[55] Kumar S, Tripathi A, Chakraborty I, Ghangrekar MM. Engineered nanomaterials for carbon capture and bioenergy production in microbial electrochemical technologies: A review. Bioresour Technol 2023; 389: 129809.
[http://dx.doi.org/10.1016/j.biortech.2023.129809] [PMID: 37797801]

[56] Mustakeem . Electrode materials for microbial fuel cells: nanomaterial approach. Mater Renew Sustain Energy 2015; 4(4): 22.
[http://dx.doi.org/10.1007/s40243-015-0063-8]

[57] Jadhav DA, Park SG, Pandit S, *et al.* Scalability of microbial electrochemical technologies: Applications and challenges. Bioresour Technol 2022; 345: 126498.
[http://dx.doi.org/10.1016/j.biortech.2021.126498] [PMID: 34890815]

[58] Logan BE, Rabaey K. Conversion of wastes into bioelectricity and chemicals by using microbial electrochemical technologies. Science 2012; 337(6095): 686-90.
[http://dx.doi.org/10.1126/science.1217412] [PMID: 22879507]

[59] Grasso G, Zane D, Dragone R. Microbial nanotechnology: Challenges and prospects for green biocatalytic synthesis of nanoscale materials for sensoristic and biomedical applications. Nanomaterials (Basel) 2019; 10(1): 11.

[http://dx.doi.org/10.3390/nano10010011] [PMID: 31861471]

[60]   Chaurasia P, Jasuja ND, Kumar S. 2022.Role of Nanotechnology in Microbial Mediated Remediation.
        [http://dx.doi.org/10.21203/rs.3.rs-1761898/v1]

[61]   Lan J, Zou J, Xin H, *et al.* Nanomedicines as disruptors or inhibitors of biofilms: opportunities in
        addressing antimicrobial resistance. J Control Release 2025; 381: 113589-9.
        [http://dx.doi.org/10.1016/j.jconrel.2025.113589]

[62]   Agarwal S, Jain H, Dhupper R, Mathur A. Optimizing methyl orange dye removal from aqueous
        solutions using white chrysanthemum floral waste-derived bioadsorbent: A study of kinetics,
        thermodynamics, isotherms, and RSM optimization. Int J Environ Res 2025; 19(3): 104.
        [http://dx.doi.org/10.1007/s41742-025-00773-z]

[63]   Cao YH, Cai WJ, He XW, *et al.* A review of advances & potential of applying nanomaterials for
        biofilm inhibition. NPJ Clean Water 2024; 7: 131-1.
        [http://dx.doi.org/10.1038/s41545-024-00423-5]

[64]   Vinci G, Savastano M, Restuccia D, Ruggeri M. Nanobiopesticides: Sustainability aspects and safety
        concerns. Environments 2025; 12(3): 74.
        [http://dx.doi.org/10.3390/environments12030074]

[65]   de Carvalho Lima EN, Souza LHM, Aguiar EM, Octaviano ALM, Justo JF, Piqueira JRC. Can
        bacteria and carbon-based nanomaterials revolutionize nanoremediation strategies for industrial
        effluents?. Frontiers in Nanotechnology 2024; 6: 1389107.
        [http://dx.doi.org/10.3389/fnano.2024.1389107]

[66]   Agarwal S, Jain H. Investigating the potential of floral waste as a vermicompost and dual-functional
        biosorbent for sustainable environmental management. Water Air Soil Pollut 2024; 235(5): 322.
        [http://dx.doi.org/10.1007/s11270-024-07144-y]

[67]   Pan J, Qian H, Sun Y, Miao Y, Zhang J, Li Y. Microbially synthesized nanomaterials: advances and
        applications in biomedicine. Precis Med Eng 2025; 2(1): 100019-9.
        [http://dx.doi.org/10.1016/j.preme.2025.100019]

[68]   lawal H, Saeed SI, Gaddafi MS, Kamaruzzaman NF. Green nanotechnology: Naturally sourced
        nanoparticles as antibiofilm and antivirulence agents against infectious diseases. Int J Microbiol 2025;
        2025(1): 8746754.
        [http://dx.doi.org/10.1155/ijm/8746754] [PMID: 40041153]

[69]   Jain H, Dhupper R, Shrivastava A, Kumari M. Enhancing groundwater remediation efficiency through
        advanced membrane and nano-enabled processes: A comparative study. Groundw Sustain Dev 2023;
        23: 100975.
        [http://dx.doi.org/10.1016/j.gsd.2023.100975]

[70]   González Cruz RA, Miranda AE, Quiroz IV. Biological treatment of wastewater: use of
        microorganisms for purification. Soil Improvement and Water Conservation Biotechnology 2025; 1:
        183-204.
        [http://dx.doi.org/10.2174/9789815322439125010012]

[71]   Tripathy DB, Chhabra P, Gupta A, Singh A. Application of nanotechnology in nutraceuticals and
        functional foods. Applications of Nanoparticles in Drug Delivery and Therapeutics 2024; 1: 46-69.
        [http://dx.doi.org/10.2174/9789815256505124010006]

[72]   Panseriya HZ, Gosai HB, Trivedi HB, Vala AK, Dave BP. Fungi and nanotechnology: History and
        scope. Mycology: Current and Future Developments 2022; 3: 1-27.
        [http://dx.doi.org/10.2174/9789815051360122030004]

CHAPTER 13

Microbial Waste Management and Resource Recovery

Mayank Chaudhary[1,*]

[1] *Ladakh Ecological Development Group (LEDeG), Leh-194101, India*

Abstract: This chapter offers a comprehensive examination of the utilisation of microbial technologies in waste management, emphasising their capacity for resource recovery, waste minimisation, and environmental sustainability. It specifically tackles critical issues in identifying appropriate waste streams, such as organic, industrial, and municipal waste, where microbial treatments provide significant benefits compared to traditional waste treatment approaches. Processes including anaerobic digestion, fermentation, and bioremediation show potential for transforming waste into valuable byproducts such as biogas, biofertilizers, and bioplastics; however, they encounter numerous obstacles that must be addressed for broad implementation. This chapter explores the fundamental mechanics of microbial waste degradation, analysing the metabolic pathways involved and their potential for effective resource recovery. It examines advanced technologies, including microbial fuel cells and biohydrogen production, assessing their benefits and drawbacks in comparison to conventional waste management techniques. A key emphasis is the construction of microbial consortia, wherein diverse microorganisms interact synergistically to enhance the degradation of particular waste types. The chapter critically assesses the environmental and economic advantages of microbial waste management, measuring reductions in greenhouse gas emissions and cost-effectiveness. It tackles the practical difficulties of expanding microbial technology, encompassing inefficiencies in severe settings and the substantial initial expenses linked to infrastructure development. Emerging disciplines like synthetic biology and systems biology are explored as prospective methods to surmount these obstacles, facilitating more efficient microbial activities and integration with other technologies. The chapter underscores the significance of microbial resource recovery in promoting sustainability objectives, especially within the framework of a circular economy and zero-waste systems. It underscores the necessity for enhanced public awareness, governmental endorsement, and investment to fully harness the potential of microbial solutions for global waste management.

Keywords: Anaerobic digestion, Microbial waste management, Microbial fuel cells, Resource recovery.

* **Corresponding author Mayank Chaudhary:** Ladakh Ecological Development Group (LEDeG), Leh-194101, India; E-mail: mayankunderscorechaudhary@gmail.com

Harshita Jain & Maulin P. Shah (Eds.)
All rights reserved-© 2025 Bentham Science Publishers

INTRODUCTION

The worldwide waste problem necessitates a fundamental change in management strategies, as conventional approaches such as landfilling and incineration are becoming progressively unproductive and detrimental to the environment. Microbial waste management provides a sustainable solution by employing microbes to digest organic waste and produce useful byproducts, including biogas, biofertilizers, and bioplastics. This chapter examines essential biochemical processes, such as hydrolysis, acidogenesis, and methanogenesis, which underpin technologies like anaerobic digestion. It additionally encompasses developing microbial technologies, such as Microbial Fuel Cells (MFCs) and biohydrogen production, which demonstrate the potential for extensive adoption [1].

Microbial waste management is very efficient for some waste streams that conventional systems do not adequately handle. These comprise food waste, sewage sludge, agricultural leftovers, and industrial effluents. Food waste is an escalating global issue; nevertheless, anaerobic digestion can transform food into biogas, yielding renewable energy and mitigating greenhouse gas emissions. Sewage sludge, typically discarded *via* expensive and ecologically detrimental techniques, can alternatively be processed using anaerobic digestion and Microbial Electrochemical Systems (MECs) to recover energy and diminish sludge volume [2].

Agricultural leftovers, often disposed of or incinerated, can be transformed into biogas and biofertilizers *via* microbial activities. This not only produces energy but also offers a sustainable substitute for chemical fertilisers. MECs have demonstrated efficacy in treating industrial effluents, including those from food processing and textiles, by decomposing organic contaminants while simultaneously generating energy. The combined capability of waste treatment and energy recovery renders MECs a promising technology for industrial applications. Microbial waste management technologies provide a revolutionary remedy for the worldwide waste dilemma [3, 4]. These technologies mitigate environmental pollution, recover energy, and shut nutrient loops by transforming waste into valuable resources, thus fostering a more sustainable and circular economy. With ongoing research and technical progress, microbial waste management will assume a progressively vital role in sustainable waste management and resource recovery [5].

The chapter primarily concentrates on the engineering of microbial consortia that improve waste degradation through the integration of several microorganisms to maximise efficiency. The environmental and economic benefits of these technologies are emphasised through lifecycle assessments, demonstrating their capacity to diminish greenhouse gas emissions, decrease waste management expenses, and reclaim resources. Challenges, including scalability and substantial initial expenses, are mitigated through advancements in synthetic biology, AI-driven optimisation, and hybrid system integration. The chapter finishes by examining how microbial waste management contributes to global sustainability, especially within the framework of a circular economy, wherein waste is converted into valuable resources.

KEY MICROBIAL MECHANISMS INVOLVED IN WASTE DEGRADATION AND RESOURCE RECOVERY

The breakdown of intricate waste materials, including food waste, sewage sludge, and agricultural leftovers, necessitates specialised microbial consortia. These waste streams consist of various organic substances, necessitating the collaborative efforts of many microbial species to decompose complicated polymers, such as lignocellulose or resistant fats. Creating efficient and customised microbial systems for these waste categories presents a considerable challenge, as individual microbes can decompose simple substances but necessitate consortia for the degradation of more complex waste [6].

Hydrolysis is the initial and crucial step in waste degradation, where complex polymers are broken down into simpler monomers by microbial enzymes. This step is often rate-limiting, especially for complex waste like lignocellulosic residues. Microorganisms such as *Clostridium cellulolyticum* and *Trichoderma reesei* produce enzymes like cellulases and hemicellulases that degrade cellulose and lignin, releasing fermentable sugars for further microbial metabolism. Research showed that microbial consortia from a biogas plant were more efficient than monocultures at hydrolyzing lignocellulosic material, highlighting the synergistic effect of diverse microbial enzymes [7].

Following hydrolysis, fermentation converts simple sugars and fatty acids into valuable products like ethanol, acids, and gases. Bacteria such as *Lactobacillus* and *Saccharomyces cerevisiae* play key roles in fermenting carbohydrates into lactic acid and ethanol, which can be further utilized in methane production or biofuel generation. It was demonstrated that fermentation of food waste by *Saccharomyces cerevisiae* produced ethanol, reducing waste volume and generating biofuels or chemicals. Methanogenesis is the conversion of fermentation products into methane by methanogens, a specialized group of archaea. This step is critical in anaerobic digestion systems for biogas production. Methanogens such as *Methanosarcina barkeri* and *Methanobacterium formicicum* utilize fermentation products like volatile fatty acids and alcohols to produce methane, a renewable energy source. The research found that combining food waste with sewage sludge increased methane production by 30%, demonstrating the synergistic effect of different organic waste types [8].

Designing microbial consortia for waste degradation requires selecting microorganisms that complement each other's metabolic functions. This can be achieved by engineering natural or synthetic consortia to optimize the efficiency of waste degradation and product recovery. Synthetic biology and metabolic engineering offer opportunities to tailor consortia for specific waste types and enhance the production of valuable byproducts such as biofuels and bioplastics. It was reported that engineered synthetic consortia convert food waste into biodegradable plastics, showing the potential of consortia engineering for both waste degradation and valuable product generation. Microbial mechanisms such as hydrolysis, fermentation, and methanogenesis are central to efficient waste degradation and resource recovery. By understanding the metabolic pathways and interactions within microbial consortia, researchers can design optimized systems that improve waste management and facilitate the recovery of valuable resources like biogas, biofuels, and biofertilizers. As research advances, engineered microbial consortia will offer sustainable, eco-friendly solutions for the growing global waste challenge [9].

CUTTING-EDGE MICROBIAL TECHNOLOGIES FOR RESOURCE RECOVERY

Transitioning microbial technologies from laboratory-scale experiments to industrial-scale applications presents significant hurdles. While technologies like MFCs, Anaerobic Digestion (AD), and biohydrogen production offer promising solutions for waste management, their scalability and economic feasibility remain major challenges. These technologies require further optimization and integration with existing systems to reduce costs, enhance efficiency, and improve resource recovery. Ongoing research is focused on overcoming these issues through new reactor designs, microbial consortia optimization, and metabolic engineering advancements. This article explores these technologies, backed by data and mathematical models from recent experiments, to better understand their mechanisms and scaling potential [10].

Microbial Fuel Cells: Harnessing Energy from Waste

MFCs offer a novel approach to converting chemical energy from organic waste into electrical energy (Fig. **1**). Electrogenic microorganisms transfer electrons to electrodes during anaerobic respiration, creating electricity. However, scaling MFCs for industrial use faces challenges like high electrode material costs, reactor efficiency, and low current densities at large scales. In a typical MFC, microorganisms like *Geobacter sulfurreducens* and *Shewanella oneidensis* metabolize organic compounds like glucose, releasing electrons that travel through an external circuit to produce electricity. Protons are transferred through a Proton Exchange Membrane (PEM), combining with oxygen at the cathode to form water [11]. A study used carbon nanotube-based electrodes to enhance electron transfer, achieving a power density of 0.3 W/m^2, a significant improvement over conventional graphite-based electrodes (0.1 W/m^2). This advancement is crucial for scaling MFCs for real-world applications. In an experiment, using wastewater treatment sludge as a substrate, the power density achieved was 0.14 W/m^2, an improvement over earlier configurations. The current efficiency was 30%, showing MFCs' potential for wastewater treatment and power generation. However, MFCs are still less efficient compared to traditional systems like anaerobic digestion. Another study proposed a model for estimating energy output in scaled MFCs. According to their calculations, scaling from a 1 m^2 to 100 m^2 system could yield approximately 300 W, sufficient for small-scale applications. However, the high cost of materials and energy recovery remains a significant barrier to large-scale implementation [12].

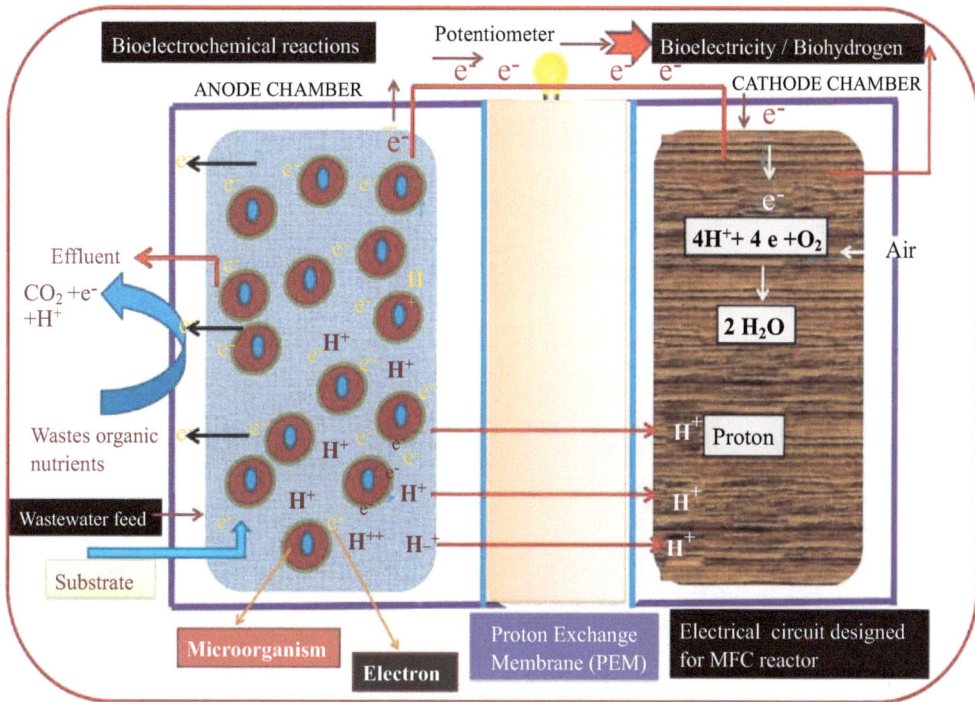

Fig. (1). Operation of MFCs with dual chambers utilising anaerobic microorganisms for the breakdown of waste nutrients in wastewater sources [13].

Anaerobic Digestion: A Proven Technology for Biogas Production

AD is a widely used microbial technology for waste treatment, breaking down organic compounds to produce methane (CH_4), a renewable energy source. AD occurs in four stages [14]:

1. **Hydrolysis**: Organic molecules are broken down into simpler monomers.
2. **Acidogenesis**: Monomers are fermented into Volatile Fatty Acids (VFAs).
3. **Acetogenesis**: VFAs are converted into acetic acid.
4. **Methanogenesis**: Methanogens convert acetic acid into methane.

A study demonstrated that thermophilic conditions (55°C) increased methane production by 40% compared to mesophilic conditions (35°C). In a study, under optimized conditions, methane yield from food waste increased from 0.15 m^3/kg to 0.35 m^3/kg of volatile solids, highlighting the benefits of controlled temperature and organic loading rates. Another study explored scaling AD for agricultural waste. Increasing the Organic Loading Rate (OLR) by 25% resulted in a 20% increase in biogas production, underscoring the importance of reactor design and operational optimization for large-scale applications [15] (Fig. **2**).

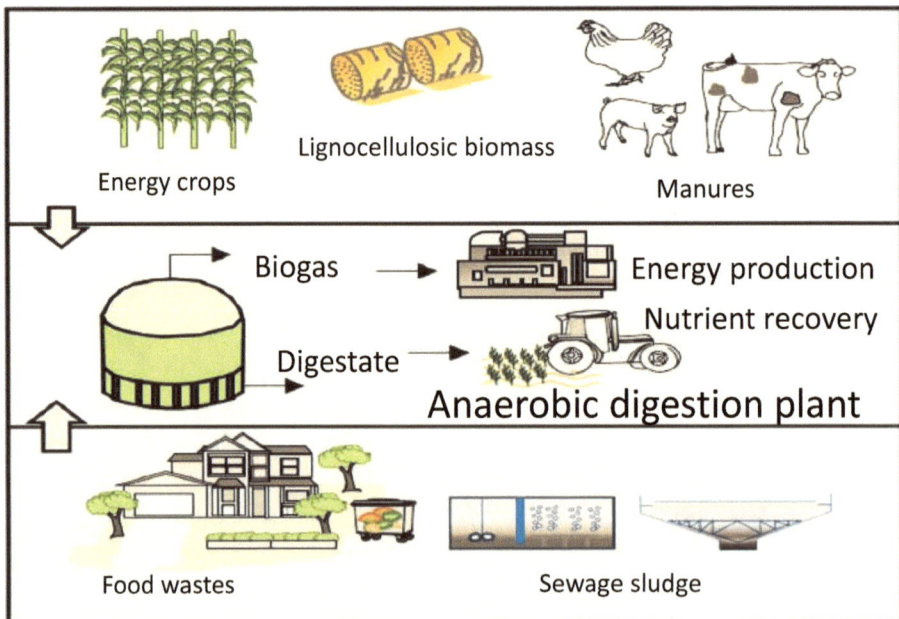

Fig. (2). Schematic depiction of various substrates appropriate for anaerobic co-digestion and the valorisation of primary process outputs [16].

Biohydrogen Production: Unlocking the Potential for Renewable Energy

Biohydrogen, produced by microbial fermentation, offers a promising renewable energy source. The process is driven by hydrogenase enzymes in bacteria and algae that catalyze the conversion of protons and electrons into molecular hydrogen (H_2). Biohydrogen can be produced through two mechanisms [17]:

1. **Dark Fermentation**: Anaerobic bacteria like *Clostridium beijerinckii* break down sugars to produce hydrogen.
2. **Photofermentation**: Photosynthetic bacteria such as *Rhodobacter sphaeroides* use light energy to produce hydrogen.

A study demonstrated that optimized dark fermentation conditions led to a hydrogen production rate of 1.8 L H_2/L/day, nearly double the previous rate. Microbial technologies, compared to traditional waste management systems like incineration or landfilling, offer the significant advantage of resource recovery (e.g., methane, hydrogen, biofertilizers). However, microbial systems still face higher capital costs and operational complexities, necessitating optimization for industrial competitiveness. Another study compared the costs of microbial technologies with incineration, finding that while microbial systems have higher initial costs, their long-term benefits, including resource recovery, make them more cost-effective over time. Microbial technologies such as MFCs, AD, and biohydrogen production show great promise for sustainable waste management and resource recovery. Despite scalability challenges, experimental advances in reactor designs, microbial consortia optimization, and metabolic engineering are driving improvements in efficiency. As these technologies continue to mature, they could revolutionize waste management and help address pressing environmental challenges. Continued research will focus on optimizing scalability, efficiency, cost, and energy recovery, making these technologies increasingly viable for industrial-scale applications [18] (Fig. 3).

Microbial Consortia in Waste Treatment

Microbial consortia, or communities of microorganisms working together, are essential in the treatment of complex organic waste. However, designing consortia for specific waste types remains challenging due to varying waste compositions and environmental conditions that affect microbial activity. The degradation of organic materials, such as Municipal Solid Waste (MSW) or food waste, involves multiple stages and requires careful optimization of microbial communities [20 - 23]. Designing microbial consortia involves understanding the dynamic relationships between species and optimizing them for efficient waste degradation. For instance, during AD, different microbial groups perform distinct metabolic functions, and fluctuations in factors like temperature, pH, or organic load can destabilize these consortia, leading to inefficient degradation or system failure [9].

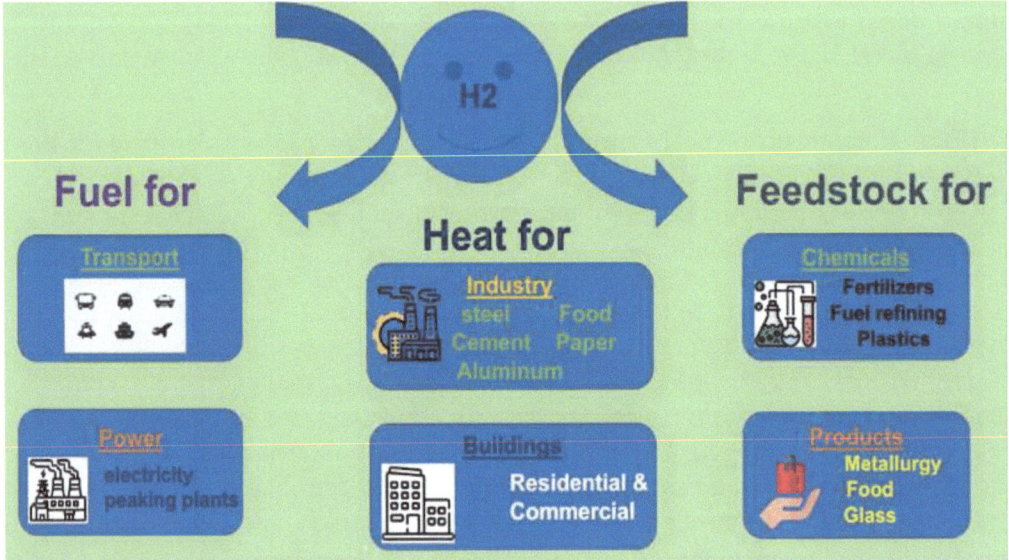

Fig. (3). Uses of green biohydrogen [19].

MAIN ENVIRONMENTAL AND ECONOMIC BENEFITS OF MICROBIAL WASTE MANAGEMENT

Microbial waste management technologies have emerged as a key solution for addressing modern waste treatment challenges. These technologies offer significant environmental and economic advantages over traditional waste disposal methods by harnessing the natural capabilities of microorganisms to decompose and convert organic waste into valuable byproducts. Below are the primary environmental and economic benefits of microbial waste management systems:

Reduction in Greenhouse Gas Emissions

One of the most profound benefits of microbial waste management is the reduction of Greenhouse Gas (GHG) emissions, particularly methane, a potent greenhouse gas that significantly contributes to climate change. Traditional methods such as landfilling result in methane release from the anaerobic decomposition of organic waste. Microbial systems, such as AD and MFCs, help mitigate this by capturing and converting methane into useful energy. This process not only captures methane but uses it as biogas for renewable energy generation. Studies show that optimizing AD can increase methane yield by up to 40%, with food waste biogas production reducing GHG emissions by 70% compared to landfilling. MFCs convert organic waste directly into electricity, reducing reliance on fossil fuels and their associated emissions. Research shows that MFCs can produce stable, low-carbon electricity, improving both wastewater treatment and energy generation [24].

Resource Recovery and Circular Economy Integration

Microbial technologies facilitate the recovery of valuable resources from waste, promoting a circular economy. Unlike traditional methods, which often waste resources, microbial systems can convert waste into renewable energy, biofertilizers, and other useful products, making waste a resource rather than a burden. AD produces both renewable energy and nutrient-rich digestate, a byproduct used as a biofertilizer. This reduces reliance on synthetic fertilizers and mitigates environmental pollution, cutting nitrogen fertilizer use by up to 40%. MFCs provide decentralized wastewater treatment, generating electricity while purifying water. The use of microorganisms like *Shewanella oneidensis* for wastewater treatment also offers efficient solutions for removing organic contaminants, pharmaceuticals, and heavy metals [25].

Economic Viability and Cost Savings

Microbial waste management technologies provide strong economic incentives by generating value from waste and reducing the costs associated with traditional waste disposal. The biogas produced in AD systems can replace expensive fossil fuels, with studies indicating that capital costs for biogas systems are recouped within 3–5 years through savings on electricity and reduced waste disposal costs. Additionally, the sale of excess energy can provide a revenue stream, particularly when connected to national or regional grids. Though still in the experimental phase, MFCs hold potential for large-scale applications, offering energy recovery while treating wastewater. Studies predict that scaling up MFCs could reduce operational costs by 20-30% [26].

Reduction in Waste Volume and Landfill Dependency

Microbial technologies like AD and composting significantly reduce the volume of waste directed to landfills, thus lowering methane emissions and conserving land for waste disposal. Studies show that microbial processes can reduce waste volume by up to 60-70%, addressing the growing global waste crisis and alleviating pressure on landfills [26].

Economic Incentives for Waste Management Companies

Governments worldwide are increasingly supporting microbial waste management technologies through subsidies, tax incentives, and feed-in tariffs for renewable energy projects, making them an attractive investment for both public and private sectors. The European Union, through its Renewable Energy Directive (RED II), offers financial incentives for biogas projects. These subsidies help waste management companies invest in AD systems, lowering the upfront costs associated with such technologies. The global market for biogas is projected to reach $50 billion by 2030, opening substantial revenue opportunities for microbial waste management systems that recover energy and resources from waste [27].

Environmental Health and Social Benefits

Beyond environmental and economic advantages, microbial waste management technologies also contribute to public health and social well-being. These technologies help reduce contamination of water and air with toxic chemicals, pathogens, and heavy metals, thereby improving environmental health. Studies on microbial bioremediation, such as those for oil spill remediation, demonstrate how bacterial consortia can effectively degrade hazardous compounds in wastewater, reducing water pollution levels. Microbial waste management technologies also contribute to job creation in the renewable energy and waste management sectors. By promoting sustainable waste practices, these technologies foster a transition towards a circular economy, reducing waste and reusing resources [21].

MICROBIAL TECHNOLOGIES AND THEIR CONTRIBUTION TO SUSTAINABILITY GOALS

Microbial technologies play a transformative role in advancing sustainability goals, particularly in the context of SDGs 6, 7, and 12. These goals are essential in addressing global challenges such as water scarcity, energy access, and responsible resource consumption. Despite the substantial potential of microbial technologies, the widespread adoption and scaling up of these solutions face significant barriers [21]. The following explores how microbial technologies contribute to sustainability and the challenges they face.

Microbial Technologies in Achieving SDG 6: Clean Water and Sanitation

SDG 6 focuses on ensuring availability and sustainable management of water and sanitation for all. Managing wastewater and providing clean water are critical challenges. Microbial technologies have proven effective in wastewater treatment, helping to improve water quality and reduce contamination through various biological processes [22]. Nitrogen is a major contaminant in wastewater, leading to eutrophication. Microbes such as *Nitrosomonas* and *Nitrobacter* perform nitrification by converting ammonia into nitrate, while denitrifiers like *Pseudomonas* convert nitrate into nitrogen gas, thereby removing nitrogen from the water. *Phosphate Accumulating Organisms* (PAOs) like *Acinetobacter* are used in wastewater treatment to remove excess phosphorus, which is another contributor to water quality degradation. Microbial processes are not only more environmentally friendly but also more energy-efficient than traditional chemical treatment methods. For example, microbial-based systems in constructed wetlands have been shown to remove up to 95% of nitrogen and 90% of phosphorus, outperforming traditional methods. A case study in Taiwan demonstrated microbial systems achieving up to 97% organic matter removal (COD) and 92% nitrogen removal. Life-Cycle Analysis (LCA) suggests microbial systems for wastewater treatment can be 30-40% more cost-effective than traditional chemical systems due to lower operational and chemical costs. Additionally, these systems reduce the need for synthetic fertilizers, leading to cost savings and reduced environmental harm [28].

Microbial Technologies in Achieving SDG 7: Affordable and Clean Energy

SDG 7 aims to ensure access to affordable, reliable, sustainable, and modern energy for all. Microbial technologies, especially through processes like biogas production and MFCs, offer innovative solutions for clean energy generation. Microbial AD processes convert organic waste into biogas, primarily methane, which can be used for energy generation. This process helps manage organic waste while producing a renewable energy source. MFCs use electroactive microorganisms to generate electricity from organic matter. Microbes such as *Geobacter sulfurreducens* transfer electrons from organic substrates to electrodes, generating an electric current that can be used for small-scale applications, such as powering sensors or wastewater treatment plants. MFCs have demonstrated power densities of 1-2 W/m^2, sufficient for powering low-energy devices. By optimizing microbial communities and electrode materials, energy recovery from wastewater and organic matter can be enhanced, contributing to sustainable, decentralized energy systems. Biogas production, as explored through anaerobic digestion, can recover significant amounts of energy from organic waste. The use of waste as a feedstock for energy production not only reduces reliance on fossil fuels but also addresses issues like waste disposal and greenhouse gas emissions [29].

Microbial Technologies in Achieving SDG 12: Responsible Consumption and Production

SDG 12 focuses on ensuring sustainable consumption and production patterns. Microbial technologies contribute to this goal by improving waste management, enhancing resource recovery, and reducing the environmental footprint of production processes. Microorganisms like *Cupriavidus necator* can produce polyhydroxyalkanoates (PHAs), biodegradable plastics derived from renewable resources. These bioplastics present an alternative to petroleum-based plastics, reducing pollution and dependency on fossil fuels. Microbial systems can convert organic waste into valuable resources such as biofertilizers, bioplastics, and bioelectricity, closing the loop on waste management and contributing to a circular economy. The production of PHAs from waste oils or agricultural byproducts can reduce the carbon footprint associated with plastic production. This contributes to reducing plastic waste in landfills and oceans, which is a significant environmental challenge. The market for bioplastics, particularly PHAs, is growing rapidly, driven by the global demand for sustainable materials. The bioplastics market is projected to exceed $44 billion by 2028, highlighting the economic potential of microbial-driven solutions [30].

Despite the proven potential of microbial technologies, several challenges hinder their broader adoption. There is a lack of public awareness about the potential of microbial technologies, which limits their acceptance and integration into existing systems. Governments often fail to provide adequate policy frameworks or incentives to support the adoption of microbial technologies. There is a need for comprehensive policy support to promote research, development, and implementation of microbial solutions. Although microbial technologies can be cost-effective, scaling them for large-scale applications remains a challenge due to the complexity of microbial systems and the need for specialized infrastructure and monitoring [31].

Microbial Technologies in Achieving SDG 7: Affordable and Clean Energy

Microbial systems are pivotal in advancing SDG 7, particularly in renewable energy production. Technologies like anaerobic digestion for biogas production and MFCs convert waste into clean energy. Anaerobic digestion transforms organic waste into methane, with efficiency determined by factors like microbial communities and substrate types. Biogas plants can produce methane yields of 0.35-0.40 m^3 per kg of volatile solids, offering significant energy production potential. MFCs leverage bacteria to directly convert organic waste into electrical energy, with studies showing power densities of 1-2 W/m^2. The economic potential for MFCs is significant, with the cost of electricity production potentially dropping to $0.12 per kWh, making them ideal for off-grid areas [32].

Microbial Technologies in Achieving SDG 12: Responsible Consumption and Production

Microbial technologies also contribute to SDG 12 by promoting sustainable resource management and waste reduction. The microbial production of bioplastics, such as polyhydroxyalkanoates (PHAs), presents an eco-friendly alternative to petroleum-based plastics. Microorganisms like *Cupriavidus necator* can produce PHAs through fermentation, achieving yields up to 90% of cell dry weight. The global bioplastics market is growing rapidly, with a projected 10.4% CAGR, reflecting a shift towards sustainable materials. Life-cycle assessments reveal that PHA production results in 30% lower CO_2 emissions compared to conventional plastics, while offering 100% biodegradability. For wider adoption, stronger policy frameworks and public education are essential. Policies should incentivize microbial technologies, support research, and enforce waste management regulations. Public awareness campaigns, like those in Europe, have shown a 30% increase in adoption, highlighting the need for education to drive change. Microbial technologies hold the key to achieving SDGs 6, 7, and 12 by enhancing wastewater treatment, clean energy production, and resource recovery. However, their full potential requires adequate support from policymakers and public education. With the right strategies, these technologies could become central to global sustainability efforts [33].

Despite the promise, scaling microbial technologies faces significant challenges, including high capital costs, complex system design, and microbial performance variability. High upfront investment is needed for large-scale systems, with capital costs ranging from $500,000 to $5 million, depending on the technology and scale. Microbial community management presents another challenge, as fluctuations in temperature and pH can lead to inconsistent performance, affecting efficiency. Scaling also introduces inefficiencies in reactor design and microbial kinetics, leading to reduced overall performance. To address these challenges, solutions like genetic engineering, improved reactor designs, and hybrid systems can enhance microbial performance. Genetic modifications in methanogenic bacteria can increase methane production by up to 60%, while innovations in reactor designs, such as fixed-film bioreactors, improve efficiency by up to 30%. Hybrid systems combining microbial and mechanical technologies, as well as decentralized modular systems, can overcome scalability barriers. Decentralized systems, for example, reduce capital and operational costs by minimizing transportation and simplifying design. Scaling microbial waste management technologies requires overcoming challenges like high capital costs, microbial variability, and system inefficiencies. However, with advancements in genetic engineering, reactor design, and hybrid systems, these barriers can be mitigated, making microbial technologies more scalable and economically viable for large-scale applications [34].

CONCLUSION

Inefficiencies in current waste management systems are often due to the challenges in effectively monitoring and optimizing microbial activity, which is essential for processes like biogas production and resource recovery. Advances in biosensors, AI-driven optimization, and real-time analytics offer promising solutions to improve microbial waste management, enhance system stability, and reduce environmental impacts. Real-time monitoring ensures microbial processes are functioning optimally, leading to better waste treatment and resource recovery. Poor monitoring can lead to inefficiencies, causing increased operational costs and reduced effectiveness in systems like anaerobic digesters or microbial fuel cells. Biosensors play a critical role in real-time microbial monitoring, offering continuous tracking of microbial activity and immediate feedback on system performance. Electrochemical biosensors, for example, detect metabolites such as methane and volatile fatty acids, enabling the optimization of anaerobic digesters, while optical biosensors track microbial degradation rates in aerobic systems. AI-driven optimization further enhances microbial performance by using advanced algorithms to predict microbial behavior and adjust operational parameters in real-time. For instance, reinforcement learning has been used to dynamically adjust parameters in anaerobic digesters, leading to improved methane yield and reduced operational costs. Real-time analytics also play a key role by enabling continuous monitoring and adaptive control of microbial waste management systems. Through the integration of various data sources, operators can make immediate adjustments to improve system efficiency and resource recovery. For example, real-time adjustments in aerobic treatment systems have led to significant improvements in waste treatment and energy efficiency. While advances in biosensors, AI, and analytics show significant potential, challenges related to scalability, data integration, and system complexity remain. Future research should focus on overcoming these barriers and integrating smart technologies to further optimize microbial waste management processes.

REFERENCES

[1] Palaniveloo K, Amran MA, Norhashim NA, *et al.* Food waste composting and microbial community structure profiling. Processes (Basel) 2020; 8(6): 723.
[http://dx.doi.org/10.3390/pr8060723]

[2] Upadhyay S K, *et al.* Transforming bio-waste into value-added products mediated microbes for enhancing soil health and crop production: perspective views on circular economy. Environ Technol Innov 2024; 34: 103573-3.
[http://dx.doi.org/10.1016/j.eti.2024.103573]

[3] Jain H, Dhupper R, Shrivastava A, Kumari M. Enhancing groundwater remediation efficiency through

advanced membrane and nano-enabled processes: a comparative study. Groundw Sustain Dev 2023; 23: 100975-5.
[http://dx.doi.org/10.1016/j.gsd.2023.100975]

[4]   Shah MP. Emerging innovative trends in the application of biological processes for industrial wastewater treatment. Elsevier 2024.

[5]   Tayou L N, *et al.* Acidogenic fermentation of food waste and sewage sludge mixture: effect of operating parameters on process performance and safety aspects. Process Saf Environ Prot 2022; 163: 158-66.
[http://dx.doi.org/10.1016/j.psep.2022.05.011]

[6]   Qaiser Z, *et al.* Microplastics in wastewaters and their potential effects on aquatic and terrestrial biota. Case Stud Chem Environ Eng 2023; 8: 100536-6.
[http://dx.doi.org/10.1016/j.cscee.2023.100536]

[7]   Gao Z, *et al.* Advances in biological techniques for sustainable lignocellulosic waste utilization in biogas production. Renew Sustain Energy Rev 2022; 170: 112995-5.
[http://dx.doi.org/10.1016/j.rser.2022.112995]

[8]   Sharma R, Garg P, Kumar P, Bhatia SK, Kulshrestha S. Microbial fermentation and its role in quality improvement of fermented foods. Fermentation (Basel) 2020; 6(4): 106.
[http://dx.doi.org/10.3390/fermentation6040106]

[9]   Cao Z, Yan W, Ding M, Yuan Y. Construction of microbial consortia for microbial degradation of complex compounds. Front Bioeng Biotechnol 2022; 10: 1051233.
[http://dx.doi.org/10.3389/fbioe.2022.1051233] [PMID: 36561050]

[10]   Sonawane A V, *et al.* A review of microbial fuel cell and its diversification in the development of green energy technology. Chemosphere 2024; 350: 141127-7.
[http://dx.doi.org/10.1016/j.chemosphere.2024.141127]

[11]   Kurniawan TA, Othman MHD, Liang X, *et al.* Microbial Fuel Cells (MFC): A potential game-changer in renewable energy development. Sustainability (Basel) 2022; 14(24): 16847.
[http://dx.doi.org/10.3390/su142416847]

[12]   Li D, *et al.* Structure evolution of air cathodes and their application in electrochemical sensor development and wastewater treatment. Sci Total Environ 2023; 869: 161689-9.
[http://dx.doi.org/10.1016/j.scitotenv.2023.161689]

[13]   Srivastava R K, Sarangi P K, Vivekanand V, Pareek N, Shaik K B, Subudhi S. Microbial fuel cells for waste nutrients minimization: recent process technologies and inputs of electrochemical active microbial system. Microbiol Res 2022; 265: 127216-6.
[http://dx.doi.org/10.1016/j.micres.2022.127216]

[14]   Laiq Ur Rehman M, Iqbal A, Chang C-C, Li W, Ju M. Anaerobic digestion. Water Environ Res. 2019; 91: pp. (10)1253-71.

[15]   Gandhi B P, *et al.* Kinetic investigations into the effect of inoculum to substrate ratio on batch anaerobic digestion of simulated food waste. Renew Energy 2022; 195: 311-21.
[http://dx.doi.org/10.1016/j.renene.2022.05.134]

[16]   González R, Peña DC, Gómez X. Anaerobic co-digestion of wastes: Reviewing current status and approaches for enhancing biogas production. Appl Sci (Basel) 2022; 12(17): 8884.
[http://dx.doi.org/10.3390/app12178884]

[17]   Ahmed SF, *et al.* Biohydrogen production from biomass sources: metabolic pathways and economic analysis. Mini Rev. 2021; 9.

[18]   Martínez-Fraile C, Muñoz R, Teresa Simorte M, Sanz I, García-Depraect O. Biohydrogen production by lactate-driven dark fermentation of real organic wastes derived from solid waste treatment plants. Bioresour Technol 2024; 403: 130846-6.
[http://dx.doi.org/10.1016/j.biortech.2024.130846]

[19] Reda B, Elzamar AA, AlFazzani S, Ezzat SM. Green hydrogen as a source of renewable energy: a step towards sustainability, an overview. Environ Dev Sustain 2024; 2024.
[http://dx.doi.org/10.1007/s10668-024-04892-z]

[20] Trends in biological processes in industrial wastewater treatment. M. P. Shah, ed.: IOP Publishing 2024.
[http://dx.doi.org/10.1088/978-0-7503-5678-7]

[21] Agarwal S, Jain H. Investigating the potential of floral waste as a vermicompost and dual-functional biosorbent for sustainable environmental management. Water, Air, & Soil Pollut 2024; 235(5): 322-2.
[http://dx.doi.org/10.1007/s11270-024-07144-y]

[22] Jain H. From pollution to progress: groundbreaking advances in clean technology unveiled. Innov Green Dev 2024; 3(2): 100143-3.
[http://dx.doi.org/10.1016/j.igd.2024.100143]

[23] Jain H. Data analytics enabled by the Internet of Things and artificial intelligence for the management of Earth's resources. In: Kumar D, Tewary T, Shekhar S, Eds. Data Analytics and Artificial Intelligence for Earth Resource Management. Elsevier 2025; pp. 19-36.
[http://dx.doi.org/10.1016/B978-0-443-23595-5.00002-4]

[24] Mundra I, Lockley A. Emergent methane mitigation and removal approaches: a review. Atmos Environ X 2024; 21: 100223-3.
[http://dx.doi.org/10.1016/j.aeaoa.2023.100223]

[25] Soo A, Kim J, Shon H K. Technologies for the wastewater circular economy – a review. Desal Water Treat 2024; 317: 100205-5.
[http://dx.doi.org/10.1016/j.dwt.2024.100205]

[26] Piadeh F, *et al.* A critical review for the impact of anaerobic digestion on the sustainable development goals. J Environ Manage 2024; 349: 119458-8.
[http://dx.doi.org/10.1016/j.jenvman.2023.119458]

[27] Malinauskaite J, *et al.* Municipal solid waste management and waste-to-energy in the context of a circular economy and energy recycling in Europe. Energy 2017; 141: 2013-44.
[http://dx.doi.org/10.1016/j.energy.2017.11.128]

[28] Holmes DE, Dang Y, Smith JA. Chapter Four - Nitrogen cycling during wastewater treatment. In: Gadd GM, Sariaslani S, Eds. Advances in Applied Microbiology. Academic Press 2019; 106: pp. 113-92.

[29] Kothari R, *et al.* MFC-mediated wastewater treatment technology and bioelectricity generation: future perspectives with SDGs 7 & 13. Process Saf Environ Prot 2024; 192: 155-76.
[http://dx.doi.org/10.1016/j.psep.2024.08.078]

[30] Jain H, Dhupper R, Shrivastava A, Kumar D, Kumari M. Leveraging machine learning algorithms for improved disaster preparedness and response through accurate weather pattern and natural disaster prediction. Rev. 2023; 11.
[http://dx.doi.org/10.3389/fenvs.2023.1194918]

[31] Salam MA, Al-Amin MY, Salam MT, *et al.* Antimicrobial resistance: A growing serious threat for global public health. Healthcare (Basel) 2023; 11(13): 1946.
[http://dx.doi.org/10.3390/healthcare11131946] [PMID: 37444780]

[32] Manikandan S, *et al.* Critical review of biochemical pathways to transformation of waste and biomass into bioenergy. Bioresour Technol 2023; 372: 128679-9.
[http://dx.doi.org/10.1016/j.biortech.2023.128679]

[33] Akinsemolu A A. Principles of green microbiology: the microbial blueprint for sustainable development. Environ Adv 2023; 14: 100440.
[http://dx.doi.org/10.1016/j.envadv.2023.100440]

[34]   Wang E, *et al.* Reviewing direct air capture startups and emerging technologies. Cell Rep Phys Sci 2024; 5(2): 101791-1.
[http://dx.doi.org/10.1016/j.xcrp.2024.101791]

SUBJECT INDEX

A

ACC deaminase 24
Acetogenesis 350, 353–354
Acinetobacter baylyi (bacterial biosensor) 292
Actinomycetes 115–116
AD (Anaerobic Digestion) 350–354, 358–359
Adsorption 160, 161–166
Advanced microscopy techniques 166
Advanced molecular techniques 265–266, 268
Affordable and Clean Energy (SDG 7) 358–359
Ag nanoparticles (Ag-NPs) 116–117
Agricultural leftovers (microbial treatment) 349–350, 354
Agroecosystems, sustainable 184
AI-driven optimization 361
Aio (Arsenite oxidase) 89–90
Air pollution 2, 5
Alcaligenes faecalis 89
Algae 114–115, 194, 333
Allophanate metabolism 66
Anaerobic degradation 64
Anaerobic digesters (real-time optimization) 361
Anaerobic microorganisms 352
Analyte detection limits 302
Antagonism 194–195
Antagonistic enzymes 194
Anthraquinones (fungal nanomaterial reduction agents) 332–333
Antibiotic resistance 9, 12
Antibiotics (produced by PGPR) 184
Applications of microbial nanomaterials 330–331
Applications of microbial nanoparticles 99, 113–117
Applications of microbial technologies 348–349
Aquifers 159, 169
Arbuscular Mycorrhizal (AM) fungi 250–251
Ars operon 88–89

Arsenate [As(V)] 88–90
Arsenic bioremediation 82–83, 88
Arsenic contamination 83–84
Arsenic detoxification 88–89
Arsenic toxicity 88–89
Arsenite [As(III)] 88–90
ATP generation via arsenite oxidation 90–91
Au nanoparticles (Au-NPs) 114–115
Auxins (phytohormone production) 239–241

B

Bacillus 52, 53, 55, 65, 90, 156, 157, 186, 241, 249, 270, 308, 311
amyloliquefaciens 249
arsenoxidans 90
cereus 249
licheniformis 249
species 156, 157, 186
spp. (heavy metal detoxification) 308, 311

subtilis 241, 270
tequilensis 241
thuringiensis 156, 157, 249
toyonensis 156, 157
Bacteria (nanomaterial synthesis) 4, 6, 8, 10, 12, 23–25, 99, 113, 183–186, 291–292, 331–333
bacteria 4, 6, 8, 10, 12, 23–25, 99, 113, 183–186
biosensors 291–292
Barriers to scaling microbial technologies 360–361
Bdellovibrio bacteriovorus (predatory bacterium) 197
Beneficial microbial interactions 185, 187
Bioaccumulation 3, 5, 10
Bioaugmentation 8, 9, 26, 52, 53, 159, 167, 311–312
Bioavailability of pollutants 288–289
Bioavailability reduction 82, 83
Bio-based technologies 16, 18

www.ingramcontent.com/pod-product-compliance
Lightning Source LLC
Chambersburg PA
CBHW050801220326
41598CB00006B/89